REFLECTION SEISMOLOGY

REFLECTION SEISMOLOGY

A Tool for
Energy Resource Exploration

KENNETH H. WATERS

Phoebe Apperson Hearst Visiting Lecturer
University of California, Berkeley
Formerly Research Fellow
Continental Oil Company, Inc.

A WILEY-INTERSCIENCE PUBLICATION
JOHN WILEY & SONS
New York · Chichester · Brisbane · Toronto

Copyright © 1978 by John Wiley & Sons, Inc.

All rights reserved. Published simultaneously in Canada.

No part of this book may be reproduced by any means,
nor transmitted, nor translated into a machine language
without the written permission of the publisher.

Library of Congress Cataloging in Publication Data:

Waters, Kenneth Harold, 1913—
   Reflection seismology.

   "A Wiley-Interscience publication."
   Includes bibliographies and index.
   1. Seismic reflection method. I. Title.

TN269.W37   622′.159   77-10837
ISBN 0-471-03186-0

Printed in the United States of America

10 9 8 7 6 5 4 3 2 1

To Frankie,

who never questioned my need to do this
and who has provided the encouragement
and renewed my persistence to see it done.

Prologue

Although the title gives the dominant theme of this book, a philosophical subtitle, suggesting the manner of treatment, might well be "The more ye know, the more ye need to know." With that as a benchmark my task will, for some readers, remain unfinished when the book has been read.

My goal has been to present an up-to-date framework of modern reflection seismology. Exploration seismology is a multidisciplinary science. A detailed description of all instruments, physical principles, geological methods, and the mathematical background is too extensive to be covered in one volume and too diverse to be adequately described by one author. At the level I have chosen I hope that it will be understandable by senior or first-year graduate students in geophysics and geology. Practicing exploration personnel who have worked for some time in their field may also find the book a useful supplement to their practical knowledge by making them aware of some of the strengths and limitations, or the doubts and hopes for the future, of reflection seismology. More advanced students may be disappointed, because the mathematical level has been truncated short of the highly useful matrix methods and a full discussion of Z-transform usage. I can only hope that this thirst for additional knowledge will be assuaged by the references given for each chapter or by a few books in the bibliography of Chapter 11.

There is an enormous literature. Two scientific journals (*Geophysics* and *Geophysical Prospecting*) devote themselves entirely to the science. Applicable papers abound in the *Journal of Geophysical Research*, the *Bulletin of the Seismological Society of America*, the *Geophysical Journal of the Royal Astronomical Society*, and many others. New material continues to pour out, ranging from highly sophisticated mathematical treatises on elastic wave propagation and data processing principles to case histories of oil field discoveries. Competition in the industry makes innovation a necessity. As time passes, it is not just one concept that has to be reexamined microscopically—and revised to meet the needs—it is all of them. The practice of reexploring the same area with different sets of instruments having diverse characteristics, over a period of several years, now involves the interpreter in a complex process of adjustment of characteristics; or it involves the exploration manager in a vexatious and costly decision to abandon some, if not all, of the previous work and reexplore the area once more with the most modern tools.

In the first six chapters the basic requirements of a structural first-look seismic section are established. In the remaining six chapters we explore the additional refinements and modifications to originally crude ideas, which are required for a more detailed attempt at the interpretation of more subtle geological events. It should be no surprise that some of the ideas offer a challenge to understanding. It is not only in the study of history that challenge has been the means by which humans advance.

The last chapter gives an indication of the direction of the most modern advancement. It is at this stage that some readers will really become fully aware of the need for additional tools to cope with the problems of finding stratigraphic oil fields or perturbations in coal seams from the surface. Appendices have been added to some chapters, usually to give more detail than seems to be necessary at first reading.

I am indebted to my present and former colleagues for many stimulating discussions over the years. Many of them had superior abilities to conquer some of the difficult problems; others had keener insight than mine into the nature of the problems. Still others have been inquisitive enough or, at times, demanding enough to require that explanations be clear and concise. In the course of their individual researches they have produced some hitherto unpublished illustrations which I am privileged to use in this book. So, even though I may be guilty of omission, I must express my admiration and my thanks to the following: D. Bahjat, L. H. Berryman, G. L. Brown, W. L. Chapman, J. T. Cherry, J. M. Crawford, E. E. Diltz, W. E. N. Doty, D. E. Dunster, E. L. Erickson, J. A. Fowler, P. L. Goupillaud, G. Harney, B. S. Heath, M. R. Lee, D. E. Miller, C. E. Payton, G. W. Rice, A. J. Scanlan, R. L. Stolt, B. J. Thomas, J. A. Ware, and M. Yancey. These colleagues, with their wide range of expertise, were all necessary in the operations and research departments with which I was associated, and I have derived great pleasure from my association with them. If this work provides help to anyone, the thanks should go to them; if it does not, then the blame is mine.

I am grateful to the Continental Oil Company not only for their very generous permission to allow me to publish this book but also for allowing me to devote a large part of my time during my final year of employment to work on it. If that had not been the case, it would have had to wait for two more years and I doubt that it would even have been finished.

My thanks are due to Mrs. Barbara Watson and to the secretarial services group for coping bravely with the unfamiliar terms and symbols, as well as my persistent requests for retypes and changes. I am also indebted to Mr. Bill Rigsby and members of his drafting staff for many illustrations.

The scientific societies and individual service companies and publishers have been generous with their permission to use some of their illustrations. That kindness is part of the atmosphere of our profession.

The original work was intended as an internal publication of the Continental Oil Company. For more general circulation a large number of revisions and modifications have had to be made. I am indebted to the Phoebe Apperson Hearst Foundation, which has provided the funds making it possible for me to accept an appointment as Visiting Lecturer in the Engineering Geoscience Section of the Material Science and Mineral Engineering Department at the University of California at Berkeley. I have spent part of the first month in the task of making this a more instructive text on reflection seismology.

KENNETH H. WATERS

Berkeley, California
October 1977

Contents

REFLECTION
SEISMOLOGY

ONE

Introduction

It is the aim of this introductory chapter to set out the goals of seismic exploration in a simple manner, uncomplicated by some of the difficulties dealt with in succeeding chapters. This has turned out to be necessary because many of the later topics are interrelated, and it is difficult to talk about the details of one topic without having a general knowledge of all. Thus to some extent the dilemma of ordering the various chapters has been resolved and, for the rest of this introduction, we deal with a simplified concept. Later, the reader can expect variations to be introduced to convert this idealistic concept into a realistic one.

The earth consists of a series of layers of rocks. In simple earthquake seismology, one can think of layers whose thicknesses are measured in terms of kilometers or hundreds and thousands of kilometers, the layers being built concentrically on a spherical inner core. The outer core, the mantle, and the crust constitute the most simple sequence. However, for considerations of hydrocarbon finding—which is assumed to be our job—there is little need to consider any layer but the crust, and then only a small portion of that. While this simplification is possible, further complications now develop, because all oil and gas accumulations are related to the sedimentary rocks where hydrocarbons are generated. These sedimentary rocks are secondary in nature and occur only in areas of the world where primary rocks have been eroded or where oceans have existed at some time in the past and the salts contained therein or minuscule fauna and flora have deposited themselves on the sea bottom. Most of these locations occur where, at some period of the earth's long history, a sequence of events took place establishing a series of rock deposits of different nature and properties, one upon the other. The rates of deposition have varied widely, as have the times during which a particular form of deposition has taken place. We look at a layered model within these sedimentary basins.

But deposition did not occur at a constant rate all over the basin, nor was the basin itself stationary during its entire history. Large-scale forces acted on basin sediments as they consolidated, causing vertical uplifts, folding due to compressional and shear stresses, and, as rock strengths were sometimes exceeded, breaks of various kinds—faulting—which interrupted the smooth change in the strata characteristics areally.

During the course of these millions of years of evolutionary history, the remains of the marine (and to a lesser extent terrestrial) fauna were acted on by pressure, heat, and bacteria so that their basic constituents were broken down first to solid prepetroleum materials (kerogen, etc.), then to a combination of liquid hydrocarbons and, in some cases, with higher temperatures, to hydrocarbon gases and to carbon dioxide.

Water and its dissolved salts were also contained in the pore spaces of rocks, sometimes tightly bound within rock particles and sometimes, when permeability existed, able to flow in response to differential hydrostatic pressure. The flow of this water out of shales into, and through, more porous rocks has been largely responsible for the movement and concentration of hydrocarbons within the pores of the rocks.

Various forms of traps exist, some of which are purely hydrostatic in nature (the oil has been pushed by the water into a position which is locally the position of minimum potential energy) and some of which are combinations of hydrostatic and permeability or porosity traps.

From a physical point of view, the task is to make use of the variation in the properties of different rocks to find these traps. Moreover, these variations must be mapped by making use of measurements made on or near, the surface. Various physical parameters have been measured in boreholes passing through rocks, and geologists are familiar with the resulting electrical logs of resistivity and spontaneous potential as well as logs made to indicate changes in density, seismic velocity, and radioactive properties. It is the variation in density and in seismic velocity that is of interest in reflection seismology. If we accept the fact that it is possible to make small shock (seismic) waves propagate through rocks, these waves must travel with a speed dependent on the nature of the rock itself and also on the nature of the possible fluid content.

When two different rocks are juxtaposed, a wave initiated at the surface, traveling through one or intercepting the boundary, is partially bounced back, or reflected, and partially transmitted so that it can travel in the second rock. These reflections eventually arrive back at the surface and, by their arrival times, convey some intelligence about how far the waves have traveled and possibly something about the rock boundaries they were reflected from. Naturally, since there are many boundaries between rocks in the earth, the intelligence conveyed is complex and the interpretation in terms of geology is not easy. Reflection seismology, which is the sole subject of this book, is therefore the science and art of initiating elastic waves and making surface measurements which are later interpreted to give information on the attitudes of, and relationships between, rock layers. In addition, whenever possible, the lateral changes in rock type and pore fluids must be inferred.

It has been established in well over 40 years of active seismic exploration that investigations have to be done so that areas as small as 1 km² (230 acres) can be investigated (oil fields of this size would be economical under the best of circumstances), and knowledge has to be gained for rock layers from a few tens of meters to over 5 km in depth. Some oil fields may exist because of the oil contained in one porous sandstone with a thickness of as little as 3 m, so the desired acuity of observation, the seismic resolution is high.

While some early oil fields were found by the so-called seismic refraction method (described at length in *Seismic Refraction Prospecting*, published by the Society of Exploration Geophysicists), these methods are not pursued further here and our emphasis remains on the use of reflections of artificially created small seismic (elastic) waves.

It should be pointed out that we are mainly concerned with the problems of obtaining and representing reflection data as free as possible of adulterating misleading information artifacts which are collected and sometimes interpreted as though they were bona fide reflections from subsurface layering—and in this volume we are able to devote only a little space to the fascinating problems of actual interpretation in terms of geology. In practice, this interpretation is rarely unique from the point of view of the physical data alone. By making sure of well-established geological principles some degree of uniqueness can be achieved. We can therefore expect to deal with the mechanics of the method. It is important to understand these basic principles thoroughly in order to appreciate later the close interplay between reflection seismology and geology, which is needed to make an accurate geological interpretation.

Figure 1.1 is a simple, diagrammatic cross section which illustrates the requirements of the technique. A method must be devised that allows a cross section of the earth to be drawn in sufficient detail to permit probable oil and gas reservoirs, such as *A, B,* or *C,* to be found. The sources and receivers of acoustic energy must reside on or close to the surface. In a manner similar to that used in radar or sonar, the method times the energy pulses from onset to time of echo. However, in contradistinction to radar and sonar, the medium through which the energy travels is heterogeneous, laminated but subject to discontinuities such as the unconformity shown or abrupt displacements due to faulting. In addition, the weathered layer is a layer of low velocity and density, of variable thickness and constitution, disconformable with the underlying consolidated rocks and even subject to seasonal fluctuations in properties.

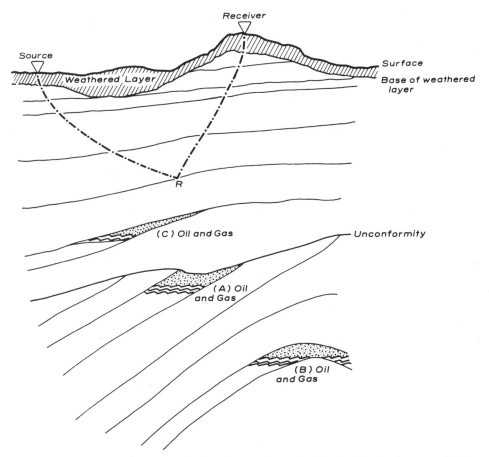

Figure 1.1 Diagrammatic cross section showing trapping modes for oil and gas in the pores of rocks.

A possible acoustic wave path is that shown from the source to R to the receiver. It is immediately obvious that not just one echo is received at each receiver from a single input pulse, since each change in the elastic properties of the rocks should give an echo—just as several radar targets in a series all give echoes from a single transmitted electro   ultiple scattering, that is, the scattered wave from o   er targets, also exist.

The   in most cases is exaggerated by having the vertica   l scale. In other words, the change in the vertica   s rapid in the horizontal direction as shown in this   prospecting problems can be looked at in the first p   the positions of rock boundaries vertically

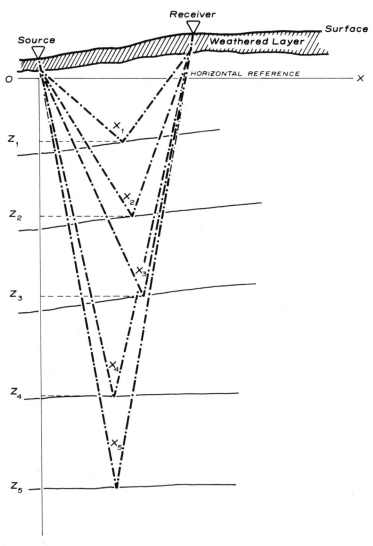

Figure 1.2 The coordinates of reflecting points to be determined by reflection seismology.

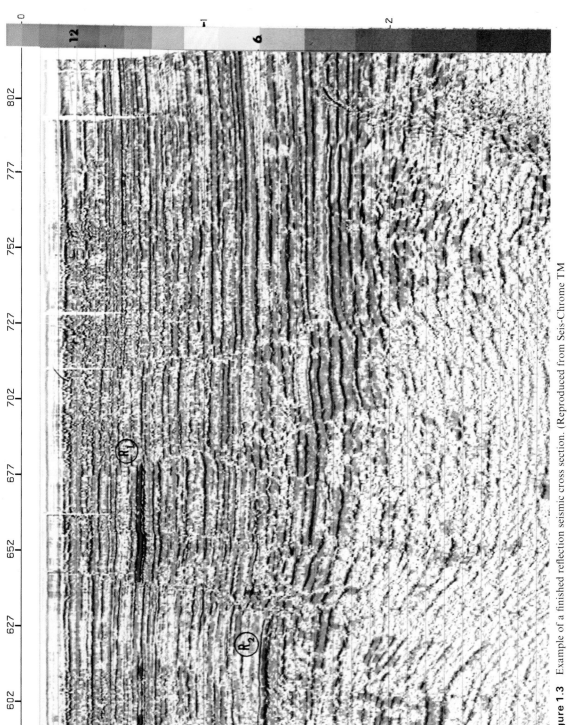

Figure 1.3 Example of a finished reflection seismic cross section. (Reproduced from Seis-Chrome TM displays with the permission of Seiscom Delta, Inc.).

beneath the experimental equipment—which can be moved along the surface from point to point, thus gradually building up the cross section from a series of similar measurements. For one of these experimental positions on the surface, the problem then reduces to that shown in Figure 1.2, determination of the coordinates (X_i, Z_i) of the successive reflecting points. This means that the raw reflection times T_i have to be corrected for elevation, for the time taken to pass through the weathered layer, and for the geometric effect introduced because the source and receiver are not at the same place on the surface. Figure 1.3 is a seismic cross section. It consists of a large number of vertical seismic traces, each of which shows the variation in received energy with time (the vertical coordinate). In the case illustrated, this variation in energy level has been converted to a color. The reflection time in seconds is given by the figures on the extreme right.

The seismic traces are very closely spaced in the horizontal direction, and the result is that the variations in energy of the individual traces, as the time of reception changes, are converted to a two-dimensional variation in color which indicates to the observer how the depth of the rock layers varies as the observation point moves along the surface. The numbers along the horizontal axis of this cross section represent the numbers of known points on the surface of the ground, which can be obtained from the land survey at any time. Thus any indication of suitable conditions for oil accumulation (say the pink reflection near R) can immediately be related to the position on the surface of the earth and steps can be taken to make appropriate tests by well drilling.

At the present time, monochromatic seismic cross sections are much more common than this polychromatic example. The reasons are discussed later. Moreover, the geological picture presented here is not yet true in all respects, and reasons for this are pointed out in subsequent chapters. Qualitatively, then, this seismic cross section is the first indication of the geological structure beneath the sequence of experimental stations. The art and science of producing these seismic cross sections (with either time or depth as the vertical variable) and proceeding from them to more exact geological information is the subject of the remainder of this book.

TWO

Basic Physical Principles:
Waves in Elastic Solids

The science of applied seismology, as it is used in the search for commercial sources of hydrocarbons, owes a debt to the parent science of earthquake seismology in that it was proved, during the latter part of the nineteenth century, that wave energy travels on the surface of, and through, the earth. Moreover, such waves were recorded and could be traced from point to point on the surface. These waves of course arose from natural causes, usually the accumulation of strain energy at particular localities in the earth until the elastic limit was passed and rupture occurred.

The waves discussed here, which have been put to work for economic purposes, are artificially generated, are much smaller in magnitude, and are of higher frequency than earthquake waves. Apart from these differences, the physics of wave propagation is the same, and the basic principles are discussed in this chapter.

2.1   WAVES IN GENERAL

A disturbance originates in some manner at a particular point A in a liquid medium and propagates outward through the medium. In Figure 2.1, one representation of such a wave is shown. This makes the assumption that a graph can be drawn which relates one property of the wave to the distance from the source. In this case, it is the displacement ξ from the rest position. On the upper line, such a wavelike form is shown, and a particular point $\bigcirc$ on the waveform arrives at B at time t_1. On the lower graph, taken at a different time t_2, we see a similar waveform reduced in amplitude but now having traveled an additional distance d. Since the waveforms are similar, it is possible to pick out a corresponding point on the waveform $\bigcirc^1$. Thus it appears that the point $\bigcirc$ has moved a distance $r_2 - r_1$ in a time $t_2 - t_1$, hence the velocity of this point must be

$$c = \frac{r_2 - r_1}{t_2 - t_1} \tag{2.1}$$

This velocity, associated with a particular feature of the waveform, such as a particular trough or a zero crossing, is called the phase velocity. As long as the wave moves without a change in form, the entire wave packet moves with this velocity, and the

6

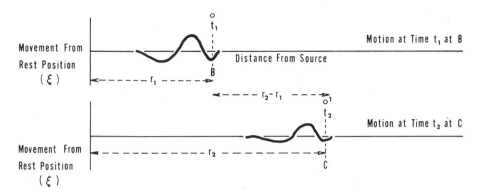

Figure 2.1 Wave profiles taken at two different positions r_1 and r_2 distant from the source.

wave is said to be nondispersive. These are the waves we concentrate on for the present.

It is also possible to plot ξ against time, while the distance remains stationary. This is shown in Figure 2.2. It is to be noted that we have now drawn the same waveform as in Figure 2.1, but the horizontal scale is now time instead of distance, and the waveform is horizontally reversed. In one figure, our frame of reference stayed fixed in space, while in the other it moved with the wave packet. It is thus possible to represent the shape of the wave packet as a function of both distance and time by the formula

$$\xi(r, t) = A(r)\xi\left(t - \frac{r}{c}\right) \tag{2.2}$$

It is seen later that the waveform ξ is related to the input stress that caused the disturbance and that $A(r)$ is a factor related to the way the medium allows the wave to spread out.

An important fact to note is that it is the abstraction "the wave" that is propagated through the material and not the material itself. Any particular small particle simply vibrates about its own rest position while the wave is passing and returns to its rest position and remains there afterward. Since motion from the rest position is a necessary characteristic of wave motion, the initial energy, which created the wave, spreads out with it. Thus, for certain rather limited special cases, it is possible to assign an approximate form to the function $A(r)$. In the body of a homogeneous liquid, the wave is generated with total energy E_0, and this is spread over the surface of a spherical

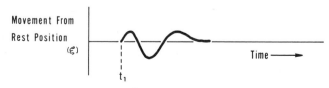

Figure 2.2 Motion of a point as a function of time.

wave (surface area $4\pi r^2$). The energy per unit area is thus $E_0/4\pi r^2$. It is seen later that the energy density is proportional to the amplitude squared.

$$\frac{E_0}{4\pi r^2} = k[A(r)]^2$$

$$A(r) = \left(\frac{E_0}{4\pi r^2 k}\right)^{1/2} = \frac{A_0}{r}$$

(2.3)

For a spherically diverging wave (2.2) therefore becomes

$$\xi(r, t) = \frac{A_0}{r} \xi\left(t - \frac{t}{c}\right)$$

(2.4)

and the amplitudes of all points on the wave packet diminish as the inverse of the distance from the source.

A second case which may be important relates to waves essentially confined to the surface of a homogeneous semi-infinite medium and, in this case, the total energy is confined to a ring of radius r. Then

$$\frac{E_0}{2\pi r} = k[A(r)]^2$$

$$A(r) = \left(\frac{E_0}{2\pi r k}\right)^{1/2} = \frac{A_0}{r^{1/2}}$$

(2.5)

For a wave spreading out over a surface (2.2) therefore becomes

$$\xi(r, t) = \frac{A_0}{r^{1/2}} \xi\left(t - \frac{r}{c}\right)$$

(2.6)

and the amplitude of the wave packet diminishes inversely in proportion to the square root of the distance. Since the inverse square roots of positive numbers decrease slower than the inverse numbers themselves as the number increases, waves confined to a plane surface (surface waves) decay in amplitude slower with respect to distance than those propagated within the body of the material (body waves).

It is sometimes convenient to consider a source of acoustic energy as being capable of emitting waves of constant frequency. In this case, it is possible to conceive of the measurement of the distance in the medium (the wavelength) between similar points of the sinusoidal waveform, and to any given frequency there corresponds a wavelength. If the frequency of the wave is n cycles per second (or hertz, abbreviated Hz) and the wavelength is λ, it is obvious that a given point on the waveform, produced at a given instant of time, travels a distance n units in 1 sec. But, by definition the distance traveled in 1 sec is the velocity c, so we have the relation $c = n\lambda$. It is not always true that waves of different frequency travel at a constant velocity but, for any given type of wave of frequency n, there is always a velocity $c(n)$ that applies. An isotropic medium is one in which $c(n)$ is independent of the direction of wave travel.

Fortunately, the waves that travel through the body of earth materials have a velocity almost independent of frequency. If this were not so, the interpretation of seismic record sections would be much more difficult. At the same time, however, waves are generated by sources of energy near the surface, which, because of rapid changes in velocity near the surface, are channeled along the surface and their velocities are strongly dependent on frequency—they are dispersive. This dispersion causes the wave packet to spread out as it propagates, and this causes the amplitude of any part of the wave packet to decay faster than simple theoretical considerations suggest.

In applied seismology, most exploration is done with instruments on the surface of the earth, and these are affected both by waves traveling along the surface and those that return to the surface having been propagated in the body of the earth. Apart from considerations of the partition of energy at the source, it is usually the case that some method of reduction of surface waves has to be used in order to avoid too large a magnitude ratio between the surface waves and the body waves, which carry most of the required information.

2.2 WAVE FRONTS AND RAYS

As a wave travels outward from the source, there is, at any given time, a surface that joins all the points where motion is just about to start. Such a surface is called a wave front. In a material with properties independent of position and direction of travel— an isotropic homogeneous material—the wave fronts are a set of concentric spheres with center at the source. However, as seen later, the form of wave fronts is determined by the wave velocity distribution. As shown in Figure 2.3, the energy can be conceived as traveling down a very large number of pyramids of infinitesimal cross section. The center line of any one pyramid is regarded as a ray, and its direction is normal to the wave front at any time. If the wave front is distorted by a nonuniform velocity distribution, the ray is bent so that it always stays perpendicular to the instantaneous wave front. In many situations, the ray convention is a convenient one to follow, since it is usually simpler than dealing with the complexities of wave front construction. However, it is merely a convenience, and it leads to difficulties whenever the normal to the wave front cannot be uniquely defined. This matter is examined in more detail

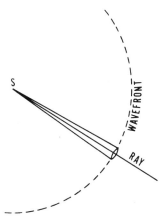

Figure 2.3   The relation between the ray and the wave front.

later. A special type of wave front is especially useful for ease in understanding additional problems. It is a *plane wave front* and, although not physically realizable, may be regarded as due to a point source of waves an infinite distance away—while still retaining finite amplitudes in the waves along the wave front. It is an artifice, but a useful one.

2.3   THE EFFECT OF BOUNDARIES BETWEEN TWO DIFFERENT MEDIA ON WAVE PROPAGATION

If we have, as in Figure 2.4, a wave front A at a particular instant of time, its position at a little later time can be predicted if we regard every point on this wave front as being a new (secondary) source of energy and draw wave fronts, a little later, for each of these new sources. This concept is known as Huyghens' principle, and it states that the new wave front B is the envelope (the common tangent surface) of all these new spherical wave fronts. In the case of a wave front in a homogeneous isotropic material, the new surface is parallel to the old one. For example, a plane wave front becomes another plane wave front parallel to the first, a spherical wave front generates another spherical wave front with the same center, and so on. The usefulness of Huyghens' principle emerges, however, when, as shown in Figure 2.5, the wave front encounters a plane boundary between two different wave velocities or in even more complex situations.

In Figure 2.5, an incident wave front I, in material 1, having a velocity C_1, propagates in a forward direction and at successive small time intervals Δt occupies positions I_2, I_3, I_4, and so on. The spacing between these positions is of course $C_1 \Delta t$. As it proceeds, a portion of the wave front reaches the boundary (at A, C, E, etc). If Huyghens' principle is used to generate new wave fronts, both forward into material 2 (velocity C_2) and backward into medium 1, it will be seen that a reflection is generated at the boundary, which sends back some of the energy into medium 1, and the angle made with the boundary is equal to the angle made with the incident wave.

The forward wave fronts travel with velocity C_2 which, in the case shown, is lower than C_1, and the angle of advancement of the wave front to the boundary is changed.

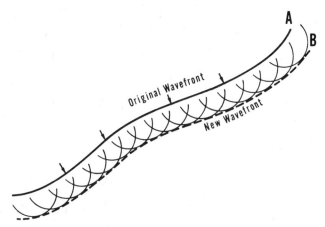

Figure 2.4   Illustrating the use of Huyghen's principle in predicting wave propagation.

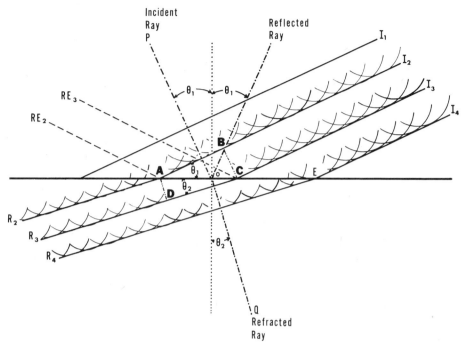

Figure 2.5 Huyghen's principle used to determine Snell's law.

This is the refracted wave. By consideration of the triangles ACD and ACB, $BC = C_1 \Delta t$, $AD = C_2 \Delta t$, and AC is common; hence

$$\frac{\sin \theta_2}{\sin \theta_1} = \frac{AD/AC}{AB/AC} = \frac{C_2}{C_1} = \mu \tag{2.7}$$

These angles are the same as those between the rays and the normal to the boundary; so Snell's laws, as these relations are called, can be stated in the form,

> When a wave crosses a boundary between two different media, both reflected waves and refracted waves are generated. The sine of the angle of incidence (between the ray and the normal to the boundary) is related to the sine of the angle of (reflection/refraction) as the ratio of the velocity of the incident wave to the velocity of the (reflected/refracted) wave.

More specification is made after the types of elastic waves have been considered.

This law does not say anything about the amplitudes of the waves transmitted (refracted) or reflected, but it does give the directions of the rays relative to the normal to the boundary. It is very general and applies to acoustic waves, to electromagnetic waves of all wavelengths, and to the two different kinds of elastic waves that can exist. In other words, it is a property of wave propagation and not of just a particular type of wave.

2.4 ELASTIC WAVES

Early in Section 2.1, the medium picked for subsequent discussion of wave propagation was a fluid. This was done to avoid the discussion of elastic waves until the
chief properties of wave propagation in general had been established. There is some
precedent for this subterfuge in that most of the exploration geophysics literature
talks about *the* seismic velocity or measurements of sonic velocity (singular) in wells.
This is not surprising, since most seismic experimentation is done with instruments
and sources that favor the reception and generation of one type of seismic wave.
The modern practices of field seismology, which use large separations between
sources and receivers, bring into question whether or not other seismic wave types
can be ignored. It is our intention to point out the characteristics of elastic waves in
general. Whenever it seems appropriate, throughout the book, dangers that may
accrue from making the single-wave-type assumption are discussed. However, it
may be profitable at times to make measurements on the other types of waves, and
these occasions are pointed out.

Solids can be distorted by external forces in two different ways. The first form involves the compression of a piece of the material without a change in shape (Figure
2.6a). The second implies a change in shape without a change in volume (Figure 2.6b).
Hence the waves are called compressional (*P*) and distortional (S). The P and S
designations, carried over from earthquake geophysics, denote *primus* and *secundus*
from their arrival sequence on earthquake records. However, the S designation is
often taken to be the initial of "shear," which is an alternate name for distortional
waves. Since fluids have no shear strength, they cannot support shear waves, hence in
these materials only one type of wave (P) has to be considered, and this type is often
called acoustic because these waves constitute sound that we hear.

We deal now, in some detail, with one type of elastic wave—in one dimension only.
The principal features of such wave propagation are derived, and then the results
extended by analogy to other wave types and more complex types of propagation.
The mathematical treatment of elastic waves in general is too advanced to be included here, but several such treatments (for those with the necessary mathematical
background) are readily available [see, for example, Ewing, Jardetsky, and Press
(1957) or Garland (1971)].

Two physical laws are invoked in discussing elastic wave propagation. Newton's
fourth law of motion states that a force applied to a body produces an acceleration
which is proportional to the force ($F = ma$). The second law required is Hooke's
law—applied to perfectly elastic materials—in which it is stated that the strain produced in a portion of an elastic medium is proportional to the stress. *Strain* is defined
as the fractional change in a geometrical property of the body (e.g., the length or the
volume), and *stress* is defined as the force applied per unit area. Stress is sometimes
referred to as *traction*.

We illustrate the use of these laws by considering a rod, having a cross-sectional
area of A and an original length L, which is held at one end and is being stretched by
the application of a force F at the other end. If the amount of stretch is ΔL, the strain
is $\Delta L/L$ and the stress is F/A, so that Hooke's law can be written as

$$K\frac{\Delta L}{L} = \frac{F}{A}$$

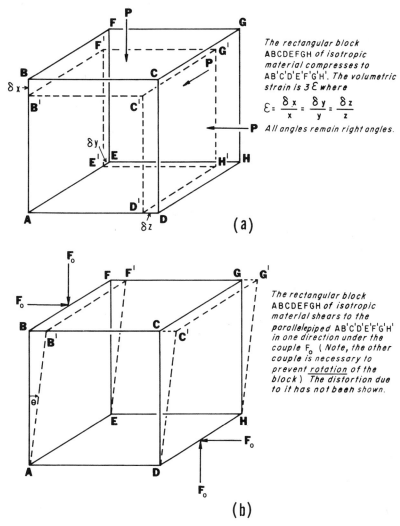

The rectangular block
ABCDEFGH *of isotropic*
material compresses to
AB'C'D'E'F'G'H'. The volumetric
strain is 3 $\mathcal{E}$ where

$$\mathcal{E} = \frac{\delta x}{x} = \frac{\delta y}{y} = \frac{\delta z}{z}$$

All angles remain right angles.

(a)

The rectangular block
ABCDEFGH *of isotropic*
material shears to the
parallelepiped AB'C'D'E'F'G'H'
in one direction under the
couple F_0 (*Note, the other*
couple is necessary to
prevent <u>rotation</u> of the
block) *The distortion due*
to it has not been shown.

(b)

Figure 2.6 Two different kinds of distortion of a rectangular block of material. (*a*) Compressive stress. (*b*) Shear stress.

where the constant K is called a modulus of elasticity. For rods for which the sides are not held the appropriate constant is called Young's modulus.

If we continue with the example of the rod, it can be shown that the displacement of one plane of cross section anywhere on the rod relative to the other cross-sectional planes (as, for example, hitting one end with a hammer) immediately causes a stress condition locally, which is propagated in both directions from the zone of the original disturbance as a stress or strain wave. The speed of propagation is established, in the next two paragraphs, as

$$c = \sqrt{\frac{K}{\rho}}$$

Note that only two different properties of the material are involved in the velocity, the proper modulus of elasticity and the density. The wave propagation process involves a continuous transfer of energy back and forth from elastic energy (compression of the material locally) to kinetic energy (motion of a small mass of material). A slightly more mathematical discussion follows.

Figure 2.7a shows a tube of constant shape and cross section S, having rigid walls and containing the material through which the elastic waves propagates. Under equilibrium conditions, two planes perpendicular to the axis are compressed or rarefied by the displacement of both boundaries in the direction of the axis, in general, by unequal amounts. At some instant of time, the first boundary is at $x_1 + u(x_1)$, and the other plane is at $x_2 + u(x_2)$. (*Note*: This is a somewhat naive way of handling materials by imagining them to be infinitely divisible, but it can be justified if we consider that we are really talking about the *average* displacement of molecules whose *average* position was x_1, x_2, etc.) Figure 2.7b shows these relationships when the distance apart is the small quantity dx.

The law of conservation of matter states that the mass of the material cannot be changed and, regarding the density ρ as constant, the mass is

$$\rho S \, dx$$

where S is the cross-sectional area of the tube.

Now, if there are unequal forces at each end of this segment, Newton's law states that the force difference is equal to the mass times the acceleration. Since u is the particle displacement from its rest position, $\partial u/\partial t$ or $\dot{u}$ is the particle velocity, and $\partial^2 u/\partial t^2$

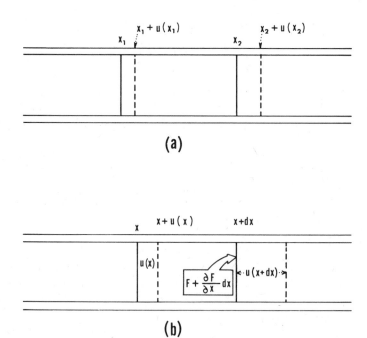

(a)

(b)

Figure 2.7 Movement of boundaries of a block of fluid in a tube of constant diameter when a wave passes through it.

(or $\ddot{u}$) is the acceleration. It is recognized that the partial derivatives are necessary, since u is a function of both time and distance. The force difference is therefore

$$\rho S \, dx \, \ddot{u}$$

It is now necessary to consider the force provided by the *elastic* nature of the material itself.

Hooke's law states that

$$\frac{\text{force}}{\text{unit area}} = \frac{F}{S} = K\left(\frac{\text{change in length}}{\text{original length}}\right) = K\frac{\partial u}{\partial x}$$

There is a force $F = KS(\partial u/\partial x)$ operating on the section at x, and an incremented force

$$F + dF = F + \frac{\partial F}{\partial x}dx = KS\frac{\partial u}{\partial x} + KS\frac{\partial^2 u}{\partial x^2}dx$$

operating on the section at $x + dx$. The net accelerating force acting on the element dx of the material is the difference between the forces acting on the two ends, or

$$KS\frac{\partial^2 u}{\partial x^2}dx$$

We can equate the two measures of force to give

$$\rho\frac{\partial^2 u}{\partial t^2} = K\frac{\partial^2 u}{\partial x^2} \qquad (2.8)$$

which is a one-dimensional partial differential equation, based on simple physical laws, which governs wave motion along the rod of material. It can be shown that it has a solution of the form

$$u(x, t) = F(x - ct) + f(x + ct) \qquad (2.9)$$

where c, the velocity of propagation of the wave, is given by

$$c = \sqrt{\frac{K}{\rho}}$$

The two terms, not necessarily of the same form, represent waves traveling in opposite directions, F in the direction of positive x and f in the opposite direction. So far, we have said little about the constant K except that it is the constant of proportionality between stress and strain in Hooke's law. Now we can particularize.

When the rod of material was specified, it was contained within a rigid tube so that it could expand or contract in one direction only. No motion across the axis could be accommodated. It can easily be seen that these are the same considerations as for plane waves passing through the body of a material, since expansion or contraction of any part of the material is incompatible with the conditions for the neighboring material. All the motion is perpendicular to the direction of wave propagation and results in compression or rarefaction. The waves are called compressional, longitudinal, or P waves (after the waves that arrive first from earthquakes). Their velocity is denoted by α and is given by

$$\alpha = \sqrt{\frac{K + 4/3\mu}{\rho}}$$

where K = bulk modulus of elasticity
$\quad\ \mu$ = modulus of rigidity
$\quad\ \rho$ = density

If these are waves of compression only, why does the modulus of rigidity enter into the velocity? The reason is that, during compression, as a plane wave passes through the material, a small portion (originally cubic in shape) becomes a parallelipiped — a change in shape as well as a change in volume, hence the modulus of rigidity, is involved.

2.5 EXTENSION BY ANALOGY

A second mode of motion may occur when the rod of material is twisted but not compressed. We let the amount of twist be $\theta(x, t)$ and, if the same arguments are followed, arrive at the same type of partial differential equation:

$$I\frac{\partial^2\theta}{\partial t^2} = Q\frac{\partial^2\theta}{\partial x^2}$$

Here the volume of any small element of the rod has not changed, but the material has been sheared. The quantities I and Q correspond, respectively, to the moment of inertia of the cross section and a constant related to the shear modulus of elasticity. Then one can simplify this to

$$\rho\frac{\partial^2\theta}{\partial t^2} = \mu\frac{\partial^2\theta}{\partial x^2} \tag{2.10}$$

which has the same *type* of solution, but the velocity, denoted by β, is given by

$$\beta = \sqrt{\frac{\mu}{\rho}} \tag{2.11}$$

Lamb (1934) showed that all plane waves in an isotropic elastic solid propagate with one or the other of these two speeds.

A third case can occur when the material rod does not have a rigid boundary but is allowed to expand or contract as necessary. Young's modulus E is then the relevant modulus of elasticity, and the velocity in the rod is

$$C_R = \sqrt{\frac{E}{\rho}} \qquad (2.12)$$

This does not have much use in seismology per se, but some models are made of rods of materials like rocks, and then E has to be used.

If we define a quantity σ, called Poisson's ratio, as the ratio

$$\frac{\text{cross section of linear dimension reduction}}{\text{longitudinal extension}}$$

there are four elastic quantities which need to be related. A further one, λ, is called Lamé's constant (Lamé used μ also), and these arise in the more formal mathematical theory of elasticity. For those who may need to read other literature the quantities are related as follows.

$$K = \lambda + \frac{2}{3\mu} = \frac{E\mu}{3(3\mu - E)} = \lambda\left(\frac{1 + \sigma}{3\sigma}\right) = \mu\frac{2(1 + \sigma)}{3(1 - 2\sigma)}$$

$$E = \frac{\mu(3\lambda + 2\mu)}{\lambda + \mu} = 9K\left(\frac{K - \lambda}{3K - \lambda}\right) = \frac{9K\mu}{3K + \mu} = \lambda\frac{(1 + \sigma)(1 - 2\sigma)}{\sigma}$$

$$\sigma = \frac{\lambda}{2(\lambda + \mu)} = \frac{\lambda}{3K - \lambda} = \frac{3K - 2\mu}{2(3 + \mu)}$$

$$\alpha = \sqrt{\frac{\lambda + 2\mu}{\rho}} = \sqrt{\frac{K + 4/3\mu}{\rho}} = \text{compression wave velocity} \qquad (2.13)$$

$$\beta = \sqrt{\frac{\mu}{\rho}} \qquad\qquad\qquad = \text{shear wave velocity}$$

$$K = \rho\left(\alpha^2 - \frac{4\beta^2}{3}\right)$$

$$\lambda = \rho(\alpha^2 - 2\beta^2)$$

Earlier we chose as our example a plane wave that propagates in a single direction. It is necessary, however, for waves to be considered that have other geometrical modes of propagation—for example, those that expand outward from a point source. If the material has properties that do not change from point to point and are not dependent on the direction of wave propagation, the symmetry is such that the waves travel outward as concentric spheres, simply expanding in radius r with time. Intuition should then tell us that, combining arguments made earlier, we can write the wave solution in the form

$$\phi_r(r, t) = \frac{1}{r}\left[F(r - \alpha t) + f(r + \alpha t)\right]$$

TABLE 2.1

SEISMIC VELOCITIES IN UNCONSOLIDATED SEDIMENTS, CONSOLIDATED SEDIMENTS, AND METAMORPHIC ROCKS

Material	Velocity (km/sec) α	β	Velocity (1000 ft/sec) α	β	Remarks
Alluvium	0.5–2.0	—	1.64–6.56	—	Near surface
	3.0–3.5	—	9.84–11.48	—	2000 m depth
Clay	1.1–2.5	—	3.61–8.20	—	
Loam	0.8–1.8	—	2.62–5.91	—	
Loess	0.3–0.6	—	0.98–1.96	—	
Sand					
Loose	0.2–2.0	—	0.66–6.56		
Loose	1.0	0.4	3.28	1.31	Above water table
Loose	1.8	0.5	5.91	1.64	Below water table
Calcareous	0.8	—	2.62	—	
Wet	0.75–1.5	—	2.46–4.92	—	
Weathered layer	0.3–0.9	—	0.98–2.95	—	
Glacial					
Till	0.43–1.04	—	1.41–3.41	—	Unsaturated
Till	1.73		5.67		Saturated
Sand and gravel	0.38–0.50	—	1.25–1.64	—	Unsaturated
Sand and gravel	1.67		5.48		Saturated
Sandstone-shale					
Tertiary	2.1–3.5		6.89–11.48		
Cretaceous	2.4–3.9		7.87–12.80		Depth range
Pennsylvanian	2.9–4.4		9.51–14.44		0.3–3.6 km
Ordovician	3.3–4.5		10.83–14.76		0.3–2.1 km
Sandstone	1.4–4.3		4.59–14.11		
Sandstone					
Conglomerate	2.4		7.87		Australia
Limestone					
Soft	1.7–4.2		5.58–13.78		
Hard	2.8–6.4		9.19–21.00		
Solenhofen	5.97	2.88	19.59	9.45	
U. S. midcontinent	—	2.75		9.02	
and Gulf Coast	3.4–6.1	—	11.15–20.01		
Argillaceous, Texas	6.03	3.03	19.78	9.94	‖ to bedding
Argillaceous, Texas	5.71	3.04	18.73	9.97	⊥ to bedding
Dolomitic, Penn.	5.97	—	19.59		
Cement rock, Penn.	7.07	—	23.20		
Crystalline, Texas,					
N.M., Okla.	5.67–6.40	—	18.60–21.00	—	
Dense, U.S.S.R.	5.90–7.00	3.03–3.59	19.36–22.97	9.94–11.78	
Salt, cornallite, sylvite	4.4–6.5	—	14.44–21.38		
Caprock, salt,					
anhydrite, gypsum,					
limestone	3.5–5.5	—	11.48–18.04		
Anhydrite,					
midcontinent and					
Gulf Coast	4.1	—	13.45		

18

Table 2.1 *Continued*

Material	Velocity (km/sec)		Velocity (1000 ft/sec)		Remarks
	α	β	α	β	
Gypsum	2.0–3.5	—	6.56–11.48		
Chalk, U.S., Germany, France, Austin, Texas	2.1–4.2	1.07 SV	6.89–13.78	3.51 SV	
	2.58	1.13 SH	8.46	3.71 SH	⊥ bedding
	3.05	—	10.01		‖ to bedding
Slate, Mass.	4.27	2.86	14.01	9.38	
Hornfels slate	3.5–4.4	—	11.61–14.44	—	
Magnetite ore	5.50	—	18.04	10.81–10.49	$V_P/V_S = 1.67$–1.72
Marble	3.75–6.94	2.02–3.86	12.30–22.77	6.63–12.66	46 samples
	5.78	3.22	18.96	10.56	Average of 46 samples
Quartzite	6.1	—	20.01	—	
Wet clay, U.S.S.R.	1.50–1.65	—	4.90–5.41	0.36–1.20	$V_P/V_S \sim 4.5$–13.7
Impermeable argillaceous clay	2.00	0.59	6.56	1.94	
Soil	0.11–0.20	—	0.36–0.66	0.59–1.27	$V_P/V_S \sim 1.7$–2.0
Tuff	2.16	0.83	7.09	2.72	

Source. These values have been selected from the compilation by Frank Press in the *Handbook of Physical Constants*, rev. ed., Memoir 97, and are printed with permission of the Geological Society of America. Copyright © 1966.

which is for compressional waves traveling outward from and inward toward the source. In most problems, but not all, the inward traveling waves can be neglected, and we have but one term:

$$\phi_r(r, t) = \frac{1}{r} F(r - \alpha t) \tag{2.14}$$

For the reason for changing from u to ϕ, see Appendix 2B.

Unfortunately, spherically traveling shear waves cannot be treated with quite the same simple type of formula, because in order to generate a shear wave shear traction must be applied to the medium—about an *axis*. This immediately involves considerations of vectors, and the displacement is no longer the same in all directions from the very small source. In the simplest case (a couple about a single axis), the results have cylindrical symmetry about this axis. However, we do not have need for such a development. Any readers interested should consult the books already cited.

This is a suitable time to remark philosophically on the role of the mathematical theory of elasticity in reflection seismology. Formal solutions to wave equations, under even simple boundary conditions, involve very sophisticated mathematical considerations. Fortunately, the solutions found for such cases (which sometimes involve only one or two boundaries) have provided guidelines as to the types of waves generated, their behavior inside the elastic media and at the boundaries, and ideas as to the distribution of energy between the various types of waves and the variation in the wave fronts. The importance of this kind of information cannot be over-emphasized, and it is given—as the need arises—without proof but usually with a reference for further study.

Field examples are usually so complex that formal solutions are impossible and only approximate answers can be found. To proceed with these approximate solutions, the field model can be simplified by (a) assuming plane parallel layering, (b) stipulating one wave type only, or (c) adopting an iterative type of computer solution which is usually limited to one or two dimensions and integrates the original differential equations (or the difference equation approximations to them). Most of these approximations are used later.

Compressional and shear wave velocities of rocks depend on many parameters. Much information on compressional wave velocities is now available for specific areas in the form of continuous velocity logs taken in drilled holes. Table 2.1 lists typical values only. It is not yet possible to obtain shear wave velocities accurately for rocks *in situ* by direct measurements in wells. The shear wave velocities, as well as most of the compressional wave velocities, listed were mostly obtained on small (several-centimeter dimensions) specimens, using very high-frequency sources and receivers, in the laboratory. The validity of these measurements is discussed later.

2.6 PARTITION OF ENERGY BETWEEN DIFFERENT WAVES AT BOUNDARIES

Now that it has been established that two different kinds of waves can be propagated in solids, we can see that the effect of boundaries on wave propagation becomes much more complex. Not only are there reflected and refracted waves generated at the boundaries, but each of these can be either compressional waves or shear waves. A generalized form of Snell's law still holds as far as the directions of the different waves are concerned, namely (see Figure 2.8),

$$\frac{\sin \phi_i}{\beta_1} = \frac{\sin \theta_i}{\alpha_1} = \frac{\sin \theta_r}{\alpha_1} = \frac{\sin \phi_r}{\beta_1} = \frac{\sin \theta_t}{\alpha_2} = \frac{\sin \phi_t}{\beta_2} \tag{2.15}$$

where subscripts 1 and 2 denote the medium and subscripts $i, r,$ and t stand for incident, reflected, and transmitted, respectively. These relations are easily proved by wave front construction near a plane boundary, as previously illustrated for a simple case in Figure 2.5.

In the figure the direction of particle motion of the shear waves is shown by an arrow with a sign. This establishes that the shear wave, which has particle motion perpendicular to the direction of propagation, has motion in the plane of the diagram. Such waves are given an auxiliary designation SV in that their particle motion lies in a plane perpendicular to the boundary and containing both the normal and the incident ray.

Shear waves with another polarity exist in which particle motion is perpendicular to the plane of the diagram. These are SH waves and, since their particle motion is parallel to the boundary, they suffer no change in type on reflection or refraction. Naturally these are the extremes, and a random shear wave striking a boundary has both SV and SH components.

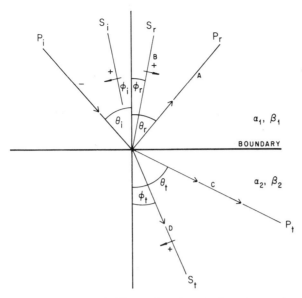

Figure 2.8 Angular relations between incident, reflected, and refracted rays when a wave encounters a boundary between two different solid media.

At perpendicular incidence, ϕ_i or $\theta_i = 0$, there is no conversion from P to SV or from SV to P but, as the angle of incidence increases, the conversion from one wave type to another becomes significant. The Zoeppritz equations determine the amplitudes of the reflected and refracted waves and are given in Appendix 2A. These equations are seldom used in calculations affecting exploration geophysics, in spite of the fact that modern seismic reflection methods use long offsets and involve significant angles of incidence. Normally, the velocity and density contrasts between layers are not large but, in special cases (e.g., high reflection coefficients caused by gas in high-porosity sands), longer offsets may be a problem.

The reflection coefficients normally used are those for perpendicular incidence. Figure 2.9 shows diagramatically the rays associated with either an S or a P wave at perpendicular incidence. The boundary conditions are simple. The boundary must move by the same amount whether it is considered the boundary of medium 1 or medium 2. (The displacement must be continuous across the boundary.) Also, the net stress on the boundary must be zero. (The boundary itself has no mass, and any stress leads to infinite acceleration.)

A plane wave traveling in the positive z direction can be represented by

$$\xi(z, t) = AF(z - ct)$$

We will not violate any rules if we choose to consider a constant frequency wave (angular frequency ω) which may be represented by

$$\xi(z, t) = A\varepsilon^{i \cdot 2\pi(kz - \omega t)}.$$

This is a commonly used form, and it is seen later that it is possible to combine sinusoidal waves of this type but having a range of frequencies and time relations to

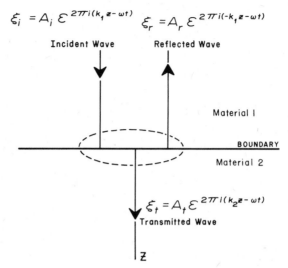

Figure 2.9 Showing the incident, reflected, and refracted rays used for calculating the reflection coefficient. All rays are actually along the normal to the surface. The incident and reflected rays are displaced from one another for clarity.

one another in order to develop any physically possible form F. k is called the wave number, and $\omega/k = c$. Since c = frequency × wavelength = $f\lambda$ and since the angular frequency (radians/sec) = $2\pi f$, we see that $k = 1/2\pi\lambda$, a scaled inverse wavelength.

In this diagram, the incident and reflected rays have for clarity been drawn separately, but they actually lie along the same normal to the boundary.

The two boundary conditions give rise to two equations:

$$A_i + A_r = A_t \tag{2.16a}$$

and

$$\rho_1 c_1 A_i - \rho_1 c_1 A_r = \rho_2 c_2 A_t \tag{2.16b}$$

making use of $F = K(\partial u/\partial x)$, and from these two it is easy to obtain

$$\frac{A_r}{A_i} = \frac{\rho_1 c_1 - \rho_2 c_2}{\rho_1 c_1 + \rho_2 c_2} = \text{reflection coefficient } R_{12} = \frac{Z_1 - Z_2}{Z_1 + Z_2} \tag{2.17}$$

where $Z_1 = \rho_1 c_1$ and $Z_2 = \rho_2 c_2$, the acoustic impedances of the two media.

Similarly, the transmission coefficient $A_t/A_i = T_{12}$ can be shown to be given by

$$T_{12} = \frac{2Z_1}{Z_1 + Z_2} \tag{2.18}$$

These simple formulas are independent of frequency, which will be evident if the two equations of continuity are derived from the expressions for the individual waveforms. If $Z_2 > Z_1$, corresponding to the wave going from a lower-velocity, lower-density rock to one with higher parameter values, the reflection coefficient—for *amplitudes*—is negative, and the transmission coefficient is positive.

Thus the displacement of the reflected wave is in a direction opposite that of the incident wave. However, the direction of travel of this wave is also reversed, so that *relative to its travel direction* the particle displacement has the same sign as the incident wave.

Since the particle velocity is a constant times the particle displacement (for a given frequency), it also has the same reflection coefficient, and it obeys the rules given above as far as particle velocity is concerned; for the situation described above, $\dot{\xi}_i = -c\xi_i$ for the downward traveling wave but $\dot{\xi}_r = c\xi_r$ for the upward traveling wave.

In an acoustic wave, the pressure obeys the same numerical formula for reflection coefficient but, since it is a number equal to $-\rho c\, \partial\xi/\partial t$, the reflected pressure is actually the negative of that given by (2.17).

In the case of reflection from the surface between water and air, the displacement has a reflection coefficient almost equal to 1 (incidence from water to air $Z_1 > Z_2$). Thus the sum of the incident and reflected displacements near such a boundary is nearly twice that of the incident wave. This is also true for particle velocities. However, the pressure of the reflected wave cancels the pressure of the incident wave. In other words, the air-water surface is a pressure-release surface. For this reason, no measurements of excess pressure are made near the surface in marine prospecting—but, provided that noise from water waves can be diminished successfully, particle velocity measurements are double those in the body of the liquid and can be used for detecting acoustic energy.

Since the energy is proportional to $\rho c\dot{\xi}^2$, it can be shown that it is conserved with these reflection and transmission coefficients.

The input energy is $\frac{1}{2}\rho_1 c_1 \dot{\xi}_i^2$ (note the analogy to $\frac{1}{2}mv^2$ for solid bodies.)

The output energy is the reflected energy $\frac{1}{2}\rho_1 c_1 \dot{\xi}_r$ plus the transmitted energy $\frac{1}{2}\rho_2 c_2 \dot{\xi}_t^2$:

$$\frac{1}{2}\rho_1 c_1 \dot{\xi}_r^2 \left(\frac{\rho_1 c_1 - \rho_2 c_2}{\rho_1 c_1 + \rho_2 c_2}\right)^2 + \frac{1}{2}\rho_2 c_2 \dot{\xi}_t^2 \left(\frac{2\rho_1 c_1}{\rho_1 c_1 + \rho_2 c_2}\right)^2 = \frac{1}{2}\rho_1 c_1 \dot{\xi}_i^2$$

It should be noted that the *amplitude* of a transmission coefficient $1 - R$ can be greater than unity when the reflection coefficient is negative. This does not violate the conservation-of-energy principle. Note also that, in *two-way* passage through a boundary, the amplitude is always diminished, since the transmission both ways must result in multiplication by $(1 - R)(1 + R) = 1 - R^2$, which is always less than unity. The implications of this are examined in more detail when transmission and reflection in a many-layered system are considered.

2.7 IMPERFECT ELASTIC MEDIA—ATTENUATION OF SEISMIC WAVES

Observations of the propagation of stress waves in solid or fluid media show that dissipation of strain energy occurs even when the waves have a small amplitude. This dissipation results from imperfections in elasticity within the body, and it occurs in *addition to* loss by radiation, geometrical spreading, or scattering.

Fortunately, the deviations from Hooke's law are usually small, and the usual approximations of elastic wave theory are sufficiently accurate for most materials if the pressure, temperature, and frequency are not varied too widely.

TABLE 2.2

INTERNAL FRICTION IN ROCKS

Material	Frequency (kHz)	dE/E	Q	Notes
Basalt	3–4	0.0112	561	20°–900° Average
Diorite (Utah)	10.7	0.035	179	Average of 2
Granite	20–200	0.0202	311	Average of 9
Granite	5.35	0.03	2.09	
Granite, Westerly, R.I.	100–800	0.09807	64.1	Average of 9 separate frequencies
Marble		0.01148	547	7 temps. 20°–1000°C
Quartzite	3–4	0.01603	392	15 temps. 20°–1000°C
Slate		0.02864	219	Average of 12 samples
Caprock	0.12–6.60	0.13375	46.98	Average of 12 samples and frequencies
Dolomite	10.44–15.00	0.03256	192.99	Average of 7 samples
Limestone	—	0.03093	203.12	Average of 12 samples
Chalk	0.002	0.046	136.6	
Hunton limestone	10.55	0.106	59.28	
	2.82	0.088	71.4	
Oolitic limestone	0.002–0.12	0.1350	46.54	Average of 11 frequencies
Shelly limestone	0.002–0.12	0.0915	68.63	Average of 11 frequencies
Solenhofen limestone	0.00389–18.6	0.00925	679.26	Average of 6 frequencies
Amherst sandstone	0.55–3.83	0.2612	24.05	Average of 10 samples
Berea sandstone	20	0.048	130.9	Average of 2 samples
Homewood sandstone	2.68–4.92	0.0909	69.11	Average of 11 samples
Old red sandstone	0.002–0.040	0.0675	93.08	Average of 2 samples
Pierre shale	0.075–0.555	0.36366	17.15	Average of 17 samples
Sylvan shale	3.36–12.8	0.08667	72.5	Average of 3 samples

Source. These values have been selected from the compilation by James J. Bradley and A. Newman Fort, Jr., in the *Handbook of Physical Constants*, rev. ed., Memoir 97, and are printed with permission of the Geological Society of America. Copyright © 1966.

The term "internal friction" is frequently used for nonelastic mechanisms that convert strain energy into heat, thereby damping or attenuating the stress waves in a medium. Table 2.2 gives values for various kinds of rocks.

Friction arising from the sliding of one grain against another along the grain boundaries, or due to imperfections in individual crystals, is widely acknowledged as the most probable mechanism of energy loss. Experiments on seismic wave attenuation in specimens of rock at different humidity levels indicate that adsorption of moisture on the grain boundaries has an important effect. Several mathematically convenient models have been proposed, but none fit the data for all frequencies. Other vibrating systems which gradually lose energy of vibration by conversion into heat are discussed extensively in textbooks on mechanical or electrical engineering. They introduce the symbol Q to denote the degree of perfection of a tuned circuit.

In discussing the gradual loss of amplitude of a seismic wave traveling through an imperfect medium it should be noted that the change in amplitude is exponential, so that each cycle bears the same ratio of amplitude to the preceding one. This ratio is called the logarithmic decrement δ. The quantity $1/Q$ has been designated the specific dissipation constant. Generally speaking, except for unconsolidated, water-saturated sediments:

1. The specific dissipation constant $1/Q$ is essentially independent of frequency.
2. It appears that $1/Q$ is substantially independent of amplitude for strain below 10^{-4}. The strain in rocks due to the passage of seismic waves used for exploration purposes is usually no more than 10^{-8}, except within a few tens of meters of the source.
3. Observations show that dissipation is less for a single crystal than for an aggregate of such crystals.
4. The rate of dissipation decreases with increased pressure.
5. The existing data suggest that dissipation is relatively independent of temperature.

One method of measurement of attenuation is the percent of stored energy lost *per cycle*. Other measures in common use are:

$1/Q$ = Specific dissipation constant. It is related to the rate at which the mechanical energy of vibration is converted irreversibly into heat energy and does not depend on the detailed mechanism by which the energy is dissipated.

δ = Logarithmic decrement—the natural logarithm of the ratio of amplitudes of two successive maxima or minima in an exponentially decaying free vibration.

a = Damping amplitude coefficient in the expression for a free vibration:

$$\varepsilon^{-at} \sin 2\pi f t$$

α = Attenuation coefficient in the expression for plane waves in an infinite medium:

$$\varepsilon^{-\alpha x} \sin 2\pi f \left(t - \frac{x}{c} \right) \qquad c = \text{wave velocity}$$

$\Delta f/f$ = Relative bandwidth of a resonance curve between the half-power or 0.707 amplitude points for a solid undergoing forced vibrations is a measure of the sharpness of the resonance curve.

$\Delta E/E$ = Fraction of strain energy lost per cycle.

Then b, the specific damping capacity, is related to the other parameters by

$$b = \frac{2\pi}{Q} = 2\delta = 2\frac{a}{f} = 2c\frac{\alpha}{f} = 2\pi\frac{\Delta f}{f} = \frac{\Delta E}{E}$$

As examples of how these attenuations affect seismic data, we make use of the formula

$$\alpha = \frac{\pi f}{Qc}$$

to find out how far a plane wave (of frequency 30 Hz) travels before it is reduced to one-tenth of its original amplitude.

In Pierre shale $Q = 17.15$, so that

$$\alpha = \frac{3.14 \times 30}{17.15 \times 7000} = 0.000785$$

Then

$$\varepsilon^{-\alpha x} = \tfrac{1}{10}$$

$$\ln \tfrac{1}{10} = -2.3 = -0.000785x$$

$$x = 2929 \text{ ft}$$

Thus, by attenuation alone, the amplitude of a plane wave drops by a factor of 10^4 in traveling to a depth of 5858 ft and back.

If the travel material is a limestone of $Q = 200$ and $c = 20,000$ ft/sec,

$$\alpha_{\text{limestone}} = \frac{3.14 \times 30}{200 \times 20,000} = 0.00002355$$

and the corresponding x is

$$x_L = 90,766 \text{ ft}$$

and the attenuation becomes an insignificant factor in amplitude decay compared with the spherical spreading.

The above is a simplistic discussion of attenuation and applies only to a single homogeneous rock body. In the practice of exploration seismology, the presence of layering has a profound effect on the change in amplitude of a signal as it propagates through the earth layers. This is particularly noticeable when the usual (reflection) observations are made of a wave that has traveled in both directions through the layering. Much more discussion of this is found in later chapters.

2.8 BOUNDARY WAVES

The discussion so far has been about waves that travel through an unbounded medium which is homogeneous and isotropic, although it may be slightly imperfect so that elastic energy is continuously lost to heat. When, as in all real systems, boundaries are present, they tend to guide waves. The first such wave to be analyzed was described by Rayleigh and is called, after him, a Rayleigh wave. He has shown that a wave is propagated along a free plane boundary with the following characteristics:

1. It has a velocity along the surface that is independent of frequency and which is somewhat less than the shear velocity for the body waves in the medium. The actual speed is dependent on the ratio V_P/V_S (or on Poisson's ratio), but for a material of $V_P/V_S = 1.732$ (or Poisson's ratio $= \frac{1}{4}$), the speed of the Rayleigh wave is $0.9194V_S$, where V_P and V_S are, respectively, the velocity of P waves and S waves in the medium.
2. The motion of a particle on the surface is not linear but an ellipse, described such that, at the top of the ellipse, the particle is traveling toward the source. Such motion is called retrograde elliptical motion. The ellipse axes are along the surface and perpendicular to it, and it can be regarded as the motion resulting from two linear motions which are 90° out of phase.
3. The amplitude of motion decreases (exponentially) as the depth below the boundary increases. At the surface the vertical motion is about 1.5 times the horizontal motion but vanishes at a depth of 0.192 wavelengths and reverses sign below this. All these parameters are dependent on the physical characteristics of the rock, being given for $\sigma = 0.25$.

The theoretical ratios of the Rayleigh wave velocity C_R and the compressional and shear velocities, α and β, of the medium are shown in Figure 2.10, taken from Knopoff (1952).

It is of course not possible, because of the invariable presence of a weathered layer, to make measurements of surface motion due to this type of a Rayleigh wave under field conditions. There the velocity increases with depth below the surface and, in

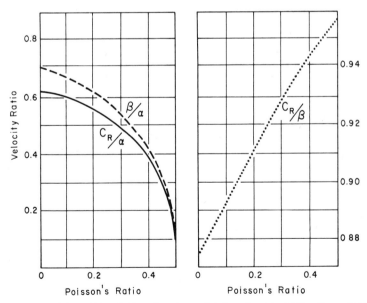

Figure 2.10 Ratios of β/α, C_R/α, C_R/β as functions of Poisson's ratio. [After Knopoff (1952), from Ewing et al., *Elastic Waves in Layered Media*. Copyright © 1957 by McGraw Hill, Inc. Used with permission of McGraw Hill Book Company.]

addition, the presence of an interface at the base of the weathered layer causes a disturbance of the simple conditions required. Dobrin, Lawrence, and Simon (1951) experimentally measured the particle trajectory variation with depth below the earth's surface for Rayleigh waves from small explosions and found that the motion was retrograde above 40 ft and prograde below. The words "retrograde" and "prograde" as used here refer to the direction of the particle motion of a surface wave. Retrograde refers to particle motion which, at the top of the ellipse, has a direction opposite that of the direction of wave travel, and prograde indicates the opposite particle motion. The crossover depth was 0.136 of a wavelength of the frequency used. These results are valid for the particular area investigated.

Figure 2.11 shows an actual particle trajectory compared with the theoretical particle trajectory and is taken from Howell (1959). The complexity is evident, but it must be remembered that the comparison is again that of an isotropic half-space with a real, complexly layered system. One reason for including this comparison is that it has been suggested that one means of canceling Rayleigh waves may be based on the elliptical retrograde motion, using a fraction of the horizontal motion suitably changed in phase to cancel the vertical motion. It seems unlikely that the real Rayleigh waves would yield to such simple treatment. This matter, however, is discussed in more detail later.

Rayleigh waves consist of a mixture of compressional and shear energy, the shear wave motion being SV, that is, polarized in a vertical plane through the line of propagation of the wave. There is no corresponding surface wave for horizontally polarized motion, that is, motion in a vertical plane perpendicular to the direction of wave propagation.

A guided wave with different characteristics is one guided by the lower boundary

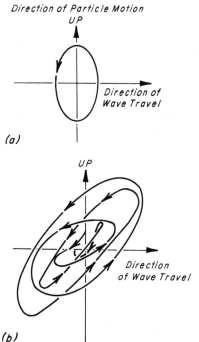

Figure 2.11   Particle trajectories from (*a*) a Rayleigh wave on a half-space of isotropic material and (*b*) a real Rayleigh wave. (After Howell, 1959.)

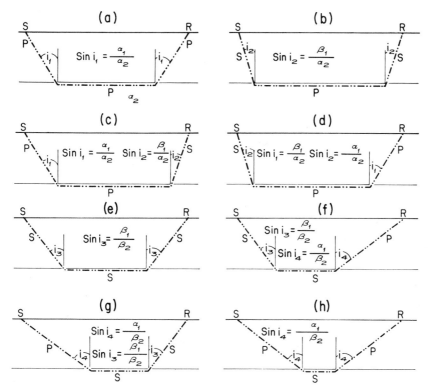

Figure 2.12 Possible refraction (or head wave) paths from the source to the receiver. Medium 1 has velocities α_1 and β_1, and medium 2, velocities α_2 and β_2. Note that, if $\alpha_1 > \beta_2$, modes (f), (g), and (h) are not possible. The cases are made more complex by a sloping lower interface or if the source is buried. (From Ewing et al., *Elastic Waves in Layered Media*. Copyright © 1957, McGraw Hill, Inc. Used with permission of McGraw Hill Book Company.)

of the weathered layer alone. In reality, there are three different types, some of which can be coproduced if the source is of the right type. They are variously called refracted waves or head waves, but in this case they travel along the boundary with the appropriate velocity (compressional or shear) for the lower medium. Their amplitudes decay exponentially with depth from the boundary, and they characteristically leak energy back into the upper layer so that they have an effect on surface measuring devices.

The paths are shown in Figure 2.12 and, as can be seen, they act as though they have been refracted into the second medium from a critically incident ray so that their paths are parallel to the boundary. An incident ray is called critical only with respect to one of the refracted (P or SV) rays. It is the ray for which the designated refracted ray has a refraction angle of 90°, in other words, it travels along the boundary. Thus, since a P wave always travels with higher speed in a medium than an SV wave, the P wave is bent more from the normal than the SV wave and the critical angle for the P wave is smaller than for the SV wave. Plane wave theory predicts zero amplitude for refracted waves traveling along the boundary. It is only when spherically diverging waves impinge on this boundary that theory shows that some energy is diffracted in the horizontal direction. Note that, to be complete, it is necessary to consider waves that change type at the boundary. Thus, for example, a wave can travel from the source to the boundary as SV, continue as P, and be reconverted to SV for upward travel to the receiver. This would be designated SPS, but SSS, PPP, SSP, PSS, PSP, SPP, and PPS are also possible, some of them obviously arriving simultaneously if the depths at the source and at the receiver are equal.

While Rayleigh waves spread out only on a surface, the amplitude can be expected to diminish as $R^{-1/2}$. Once the refracted waves reach the refracting surface, they too spread out over a surface. However, contrary to expectation, it can be shown (Heelan, 1953) that their amplitudes decay as R^{-2}, so they rapidly die out—not sufficiently rapidly, however, that they are not a considerable nuisance when reflection waves are being recorded. This comparatively rapid rate of decay is accounted for by the fact that the refractor continually broadcasts energy into the upper formation.

2.9 SURFACE WAVES RESULTING FROM ENERGY TRAPPED IN THE NEAR-SURFACE LAYER—LOVE WAVES

Because of their importance in seismic reflection recording, we show some of the physical characteristics of waves that are one stage more complex than those in the previous two classes. For a full mathematical treatise on these waves, which are Rayleigh and Love waves formed by energy trapped in a surface layer, the reader is advised to consult Ewing, Jardetsky, and Press (1957). The treatment here is of the physical plausibility of the existence of such waves, together with illustrations of waves actually encountered in exploration practice.

It is easy to see that, in a layered situation, such as that shown in Figure 2.13a, rays emanating from source S are refracted into the lower half-space, and thus a proportion of the energy is lost from the layer if a ray strikes the lower boundary at less than the critical angle. For an incident SH wave, which is not converted at the boundary into any other type, the critical angle is given by

$$\theta_c = \text{arc} \sin \frac{\beta_1}{\beta_2}$$

For an incident P wave, there is still leakage of shear energy until

$$\theta_c = \text{arc} \sin \frac{\alpha_1}{\beta_2}$$

And for an incident SV wave,

$$\theta_c = \text{arc} \sin \frac{\beta_1}{\beta_2}$$

Since such waves, which are incident at angles less than the relevant critical angle, lose energy, multiple reflections such as those shown in Figure 2.13b gradually decay in energy, and a true trapped wave does not exist. All waves in the above category are classed as "leaky" modes and of course contribute to the energy available for reflection work.

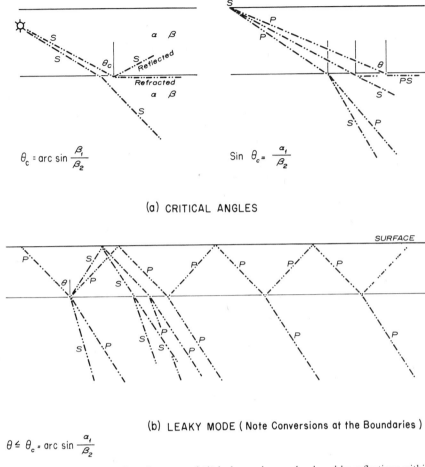

$$\theta_c = \text{arc} \sin \frac{\beta_1}{\beta_2}$$

$$\sin \theta_c = \frac{\alpha_1}{\beta_2}$$

(a) CRITICAL ANGLES

(b) LEAKY MODE (Note Conversions at the Boundaries)

$$\theta \leqq \theta_c = \text{arc} \sin \frac{\alpha_1}{\beta_2}$$

Figure 2.13 (*a*) Critically reflected waves and (*b*) leaky modes are developed by reflections within the layer at angles less than the critical angle. The latter case is not completely developed to preserve some clarity; however, all reflections from the boundaries should be treated in the same manner.

However, at angles beyond the critical angle and at distances from the source such that the wave fronts are appreciably plane, an interesting phenomenon of interference takes place, which is sensitive to frequency. The obvious complication introduced by the conversion from P partially to S at each reflection and from S partially back to P can be deferred if we look first at the simpler case of SH waves—in which case, the trapped waves are called Love waves. These are the large-amplitude surface waves whose presence makes shear wave (SH) reflection records difficult to obtain.

In Figure 2.14, a multiply reflected ray *ABCD* is drawn for which the angle of incidence θ is greater than the critical angle. It is postulated that we are looking at this ray at a distance from the source such that the wave fronts are essentially plane, and we can disregard amplitude change of the wave by spherical divergence (or by other forms of attenuation). Only one ray has been selected to look at, but it must be borne in mind that there is an infinite number of them, parallel to one another. *E* then represents one position of a wave front on the ray *AB* and, as *E* moves, *A* sweeps along the

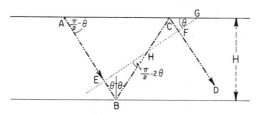

Figure 2.14 Illustrating conditions for generating a Love wave in a surface layer by multiple reflections of SH waves.

surface. It is easy to see that, if the wave moves through the upper layer with velocity β_1, the wave front moves along the surface with the (faster) phase velocity

$$C_s = \frac{\beta_1}{\cos(\pi/2 - \theta)} = \frac{\beta_1}{\sin\theta}$$

For waves of constant frequency, there are points E and F (on the same wave front) which interfere constructively if the distance $EBCF$ is an integral multiple of a wavelength (any phase changes at boundaries must be included in this assessment). For SH waves, there is no change in phase at the free surface.

The path length $EBCF$ can therefore be written in terms of the thickness of the layer, the angle θ, and an equivalent length due to the phase change at the lower boundary $L_0(\theta)$:

$$EBCF = \frac{H}{\tan\theta}\left[1 + \sin\left(\frac{\pi}{2} - 2\theta\right)\right] + L_0(\theta)$$

$$n\lambda = \frac{H}{\tan\theta}(1 + \cos 2\theta) + L_0(\theta)$$

$$\frac{2\pi n\beta_1}{\omega} = \frac{H}{\tan\theta}(1 + \cos 2\theta) + L_0(\theta)$$

The value of $L_0(\theta)$ can be found by application of the boundary conditions at the lower surface of the layer, hence various values of θ can be obtained that satisfy this equation for different values of n (the mode number) and ω (the angular frequency).

As the values of θ vary with ω, it follows that the phase velocity $C_s = \beta_1/\sin\theta$ also changes with the frequency. This is the first characteristic of surface layer-guided waves of all types. It is seen later that this leads to dispersion, that is, broadening of a pulse so that it becomes a long train of waves after traveling some distance.

A second characteristic is that constant-frequency waves set up nodal planes within the layer (i.e., they are standing waves as far as the vertical direction through the layer is concerned). This is obvious from the fact that, if the paths are such that the effective length $BCDE$ is a full wavelength, there is some point along this length where there is a null point. The actual point depends on the phase change at the lower boundary, but the situation is similar to that of an open organ pipe (except that the closed opposite end is replaced by a partial opening). In Figure 2.15, the amplitudes of motion are shown for various modes.

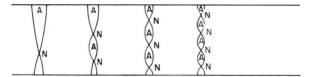

Figure 2.15 Possible distributions in a vertical direction of the amplitude of a Love wave. The actual position of the node is dependent on the mode frequency and wave number, since the phase change at the lower boundary changes with frequency. All modes have antinodes at the free surface and can therefore be generated efficiently by surface sources.

Additional discussion of the Love wave case is necessary in order to firm up some ideas with respect to the significance of the terms "phase velocity" and "group velocity." If a pulse travels in a manner that allows it to maintain its shape, one can associate a velocity with it which applies to both the velocity of motion of any particular part of the pulse (say, a particular peak) and also to the velocity of the energy through the medium. However, for a medium that allows dispersive propagation, the different constituent frequencies travel with different speeds. The term "phase velocity," as used here, really refers to the effective velocity along the ground surface of a particular frequency. If observations are made at points sufficiently together, the velocity is simply the change in time of arrival of a particular phase divided into the distance between observation points:

$$c = \frac{d}{\Delta t} = \frac{2\pi f d}{\theta} = \frac{\omega d}{\theta}$$

where θ = phase angle measured for the frequency f.

In a pulse, this phase angle as a function of frequency is impossible to see but, in the case of surface waves, the dispersion often results in long trains of waves, and a portion of the wave train can often be selected and the correlation at two or more sampling points measures the phase velocity. Moreover, when sources of constant frequency (vibrators) are available, the comparison of phases at two points is often a convenient method of measuring phase velocity.

"Group velocity" can be defined as the velocity with which the energy associated with a narrow band of frequencies has traveled. Obviously, if, as in explosion seismology, the time of energy release at the source is known and if the arrival time of the band of frequencies of interest at the receiver is known, the group velocity is just the distance from the source divided by the time of arrival. It is also a function of frequency, the velocity of the media, and the thickness of the formation. The group velocity u is related to the phase velocity c by

$$U = c + k\frac{dc}{dk}$$

where k = wave number = 1/wavelength λ.

Thus if the slope of the phase velocity–wave number curve is negative, the group velocity for a given wave number will be less than the phase velocity. The group

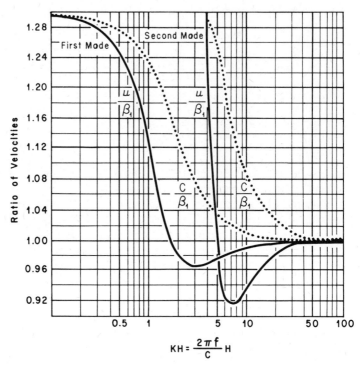

Figure 2.16 Phase and group velocity curves for first- and second-mode Love waves for the case $\beta_2/\beta_1 = 1.297$ and $\mu_2/\mu_1 = 2.159$. (From Ewing et al. *Elastic Waves in Layered Media.* Copyright © 1957 by McGraw Hill, Inc. Used with permission of McGraw Hill Book Company.)

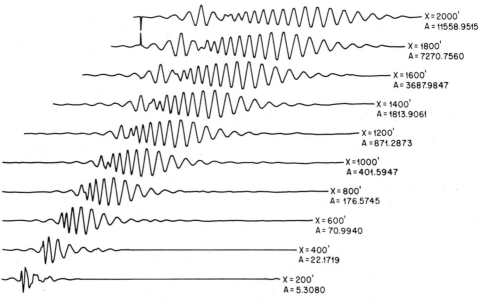

Figure 2.17 Love waves generated by a computer program which adds together all possible critically reflected rays. The dispersion with increasing offset is strikingly shown. By this method attenuation in the layer can be included. In this case $\beta_1 = 1400$ ft/sec, $\beta_2 = 4000$ ft/sec, $\rho_1 = \rho_2 = 2.5$, and $Q = 14$.

34

velocity sometimes exhibits a minimum, which means that, over a certain range of wave numbers, the waves all travel with the same velocity and arrive at the same time. From an energy point of view this is important, since it is a time of maximum surface wave amplitude. It is called, in earthquake seismology, the Airy phase.

Figure 2.16 shows some typical phase and group velocity curves for first- and seond-mode Love waves for a selection of velocities and elastic parameters. The vertical scale is a ratio of the velocity concerned to the shear wave velocity in the layer, while the horizontal scale is a logarithmic measure of the dimensionless parameter kH.

With the use of a digital computer, it is possible to program the amplitude and phase of each individual multiple reflection frequency arriving at a particular receiving point. Then, by a summation procedure, the total waveform of the Love wave can be synthesized and compared with that derived from normal mode analysis. The two compare very favorably. An example of a sequence of Love waveforms at different distances is given in Figure 2.17. In addition to normal considerations, it is possible this way to apply attenuation to each ray, both spherical divergence and change into heat. Attenuation with $Q = 20$ or less very effectively damps out the high frequencies and makes the true character of the Love wave easier to see.

2.10 SURFACE WAVES RESULTING FROM ENERGY TRAPPED IN THE NEAR-SURFACE LAYER (PSEUDO-RAYLEIGH WAVES)

The analysis of waves that propagate along the surface of an ideal earth—first made by Rayleigh—shows that such waves maintain the same pulse shape, only diminishing in amplitude with travel distance as a result of cylindrical spreading of the energy. However, when an imperfect earth is considered, the existence of one, or more, near surface layers having lower-than-normal velocities must be taken into account. As far as exploration seismology is concerned, the most important low-velocity layer is the weathered layer, which may include recently deposited alluvium or may be due to the alteration of previously competent rock by the action of rain or other agents which leach out soluble compounds and cause partial disintegration of the con-solidated rock.

The presence of a low-velocity layer profoundly influences the transmission of surface waves. Because some of the characteristics are changed (for instance, dis-persion, or broadening of the pulse with distance is now introduced) the surface waves originating from a compressional wave source are now called pseudo-Rayleigh waves.

Although the discussion of pseudo-Rayleigh waves follows the same pattern as that of Love waves, the added complication of conversion of waves between P and S at each boundary incidence now exists. Figure 2.18 shows two different path types which give rise to guided waves. The conditions always have to be such that the angles of incidence and reflection are such that all the energy is contained within the layer. There is no refracted energy, since the assumption is that these are plane waves—it does not hold close to the source, but experience has shown that, at offsets of five or six times the depth of the layer, the guided wave has already become stabilized.

Reflection of both SV and P waves from the free surface introduces a change in phase (in distinction from the SH case) which must be taken into account. Given the

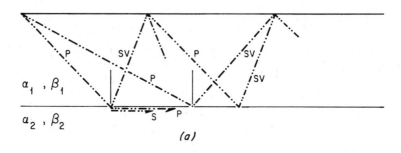

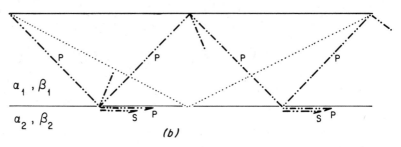

Figure 2.18 Illustrating some of the many ray paths and conversions that allow the formation of a pseudo-Rayleigh wave in a surface layer. (*a*) Conversion from P to SV or SV to P. (*b*) No conversion, all energy contained in the layer.

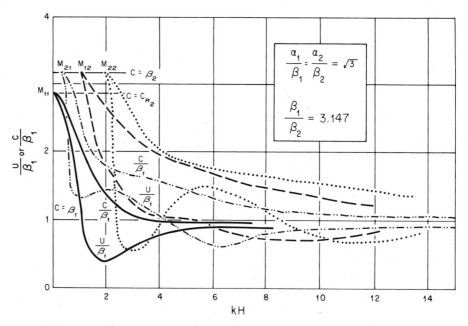

Figure 2.19 Dispersion curves for an elastic layer over a semi-infinite solid. The modes M_1 and M_2 correspond to the symmetrical and antisymmetrical modes of a plate, as modified by the attachment to a seimi-infinite medium. (From Ewing et al., *Elastic Waves in Layered Media*. Copyright © 1957 by McGraw Hill, Inc. Used with permission of McGraw Hill Book Company.)

relevant parameters for the layer and the half-space, dispersion curves can be drawn. There are now, however, more possibilities, since the conversion from compression to shear waves (or the reverse) must be considered and the curves become more complex, with many modes and branches—similar to those shown in Figure 2.19. The technique employed in the interpretation of surface waves is to measure the phase velocity as a function of frequency and to compare and match this experimental curve as nearly as possible to one or more theoretical curves. The number of variables involved makes this a laborious task and, as yet, there does not appear to be any way of working directly back from the dispersion curve(s) to a unique two-layer system. In the case of exploration technology, the surface waves can be only approximately accounted for by a two-layer system. Except for special cases, the parts of the dispersion curve of diagnostic interest are at very low frequencies $(1.0 \rightarrow 10\ Hz)$. Attempts have been made, without too much success, to use surface wave measurements for weathering calculations. There are several facts that make this process frustrating. Weathering materials are not constant over the distances needed, nor is the depth of the weathering, since more often than not the weathered layer cannot even approximately be regarded as a single layer.

For some exploration purposes, surface waves must be rejected to allow proper visibility of reflected waves. The fact that they usually have frequencies below the band (8 to 60 Hz) used for reflections has been used for filtering—either by employing

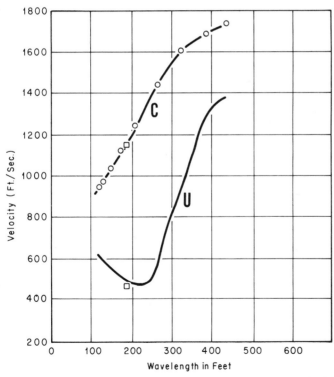

Figure 2.20 Group velocity (U) and phase velocity (C) of Rayleigh waves obtained from special exploration seismograms. The square points are data from air shots. (After Dobrin et al., from Ewing et al., *Elastic Waves in Layered Media.* Copyright © 1957 by McGraw Hill, Inc.

geophones which have a low response for such frequencies (the old approach) or by velocity filtering (q.v.) during data processing.

Finally, it must be said that the normal mode theory shows that the same kind of standing waves (in a vertical direction) are set up for pseudo-Rayleigh waves as for Love waves but, with the added complication of conversion between wave types, the nodes for different modes, branches, and frequencies are scattered throughout the layer. Any attempt to generate seismic waves when the source is in the weathered layer—or on the surface—results in the generation of large-amplitude guided waves as well as body waves used for reflection purposes.

Dobrin, Simon, and Lawrence (1951) applied the theory of Rayleigh wave dispersion using values for velocities and layer thickness found in borehole surveys. Figure 2.20 shows that fairly good agreement was found between the theoretical and experimental points.

2.11 SUMMARY

If a mechanical disturbance is caused in a body of ideal rocklike material, energy travels from the source throughout the medium in the form of waves. Two types of waves, compressional (P) and shear (S), can exist. They travel with different velocities corresponding to the appropriate elastic constants.

At boundaries between different types of rock, a generalized form of Snell's law applies, which regulates the direction of travel; and partial conversion, in both reflection and refraction, occurs between one wave type and another.

As the wave travels outward from the point source, the amplitude of the particle motion diminishes, since the energy available is spread out over the entire wave front. It is not necessary, however, that the energy density be constant over the wave front.

In imperfect rocklike materials, an additional loss of amplitude of the wave occurs as a result of the loss of energy due to heat (generated at imperfections in the medium). The specific dissipation constant $1/Q$ appears to be independent of frequency. The wave amplitude decreases exponentially with distance of travel, and the rate of attenuation is roughly proportional to the frequency in most rocklike materials.

In addition to body waves, it is possible for a Rayleigh wave to be generated, which is guided by a boundary between the earth and the air. This form of Rayleigh wave is distinguished by retrograde elliptical particle motion, constancy of pulse shape with distance of travel, exponential decay of amplitude below the surface, an amplitude decay with distance as $R^{-1/2}$, and a velocity slightly less than S velocity.

If a low-velocity layer is present, superimposed on a more consolidated rock half-space, guided waves are possible, the type depending on the mode of initiation of the energy. A Love wave is a result of the interference of different orders of multiple reflections whenever the latter strike the lower boundary at angles greater than the critical angle. Such a wave is generated by a source of SH motion and generates SH motion on the surface; the energy decays as $R^{-1/2}$ (because it is confined to a layer) but, since it is dispersive, the amplitudes of individual portions of the way decay faster than $R^{-1/2}$ (more as R^{-1}). The characteristics of these waves are summarized by the dispersion curves relating phase and group velocities to the wave number and layer thickness product.

Pseudo-Rayleigh waves exist in the layer when the excitation produces P or SV waves. The interference of multiply reflected waves (at greater than critical incidence) forms different modes, each of which has characteristic dispersion curves associated with it, giving the phase and group velocities as functions of the mode and branch number, the wavelength, the layer thickness, the P and S velocities, and the densities.

These pseudo-Rayleigh waves or ground roll constitute one of the main difficulties in the recording of seismic reflections, since they are usually of very large amplitude compared with the reflections. Since they are usually low-frequency and have low velocities along the surface, frequency filtering or other special forms of filtering can be used to reject them from the reflection energy.

APPENDIX 2A: ZOEPPRITZ EQUATIONS

Using the notation introduced in Figure 2.8, the equations for determining the amplitudes of reflected and refracted waves (both P and SV) from an incident P wave are:

1. Snell's law:

$$\frac{\sin \theta_1}{\alpha_1} = \frac{\sin \phi_1}{\beta_1} = \frac{\sin \theta_2}{\alpha_1} = \frac{\sin \phi_2}{\beta_2} \tag{2A.1}$$

2. Continuity of tangential displacement:

$$\sin \theta_1 + A \sin \theta_1 + B \cos \phi_1 = C \sin \theta_2 - D \cos \phi_2 \tag{2A.2}$$

where the P-wave displacement (A and C) is positive in the direction of propagation, and the S-wave displacement is positive to the right of the direction of propagation.

3. Continuity of normal displacement (direction):

$$\cos \theta_1 - A \cos \theta_1 + B \sin \phi_1 = C \cos \theta_2 + D \sin \phi_2 \tag{2A.3}$$

4. Continuity of normal and tangential stress yields two equations:

$$-\sin 2\theta_1 + A \sin 2\theta_1 + \frac{\alpha_1}{\beta_1} B \cos 2\phi_1 = \frac{-\rho_2 \beta_2^2 \alpha_1}{\rho_1 \beta_1^2 \alpha_2} C \sin 2\theta_2 + \frac{\rho_2 \beta_2 \alpha_1}{\rho_1 \beta_1^2} D \cos 2\phi_2 \tag{2A.4}$$

$$\cos 2\phi_1 + A \cos 2\phi_1 - \frac{\beta_1}{\alpha_1} B \sin 2\phi_1 = \frac{\rho_2 \alpha_2}{\rho_1 \alpha_1} C \cos 2\phi_2 + \frac{\rho_2 \beta_2}{\rho_1 \alpha_1} D \sin 2\phi_2 \tag{2A.5}$$

where A = amplitude of reflected P wave
C = amplitude of refracted P wave
B = amplitude of reflected SV wave
D = amplitude of refracted SV wave

These equations can be put in matrix form which is more suitable for computer solution:

$$
\begin{vmatrix}
\sin\theta_1 & \cos\phi_1 & -\sin\theta_2 & \cos\phi_2 \\
-\cos\theta_1 & \sin\phi_1 & -\cos\theta_2 & -\sin\phi_2 \\
\sin 2\theta_1 & \dfrac{\alpha_1}{\beta_1}\cos 2\phi_1 & -\dfrac{\rho_2\beta_2^2\alpha_1}{\rho_1\beta_1^2\alpha_2}\sin 2\theta_2 & \dfrac{\rho_2\beta_2\alpha_1}{\rho_1\beta_1^2}\cos 2\phi_2 \\
\cos 2\phi_1 & -\dfrac{\beta_1}{\alpha_1}\sin 2\phi_1 & -\dfrac{\rho_2\alpha_2}{\rho_1\alpha_1}\cos 2\phi_2 & -\dfrac{\rho_2\beta_2}{\rho_1\alpha_1}\sin 2\phi_2
\end{vmatrix}
\begin{vmatrix} A \\ B \\ C \\ D \end{vmatrix}
=
\begin{vmatrix} -\sin\theta_1 \\ -\cos\theta_1 \\ \sin 2\theta_1 \\ -\cos 2\phi_1 \end{vmatrix}
$$

$$\mathbf{P} \qquad\qquad\qquad \mathbf{Q} \;=\; \mathbf{R}$$

and the solution is

$$\mathbf{Q} = \mathbf{P}^{-1}\mathbf{R} \tag{2A.6}$$

All these expressions are relative to an input P-wave amplitude of unity and are given in R. E. Sheriff, "Encyclopedia Dictionary of Exploration Geophysics" as amended in *Geophysics*, Supplement to Vol. 40, No. 2.

Tooley, Spencer, and Sagoci (1965) have given curves for the energies associated with the various reflected and refracted waves for all angles of incidence and various material contrasts.

Beyond the critical angles for P and S waves, the respective refracted waves vanish. (These formulas hold only for plane waves.) Therefore they do not allow calculation of head wave amplitudes which arise from spherical waves incident on a boundary. Such refracted energy is explainable only by the decomposition of spherical waves into plane waves with *both real and complex angles of incidence*, or by a formal solution to the spherical wave problem.

The increase in reflection energy near the critical angle is sometimes exploited in seismic surveying. However, beyond the critical angle, a phase shift is introduced in the reflected wave, which must be taken into account. Similar equations exist for incident SV waves. This subject has been extensively treated by Tooley, Spencer, and Sagoci (1965) and in other literature. Unfortunately, sign conventions differ, and many mistakes in sign have been made so care has to be taken in selecting *results* without checking. Probably the most satisfactory way to check is to take a few special cases for which the results are known and use the Zoeppritz equations to predict these results. Ewing, Press and Jardetsky (1957) obtains the general equations on p. 76, unfortunately with a different set of axes and a different notation.

This appendix has been included mainly to show how complex the reflection process becomes for large angles of incidence. Remember that these relations apply at every interface encountered by a wave as it propagates through a medium. With the tendency, which is seen later, for some reflection energy to be used that has been reflected at large angles of incidence, some attempt should be made to assess the importance of this continual conversion from one wave type to another.

APPENDIX 2B: SPHERICALLY SYMMETRICAL ACOUSTIC WAVES

Earlier in this chapter, in (2.14), it was noted that, when dealing with spherically diverging or converging acoustic waves, it is necessary to change from the designation u—a particle displacement which obeys the plane wave equation—to a different quantity, ϕ. It should be noted that, for plane waves, other quantities that obey the same equation are u (the particle velocity), p (the excess pressure), and ϕ (the velocity potential).

ϕ is related to the particle displacement by the relations

$$u = \frac{\partial \phi}{\partial x} \qquad \text{(displacement in the } x \text{ direction)}$$

$$v = \frac{\partial \phi}{\partial y} \qquad \text{(displacement in the } y \text{ direction)}$$

$$w = \frac{\partial \phi}{\partial z} \qquad \text{(displacement in the } z \text{ direction)}$$

or, for spherical symmetry,

$$u_r = \frac{\partial \phi}{\partial r} \tag{2B.1}$$

where r = radial distance

Now it can be shown that $r\phi$ satisfies the acoustic wave equation in spherical coordinates so that

$$u_r = \frac{\partial \phi}{\partial r} = \frac{\partial}{\partial r}\left[\frac{1}{r}F(r - ct)\right]$$

for outward traveling waves from a spherically symmetrical source. Pursuing this one stage further

$$u_r = \frac{F'(r - ct)}{r} - \frac{F(r - ct)}{r^2} \tag{2B.2}$$

There are therefore two different terms of which the first is important at long distances and the other is more important close to the source. These two forms cross over in importance at a distance about equal to the wavelength of the acoustic wave. Note also that the first term $F'(r - ct)$ is differentiated with respect to distance, while the second is not. This action of taking the derivative causes the higher frequencies to be accentuated in solutions further from the source.

Wave propagation for elastic waves is much more complex and involves a vector potential ψ as well as a scalar potential ϕ. Details of the complete treatment should be sought in Ewing, Jardetsky, and Press (1957).

In reflection seismology, it is usually the pulse shape at long distances from the source that is important. At the same time, it is usually more convenient (as in the case of marine exploration) to make measurements on the pulse shape within the water layer and close to the source. The pulse measured should therefore not be construed as being the same as the pulse shape involved in deep reflections, unless a suitable transformation is applied.

REFERENCES

Dix, C. H. (1955), "The Mechanism of Generation of Long Waves from Explosions," *Geophysics*, Vol. 20, No. 1, pp. 87–103.

Dobrin, M. B., Simon, R. F., and Lawrence, P. L. (1951), "Rayleigh Waves from Small Explosions," *Transactions of the American Geophysical Union*, Vol. 32, pp. 822–832.

Ewing, W. M., Jardetsky, W. S., and Press, F. (1957), *Elastic Waves in Layered Media*, McGraw-Hill, New York.

Garland, G. D. (1971), *Introduction to Geophysics, Mantle, Core and Crust*, W. B. Saunders, Toronto.

Heelan, P. A. (1953). "Radiation from a Cylindrical Source of Finite Length." *Geophysics*, Vol. 18, pp. 685–696.

Heelan, P. A. (1953), "On the Theory of Head Waves," *Geophysics*, Vol. 18, pp. 871–893.

Howell, B. (1959), *Introduction to Geophysics*, McGraw-Hill, New York.

Knopoff, L. (1952), "On Rayleigh Wave Velocities," *Bulletin of the Seismological Society of America*, Vol. 42, pp. 307–308.

Lamb, H. (1934), *The Mathematical Theory of Elasticity*, Cambridge University Press, Cambridge.

Tooley, R. D., Spencer, T. W., and Sagoci, H. F. (1965), "Reflection and Transmission of Plane Compressional Waves," *Geophysics*, Vol. 30, No. 4, pp. 552–570.

Tolstoy, I., and Usdin, E. (1953), "Dispersive Properties of Stratified Elastic and Liquid Media: A Ray Theory," *Geophysics*, Vol. 18, No. 4, pp. 844–870.

THREE

Sources and Receivers

3.1 GENERAL TYPES OF SOURCES OF ELASTIC ENERGY

In nature, elastic waves are generated whenever a sudden change in stress occurs in the earth—usually in the form of a stress change due to strain relief by faulting. The tectonic forces acting on the earth's crust build up strain energy, usually both compressive and shear, until the elastic limit of the material is exceeded. The change in strain energy usually implies a change both in the gravitational potential energy of the rock masses as well as in the heat energy supplied to the rock in the vicinity of the fault and energy radiated in the form of various kinds of elastic waves.

While the use of these natural sources of elastic energy supplies some general knowledge of the earth's structure, such sources are generally unsuitable for more detailed investigations. Some of these detailed investigations have to take place beneath the land, and others below the shallow seas overlying the continental shelves. This is no artificial division, as it turns out, because exploration seismic methods are considerably different for the two different environments. We deal first with exploration on land.

Explosions of various types also give rise to radiated seismic waves; the distribution of energy into different types of elastic waves varies considerably, depending on the location of the explosion with respect to significant rock boundaries, as well as on the type and strength of the rock. The historical (and present-day) use of explosions to generate suitable waves for exploration makes use of the very compact chemical energy available in high explosives. Although explosives have been used in the air to generate seismic waves in the earth (Poulter, 1950) for exploration purposes, most of the work has been done with buried, tamped charges. A comprehensive account of the seismic signals generated by explosions has been given by Kisslinger (1963). This reference is recommended for a more detailed account than can be given here. Figure 3.1 is taken from this work and gives a comparison of the relative amplitudes of seismic signals (body waves) as a function of height above or depth below the ground-air interface.

In practice, charges are buried by drilling a suitable shothole 10 to 15 cm (4 to 6 in.) in diameter so that an explosive charge of 0.05 to 100 kg ($\frac{1}{8}$ to 200 lb) can be lowered below the weathered layer. These holes range in depth from 10 to 100 m (30 to 300 ft). After the explosive has been loaded, the hole is tamped by filling at least the next 3 m (10 ft) with dirt or snow, or sometimes by filling the hole completely with water. The charge is detonated electrically. It is usual to record the time of the explosion by

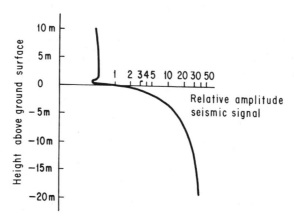

Figure 3.1   Relative amplitude of the seismic signal as a function of depth or altitude of the shot, with reference to the ground surface. [After Rocard–Kisslinger (1963).]

an impulse derived from the breaking of the firing current when the detonator fires. Very high-voltage blasters are used, and the time of application of the high voltage and the time of rupture of the bridge wire in the cap are, for normal use, simultaneous.

The location of the charge (or any other source) is quite important in order to minimize the generation of unwanted waves, while simultaneously satisfying the need for the generation and transmission of as complete a spectrum of frequencies as possible. The phrase "spectrum of frequencies" or, more commonly, "spectrum" is used at this point to denote the amplitudes of waves of various frequencies of waves produced. A plot can be made which shows amplitude as a function of frequency, and this is called the amplitude spectrum. This takes for granted that an impulsive wave can be regarded as the sum of waves of different harmonically related frequencies. A more formal treatment of this concept is presented in Chapter 4 [see also, for example, Bracewell (1965)]. A simple case of initiation of vibration can be examined to see the general physical principles involved. Figure 3.2 shows a string supported between two rigid supports. It is well known that such a string can be made to vibrate by an impulsive transverse force at some point along its length. Now the string must vibrate in such a manner that the boundary conditions (no motion at either support) are satisfied at all times; and one way of expressing this is to say that the string has a series of (standing wave) vibrational modes which in this case are integral multiples of a fundamental mode whose wavelength is twice the length of the string. The points marked A are antinodes (points of maximum displacement), and those marked N are nodes (points of zero displacement), when any of the pure harmonics are excited. It is also known that these harmonics cannot be generated when the string is struck at the corresponding nodal points. When any point is selected at random, it is generally possible to excite several modes of vibration but not the mode for which the selected point is a nodal point.

The degree of excitation is again a matter of fitting the initial conditions of the problem, and this requires that the amplitude, velocity, and acceleration of the string at the point in question, as well as at all other points on the string, agree with the sum of the amplitudes, velocities, and accelerations of the individual harmonics and with the impulse supplied.

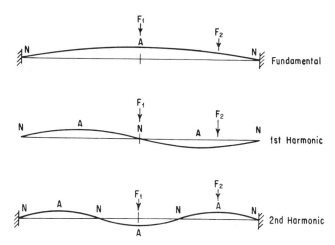

Figure 3.2 Standing waves possible for a string vibrating between two rigid supports. Force F_1 generates only modes of vibration for which it does not occur at a node, that is, the fundamental and second harmonic in the case illustrated. Force F_2 generates all three modes.

This general approach can be carried over to three-dimensional models. For example, it has been shown both mathematically and by the analysis of earthquake records that the earth as a whole has several modes of vibration which can be excited by a sudden rupture of the earth and the release of stored elastic energy. The types of vibration are independent of each other, but all are dependent on the near-spherical form of the earth and on the distribution of elastic moduli and density with depth into the earth. These are called the free modes of vibration of the earth.

In the same way, the near surface layers of the earth exhibit independent modes of vibration, which can be excited by a suitably placed explosion or other source of seismic energy.

In principle therefore it is necessary to explode the dynamite, or apply the time-varying force, at a position that is a node of any possible standing wave for the weathering and also sufficiently far below the surface that the Rayleigh waves (exponentially diminishing in amplitude from the surface downward) are not appreciably excited in the frequency band desired. Since all the standing waves (or normal modes) are frequency-dependent, their nodes occur at different depths and it is impossible to satisfy all the requirements simultaneously. The practice of taking several shots, at different depths *below* the base of the weathered layer, generates waves of different forms. The major factor contributing to this is then the presence of a reflected wave from the base of the weathering. While this is a nuisance and tends to introduce unwanted variations in the spectrum of the downward traveling composite pulse, the standing waves in the weathered layer are not initiated to anything like the extent they would be if the shot were in the weathered layer. The near constancy of the reflection coefficient of the base of the weathering then gives generated pulses which, while complex, can be held nearly consistent in waveform.

Aside from the position of the explosion in the earth, the frequency content or form of the waves generated is affected by several other factors. If the assumption is made that the explosion suddenly increases the ambient pressure in a spherical cavity of radius r and the elastic limits of the material surrounding the cavity are not exceeded, Sharpe (1942) showed that the shape of the elastic pulse generated, a highly damped sinusoid, is determined both by the radius of the cavity and the elastic constants of the material. It has been suggested that an equivalent cavity can always be found outside which the elastic limits are not exceeded. A high-frequency pulse is generated when the cavity radius is small. While such an explanation carries elements of truth, the actual mechanism is much more complex, since almost always crushing of the rock takes place until the shock wave reaches a point where the behavior of the rock is elastic. This theory does not explain fully the number of relatively low frequencies (5 to 100 Hz) generated in the earth.

There is no question that the use of explosives in drilled holes is an inconvenience and that this practice is at an economic disadvantage compared with the more modern practice of surface sources. This inconvenience and expense are, however, mitigated by the following advantages:

1. Relative freedom from surface waves.
2. The possibility of measuring directly the time required for the wave to pass through the weathered layer.

To quote Kisslinger (1963) in his "Summary and Conclusions: The State of the Art":

> The study of explosion-generated seismic waves is still primarily an empirical science. In the absence of a complete theory of the processes that result in the radiated wave, progress in further understanding of the subject is dependent on well designed experiments. Unfortunately, the complex properties of earth materials and geologic structure in which full-scale tests must be carried out make the design of experiments difficult.

3.2 DIRECTIONALITY

In the discussion of explosive and other types of source, we talk about the energy (or amplitude) of the wave not only as a function of distance but also as a function of direction. The general plane wave solution of the wave equation in Cartesian (x, y, z) coordinates is

$$\xi(x, y, z) = F[(lx + my + nz) - ct] \tag{3.1}$$

where l, m, and n = cosines of the ray path with the x, y, and z coordinate axes

For the present purpose, it is more convenient to express the solution to the wave equation in spherical coordinates:

$$\xi(\theta, \phi, R) = \frac{1}{R} P(\theta)Q(\phi)F(R - ct) \tag{3.2}$$

where $P(\theta)$ = a function of the azimuth θ only
$Q(\phi)$ = a function of the vertical ϕ only
$F(R - ct)$ = a function of R and t only

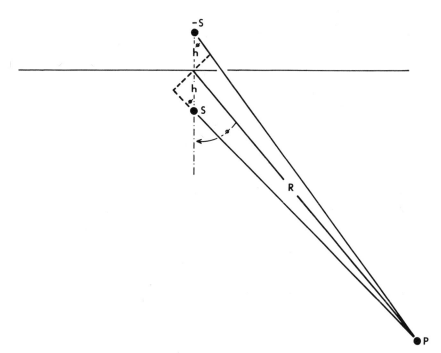

Figure 3.3 Geometry for calculation of amplitude as a function of the vertical angle θ for a source below the free surface.

For an explosion in the body of an infinite isotropic medium, the wave has no way of showing a preference as to direction, so *by symmetry* we put $P(\theta)$ and $Q(\phi)$ equal to constants. This leaves the wave amplitude ξ as a function of distance only.

A single spherically symmetrical source occurring in an infinite medium is, however, an abstraction; and in any real situation there are one or more boundaries near it. A relatively simple condition is shown in Figure 3.3 where the spherically symmetrical source is buried at a small distance h below the free surface. If the measurement of amplitude is made at a point P where the distance R is very much larger than h, we can make use of the method of images to satisfy the (free) boundary conditions at the surface. If only P waves are considered (the entire medium is a fluid), a *negative* image of the same strength S as the original source is assumed to be present at a height h above the boundary. By so doing, the surface boundary conditions are satisfied. Since both the real and virtual sources lie on the vertical axis, this source combination gives waves that are independent of the azimuth. The variation with the vertical angle ϕ is calculated as follows.

The wave amplitude at a distance R from any single source *of frequency* ω is given by

$$\xi_R = \frac{A}{R}\left[\cos\omega(t-t') - \cos\omega(t+t')\right]$$

where t' = time difference due to path difference $h\cos\phi$

$$\omega t' = \frac{\omega h\cos\phi}{c} = \frac{2\pi h}{\lambda}\cos\phi$$

$\dfrac{h}{\lambda} = .01$

(a) Maximum amplitude is down (= .0627)
Variation near as cos ∅. All values positive.

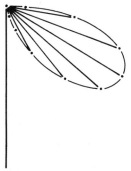

$\dfrac{h}{\lambda} = 0.5$

(b) Maximum amplitude is ∅ = 60° (1.00)
All values positive.

$\dfrac{h}{\lambda} = 1.0$

(c) Maximum amplitude is − .992
Values cycle from negative to positive.

Figure 3.4 Directivity functions for a constant-frequency P-wave source located at different distances below a free boundary. λ is the wavelength of the signal, and h is the source depth.

Therefore

$$\xi_R = \frac{A}{R} (\sin \omega t) \sin \left(\frac{2\pi h}{\lambda} \cos \phi \right) \qquad (3.3)$$

The amplitude at distance therefore vibrates 90° out of phase with the individual source standard phase (accounting for travel time) but varies with ϕ. The manner of variation is much controlled by the factor $2\pi h/\lambda$, that is, the ratio of the depth of the source to the wavelength of the sound in the medium. Figure 3.4 shows three cases.

a. If the depth below the surface is very small compared with the wavelength (the usual case for seismic exploration), the variation is nearly proportional to $\cos \phi$. (In polar diagrams, the length of the line from the origin is proportional to the amplitude at that angle from the vertical.)
b. For $h = \frac{1}{2}\lambda$, the variation is determined by both the angle and the phase difference between the two sources for the frequency concerned ($f = c/2h$).
c. For $h = \lambda$, the variation with angle becomes more rapid, going through two complete half-cycles. Note now that the net amplitude is negative over part of the range of angles.

Even for this simple combination of constant-frequency source and image, the resultant directivity pattern can be a complicated function of the vertical angle. More complex cases of patterns of sources are dealt with later in this chapter. In the case of impulsive sources, the method of combining the outputs for constant-frequency sources will be evident after Fourier analysis and synthesis are discussed (Chapter 4).

Although these spherically symmetrical sources generate no shear waves by themselves, the combination of source and image, when considered together, can yield SV waves when the combined effect is measured close to the source. It is necessary first to draw the lines of equal phase (wave fronts) for the combination and then to calculate the resultant amplitude and direction to see that two components, one along and one perpendicular to the ray, are necessary to explain the resultant motion. Of course, in an elastic medium, a particular type of wave impinging on a boundary always undergoes partial conversion of type. This is, qualitatively, the way the Rayleigh wave is set up. The generation of Rayleigh waves from an explosive, buried shot has been the subject of numerous investigations by Dix (1955) and others.

3.3 SURFACE SOURCES

The inconvenience and expense of drilling holes, the inconsistency of the seismic pulse generated, and, to some extent, the continued existence of areas in which no reflection records could be obtained led to experimentation with other types of seismic sources. While some appeared temporarily to offer an advantage over others, over the course of time only a few types remained. Poulter (1950) showed that reflection records could sometimes be obtained by arrays of explosive charges fired in the air a few meters above the surface of the ground. Generally, these arrays were about 100 m in diameter, with 7 to 13 charges of $2\frac{1}{2}$ kg (5 lb) of dynamite mounted on poles 2 to 3 m (6 to 10 ft) high. Interconnections between the charges were usually

made by Primacord, although separate electric blasting caps were sometimes employed. Although used for oil prospecting for a short time, the disadvantages out-
weighed any (dubious) advantages and the method succumbed to progress.

The use of arrays of small (2- to 5-kg) explosive charges in multiple holes a few
feet deep gained some acceptance, largely because of the advantages of arrays over
single charges in diminishing the size of surface waves and increasing the downward
traveling energy. The drilling and tamping of 2- to 5-m holes, however, were a deterrent. While this method gradually died out, the concept of arrays or patterns of sources
and receivers showed promise provided that a suitable unit source could be found.

The advent of the weight drop method issued in the era of surface sources. In this
method, a large mass weighing 1500 to 2000 kg (3000 to 4500 lb) was dropped from a
height of about 2 to 3 m (6 to 10 ft) onto the earth, the time of initiation of the seismic
wave being derived from an accelerometer contained in the concrete block. The
weight could be raised easily, the position of the truck changed quickly, and the
generation of source patterns was facilitated.

When the mass always fell flat on the earth and the latter had consistent elastic
and density characteristics, a step function pressure was formed which was responsible
for generating both surface waves and body waves. A later discussion shows that the
amount of energy diverted into surface waves was probably 10 times that for compressional waves. A typical pulse shape generated was given by Neitzel (1958) (see
Figure 3.5). The difficulties experienced were largely due to inconsistencies in the soil
and near-surface parameters but also to the predominance of low frequencies (5 to
20 Hz) generated. As can be seen from the examples given by Neitzel, the pulse shape
changes, even when the weight is dropped many times in a single location. The result
is a variation in the effective initiation time of the low-frequency spectrum. In the later
process of summation of the results of multiple drops, only the low-frequency reflections survive, and even their phase (or initiation time) is in question by an undesirable amount. Because of the relative simplicity of the equipment and the fact that
that lower frequencies often have an advantage in prospecting for deeper geological
structures (lack of attenuation, and the tendency for geological formations to be
thicker in older sediments), this method has continued to be used but now appears
to be losing ground to newer methods. The advantage (or necessity) of using arrays of
surface sources and geophones was nevertheless firmly established.

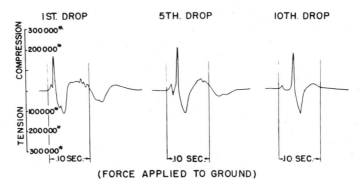

Figure 3.5 Force-time curves at the bottom of a concrete mass. Approximate value for the first peak
is 90,700 kg (200,000 lb). [After Neitzel (1958). Reprinted with permission from *Geophysics*.]

The Dinoseis® (a registered trademark of the Atlantic-Richfield Company, Inc.) method substitutes a controlled, contained explosion of a propane-air mixture to create a transient pressure on the surface of the earth. The explosion is contained in a chamber with one flexible side. It is pressed against the earth by the weight of the vehicle. A fixed amount of energy is generated each time, and the method is sufficiently mobile that patterns or arrays of sources can easily be used. If the surface soil or rock had consistent properties, consistent pulses resulted; however, this again is not realistic, and the method suffers from some of the disadvantages of the weight drop method. Its ease of use in generating arrays of sources has gained it some popularity for land use. It does not involve the problem of interaction of the mass of the concrete block dropped with the spring constant of the earth, since the diaphragm of the Dinoseis® method has virtually zero mass, hence the resonant frequency is very high—well outside the normal seismic frequency range. Thus the prevalence of very low-frequency generation in the spectrum of the weight drop pulse is not carried over to the Dinoseis® pulse. The method is therefore more suitable for the generation of higher-frequency reflections and, of some consequence, has less of a tendency to generate ground roll.

While the Vibroseis® (a registered trademark of Continental Oil Company, Inc.) method, in its full implementation (described in detail in Appendix 3B), is much more complex, looked at from the view point of the required source, it has the advantage that it is possible to exert some control over the frequency spectrum of the energy introduced into the earth. By a feedback system using a monitor signal from the baseplate resting on the earth, the force applied can be varied to accommodate phase changes brought about by inconsistencies in earth materials. At present, the amplitude is not controlled at the source. All this control results from the nature of the method itself, which is described by Crawford, Doty, and Lee (1960).

For the purposes of this chapter, the signal is introduced into the earth in the form of a long train of waves of gradually changing frequency. It is a signal of the general form

$$S(t) = A(t) \sin\left(at + \frac{b}{2}t^2\right) \qquad 0 < t < T$$

where $A(t)$ is a relatively slowly varying envelope function and a and b are constants relating to the initial frequency and the rate of change in frequency, respectively. As far as the source design is concerned, the important point is that the frequencies are introduced in a slowly varying manner and feedback methods can be employed in a phase compensation circuit. The baseplate motion is used as a measure of the earth motion, and a reference signal is provided by a sweep generator either in the vibrator or radioed in from the recording truck.

Diagrammatically, Figure 3.6 shows the mode of operation. A central cylinder is bored through a large steel mass which can be sealed by sleeves incorporating O-ring seals. A piston is formed by steel ridges on a central steel bar in which is drilled several holes which can convey oil under pressure to two enclosures, 30 and 32, and which can exhaust oil to an accumulator (not shown). Enclosures 30 and 32 are alternately connected to a high-pressure, servocontrolled hydraulic system. Also, 30 and 32 are alternately connected to the return hydraulic line.

Under the action of these pressures in the enclosures, the mass moves up and down.

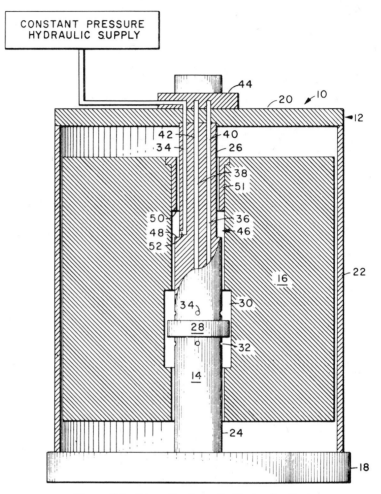

Figure 3.6 Schematic of a vertical hydraulic vibrator.

The reaction force on the piston moves the baseplate, thus inducing seismic energy into the earth. The baseplate is held down by the weight of the vehicle through decoupling members (very low-pass frequency filters) such as air bags or springs. The hold-down weight must be greater than the peak dynamic force supplied by the oil pressure in the cylinder-piston assembly. Peak forces up to 16,300 kg (36,000 lb) have been obtained. The rigid baseplate and piston assembly have a mass which is coupled to the impedance of the earth. For the seismic frequency range, the earth acts approximately as a spring, although the action is more complicated because of damping (both by radiation of energy and by heat loss) and because of the nonlinear behavior of the baseplate-earth system. As pointed out earlier when the weight drop method was being considered, the earth consolidates under the action of successive impacts and so presents a varying load to the force being impressed on it. In the case of the vibrator, the behavior of the earth under the baseplate is more complex, because the zone is not in contact with the air between successive force peaks and the heat generated in crushing can change the nature of the soil. From the point of view of signal shape,

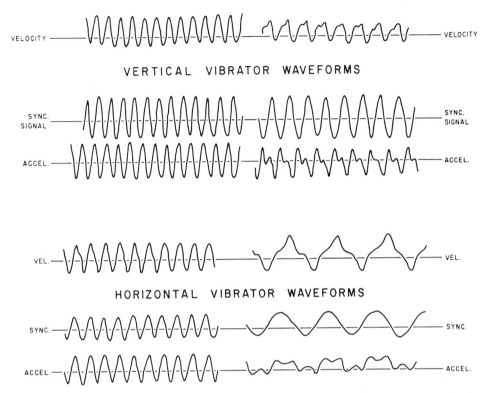

VELOCITY ———— ⋀⋀⋀⋀⋀⋀⋀⋀⋀⋀——⋀⋀⋀⋀⋀⋀—— VELOCITY

VERTICAL   VIBRATOR   WAVEFORMS

SYNC.
SIGNAL ————— ⋀⋀⋀⋀⋀⋀⋀⋀⋀⋀——⋀⋀⋀⋀⋀⋀—— SYNC.
SIGNAL

ACCEL. ————— ⋀⋀⋀⋀⋀⋀⋀⋀⋀⋀——⋀⋀⋀⋀⋀⋀—— ACCEL.

VEL. —— ⋀⋀⋀⋀⋀⋀⋀⋀——⋀⋀⋀⋀—— VEL.

HORIZONTAL   VIBRATOR   WAVEFORMS

SYNC. —— ⋀⋀⋀⋀⋀⋀⋀⋀——⋀⋀⋀⋀—— SYNC.

ACCEL. —— ⋀⋀⋀⋀⋀⋀⋀⋀——⋀⋀⋀⋀—— ACCEL.

Figure 3.7   The vertical and horizontal vibrator baseplate motions corresponding to a given synchronizing electric signal. High frequencies are on the left, and low frequencies on the right. Development of the double-frequency components as the frequency is lowered is evident.

the effect of the nonlinear behavior is to introduce harmonics (see Figure 3.7) into the low-frequency signal generated, sometimes almost to the extinction of the desired fundamental. This effect is considered in more detail when the full Vibroseis® system is considered, since these harmonics may be detrimental to the quality of the seismograms eventually produced. In the first few cycles of vibration at a particular position on the earth, much energy is used in crushing and consolidating the soil. It is for this reason that no serious attempts have been made to establish a continuously moving Vibroseis® source except in a marine environment.

Recent developments in land-based seismic sources have included a vibrator and an impulsive source based on a surface-effect vehicle (or Hovercraft). As is well known, these vehicles ride on a cushion of air continuously provided by an air compressor. A flexible skirt forms a partial seal against air leakage near the earth, and the vehicle platform rises to a height automatically determined by the available airflow, air pressure, and the quality of the seal.

Independently, R. A. Broding and D. E. Miller suggested that such a vehicle could be made into a seismic source either by (a) vibrating the platform with a reaction-type vibrator or (b) modulating the air stream. Both modes of operation do in fact work, but a larger force can more conveniently be obtained by using the second method. In Figure 3.8, a schematic of such a vibrator is shown. The action of the

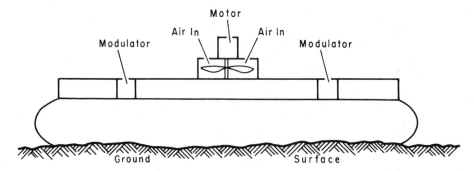

Figure 3.8 Schematic of a vibrator based on a Hovercraft. The two air modulators are synchronized with a sweep signal. A shutter is open, allowing airflow only during the linear increase in frequency, and is closed for a return to the original starting frequency.

modulators causes the pressure inside the skirt to be a combination of dc and ac air pressure—the former holding the vehicle near an equilibrium position while the latter gives an alternating pressure on the ground. The alternating pressure is relatively small [about 0.035 to 0.07 kg/cm² (0.5 to 1.0 lb/in²)], but by using a large (3.04 × 6.08 m) vehicle the total force [about 0.65 to 1.30 × 10⁴ kg (20,000 lb)] can be made as large as that of a large conventional field vibrator. Several advantages can accrue from this arrangement:

1. Coupling can be made to a rocky surface.
2. Easily damaged ground (permafrost) can be traversed without environmental damage.
3. Pressure can be provided on the ground that does not exceed the elastic limit of the soil.
4. Work can be carried across tidal flats and other marginally accessible areas.

Such a vehicle has been built and tested. In addition, a similar concept has been used in which the air is suddenly released inside the air cushion to give a sudden increase in pressure—which of course decays rapidly because of the leak at the ground surface.

An air cushion vehicle is resonant at a very low frequency determined by the mass of the platform and the volume of air in the cushion (see Appendix 3A). It is of course damped because of the air leak at the ground surface. It is therefore predominantly a low-frequency vibrator whose output for constant peak input pressure variation decays by a factor of 2 each time the frequency is doubled (an octave). Expressed in the usual way the rate of decay is 6 dB per octave. [The *decibel change* is given by $20 \log_{10} (A_2/A_1)$, where A_2 and A_1 are the *amplitudes* of the measured signals.]

3.4 MARINE SEISMIC SOURCES

In seismic exploration confined to water-covered areas such as the continental shelves, the early source was always a form of unconfined explosion. However, environmental damage was recognized here, perhaps earlier than the damage done on land, since the killing of fish was an immediate economic threat to the livelihood of fishermen. There existed therefore immediate pressure to devise means of creating, in water, seismic waves that did not have sufficiently high peak pressures to cause damage to marine fauna. There was moreover an additional reason arising from the so-called bubble effect. As long as the bubble of gas formed immediately after the explosion is spherical, there is a compression of the water and this continues, as a result of inertia, for a short time after the average pressure within the bubble becomes equal to the water pressure. For a 50-lb charge of TNT, the outward expansion is brought to a halt after 200 msec, and bubble contraction starts. An implosion results, since the water converges into the limited spherical volume with higher and higher velocity until suddenly there is no space left. The pressure rapidly increases, overshoots, and then the expansion starts again. This process continues, with less and less amplitude (since energy is lost each time and some heat is lost to the water). Eventually, the bubble breaks the water surface, the spherical symmetry is broken, and the process stops (see Figure 3.9).

As far as the production of seismic waves is concerned, a series of nonuniformly spaced pulses is obtained which decay in amplitude but, at the same time, give rise to a complex source pulse (or signature) which is far from predictable. To avoid this

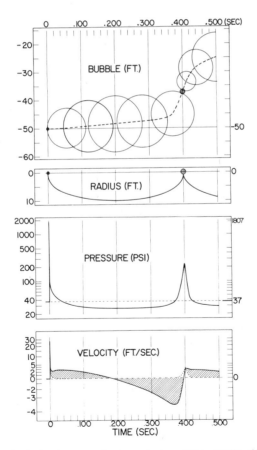

Figure 3.9 Pressure and velocity signatures for the explosion of 22 kg (50 lb) of TNT taken at a distance of 10 m (33 ft). The depth below the water surface was 16 m (50 ft). [After Seismic-Energy Sources, United Bendix (1968).]

complex signature the explosive charges were used very inefficiently at depths sufficiently low that the first bubble broke the surface.

Another predictable consequence was that other types of source were developed. There were analogs of the weight drop, Dinoseis® and Vibroseis® sources, and for specialized purposes these still are being used. However, largely because of the consistency of the water medium, it was found possible to generate energy within the frequency band used in seismic exploration by a more controlled release of gas pressure or by other means of producing a sudden volume increase within the water while at the same time minimizing the bubble effect. One of these, widely used, is the Aquapulse source in which a propane-oxygen mixture is exploded inside a steel mesh-supported cylindrical rubber sleeve. After detonation, the sleeve expands from its initial cylindrical form but retains its cylindrical symmetry. This effectively prevents the bubble effect. Later, the explosion-generated gases are expelled, and a new charge enters. The cycle time is about 8 sec, which is adequate for most marine exploration. Another type of source, the air gun, uses compressed air which is suddenly released into a chamber whose volume controls the frequency content of the pressure waveform. In many operational arrangements, several air guns of different chamber volumes are used to create an impulse in the water having a wide frequency spectrum. Other sources are electromagnetic, in which the repulsion between two metal coils, one carrying a large transient current obtained from the discharge of a capacitor bank, is exploited to create a sudden volume change in the water. The mutual inductance between the two coils (or simply between one coil and a highly conductive metal plate) causes a large secondary current to flow in the second conductor—hence the repulsion effect.

Finally, spark sources rely on the creation of an acoustic wave in the air, and arc sources achieve their acoustic output by the ohmic heating of conducting water between two renewable electrodes. The high temperatures cause a steam bubble to be formed suddenly, which expands to give the seismic wave. Most of these marine sources have been treated at length in the excellent tutorial booklet "Seismic Energy Sources 1968 Handbook" published by the United Geophysical Corporation. Reference to this publication should be made for additional details.

The requirements for a marine seismic source are:

1. Ability to generate a discrete powerful pulse or a signal which can be subjected to later compression in time such as in the Vibroseis® system.
2. A rechargeable or repeatable system which can be used in a sequence of operations at short intervals of time (10 sec or so).
3. A relatively simple system which will operate consistently, be trouble-free, and have a long life between overhauls.
4. A system that can be used at constant depth below the water surface and results in a minimum drag on the vessel carrying it.
5. A system that does not injure marine life.
6. A system that minimizes the bubble effect.

Most of the successful sources have been air guns or gas guns for deep seismic prospecting and the sparker or Boomer for shallower, higher frequency work. In Figure 3.10 is shown a comparison chart which relates the strength and other characteristics of these various sources.

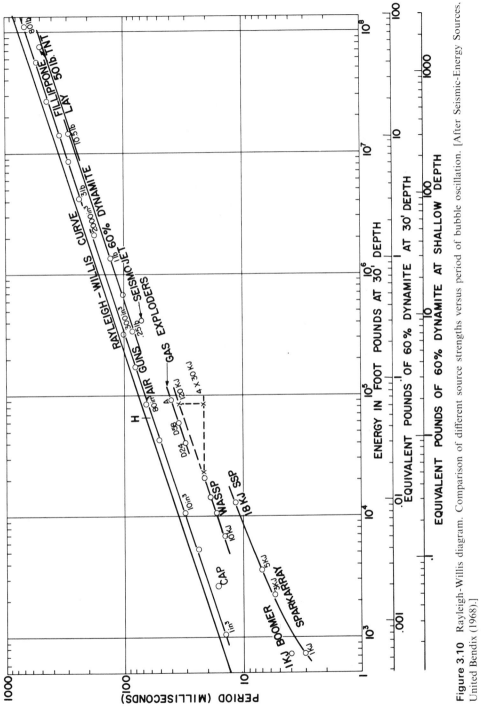

Figure 3.10 Rayleigh-Willis diagram. Comparison of different source strengths versus period of bubble oscillation. [After Seismic-Energy Sources, United Bendix (1968).]

Rayleigh showed (1917) that the period of spherical bubble oscillation is related to other parameters by the equation

$$T = 1.83 A_m \sqrt{\frac{\rho}{P_0}}$$

where T = period of bubble oscillation (sec)

A_m = maximum radius of the bubble (cm)

ρ = specific gravity of the fluid (gm/cc)

P_0 = ambient absolute hydrostatic pressure (dyn/cm^2)

Willis (1941) combined this relationship with the formula for the potential energy Q of a bubble of radius A_m:

$$Q = \tfrac{4}{3}\pi A_m^3 P_0$$

to give

$$T = 1.14 \rho^{1/2} P_0^{-5/6} (KQ)^{1/3} \qquad (K = 1 \text{ when cgs units are used})$$

which is known as the Rayleigh-Willis formula. The diagram shows this relation on a log-log scale.

For prospecting purposes, the source is dragged through the water and the individual bursts of energy are summed to make a composite source. If we assume that a composite source is available on the surface every 61 m (200 ft) and if the speed of the vessel is (say) 11 km/h (6 knots), the coverage is 11 km/h or 183 m/min. Every 61 m therefore takes 20 sec, and the source must be strong enough that one is sufficient if the recovery time is 20 sec, or two if the recovery time is 10 sec. The latter time is

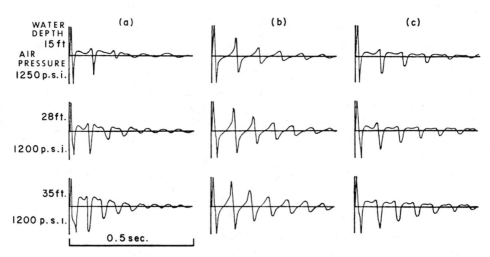

Figure 3.11 Comparison of waveforms generated by a 2.62-liter (160-in.^{3}) air gun. Measured waveform (*a*), theoretical waveform (*b*), and theoretical waveform (*c*) after correction for filtering by the recording system. [After Smith, *Geophysical Journal of the Royal Astronomical Society* (1975), Vol. 42, pp. 273–280.]

approximately correct for the Aquapulse source. Enough units are fired in parallel (simultaneously) so that adequate strength is obtained.

For precise work, the signature (the pulse shape versus time) of the source must be known exactly, and it is now customary to measure it at a given depth below the surface by using a hydrophone of known characteristics at distances sufficiently far below the source known to be in the far field. Usually the distance involved must be comparable with a quarter-wavelength or more of the longest wavelength required. Figure 3.11 shows an example of such a measurement.

Even though the sources are designed for maximum pulse discreteness, that is, the amplitude of all events after the initial pulse are reduced to a minimum, the after-pulses, whatever their cause, are still important and must be reduced by later processing steps. This matter is considered in more detail in later discussions of deconvolution (Section 6.10).

3.5 EARTH RESPONSE TO A DRIVING FORCE DISTRIBUTED OVER AN AREA OF THE SURFACE

The mathematical treatment of the production of elastic waves due to prescribed, periodic stresses on the surface of a semi-infinite isotropic solid has been provided in two important papers by Miller and Pursey (1954, 1956) entitled "The Field and Radiation Impedance of Mechanical Radiators on the Free Surface of a Semi-infinite Isotropic Solid" and "On the Partition of Energy between Elastic Waves in a Semi-infinite Solid." Only a few results are quoted here. Different types of motion on the surface were considered:

a. An infinitely long strip of finite width vibrating in a direction normal to the surface of the medium.
b. An infinitely long strip of finite width vibrating tangentially to the surface and normally to the axis of the strip.
c. A circular disk of finite radius vibrating normally to the surface of the medium (a model for a vertical vibrator).
d. A torsional radiator in the form of a disc of finite radius performing rotational oscillations about its center. (a model for a torsional vibrator).

The directivity functions in Figure 3.12 are taken directly from that paper, but note that the motions are given as functions of R (distance from the origin) and θ (the vertical angle used previously). It should be noted that the radial motion U_R is almost proportional to $\cos \theta$ and varies only a small amount with a change in Poisson's ratio. For most work in sedimentary sections $\sigma = \frac{1}{3}$ (or $V_p/V_s = 2$) fits the field data closely.

The SV motion (perpendicular to the direction of travel and in a vertical plane) is a much more complex function. The directivity function has two loops, with a zero in between. Actually, the phase changes by 180° between the two loops, thus making the zero a natural consequence of the (sudden) phase change. There is no SV motion down the axis (by symmetry this must be the case), and the strongest SV motion is developed at an angle near 30° from the vertical axis.

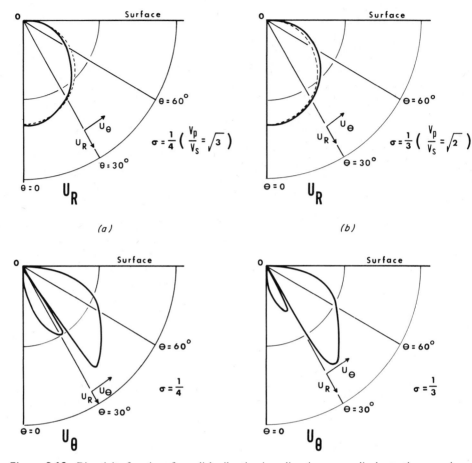

Figure 3.12 Directivity functions for a disk vibrating in a direction perpendicular to the ground surface. Two cases are shown: (a) Poisson's Ratio $\sigma = 0.25$, $V_s/V_p = 0.577$; (b) $\sigma = 0.33$, $V_s/V_p = 0.5$. [After Miller and Pursey (1954).]

The distribution of the available energy from the vibrating source (for the particular case $\sigma = \frac{1}{4}$) has been evaluated in Miller and Pursey (1956) as:

$$\text{Compressional energy} = W_c = 0.333 \frac{\pi^3 v^2 a^4 P_0^2}{\rho V_c^3}$$

$$\text{Shear energy} = W_{SR} = 1.246 \frac{\pi^3 v^2 a^4 P_0^2}{\rho V_c}$$

$$\text{Surface wave (Rayleigh) energy} = W_{Su} = 3.257 \frac{\pi^3 v^2 a^4 P_0^2}{\rho V_c^3}$$

where v = frequency
$\quad\quad a$ = radius of disk
$\quad\quad P_0$ = stress under disk (peak)
$\quad\quad \rho$ = medium density
$\quad\quad V_c$ = medium compressional wave velocity

Thus the distribution of energy is:

Compressional energy   6.9%
Shear energy   25.8%
Surface wave energy   67.3%

In practice, the body wave energy figures appear to be optimistic. The presence of a surface layer causes much of the body wave energy radiated at high angles to be multiply reflected inside the surface layer and thus become additional surface wave energy. However, the effect of the surface layer on the mechanical efficiency of the source does not appear to have been investigated, and the above conclusions may not be valid.

As with the explosive sources, we deal later with the pattern effect of multiple surface sources. This pattern effect is obtained whether the vibrators are used simultaneously or sequentially, and the results are summed afterward. There may, however, be some question concerning the mutual effect between simultaneously operated mechanical sources situated close together so that the motion due to one source changes the radiation impedance (or efficiency) of another. Physically, this nonlinear effect must be present. Results from finite elastic modeling show that the surface motion from one vibrator dies off so rapidly with distance that—for all practical distances (say, greater than 5 m)—one vibrator cannot "feel" the effect of another in any significant amount. This matter is also discussed (mathematically) in Miller and Pursey (1956). Experimental evidence has also been obtained that, in a real situation, the influence of one vibrator on the efficiency of a nearby one is negligible. Multiple, simultaneously operating vibrators are used in field practice, but for considerations of signal/noise ratio and economics rather than any gain in efficiency due to interaction between vibrators. For a given random noise power, four vibrators operating simultaneously increase the signal/noise level four times over the action of a single vibrator. However, if four single-vibrator traces are taken at different times, the noise samples must be assumed to be different and the signal/noise ratio increases only as the square root of the number of trials, or 2:1 in this case. Thus it is always more beneficial to increase the signal strength than to try to increase the number of samples by the same proportion, and in fact this increase in sampling brings diminishing returns as the number is increased and eventually becomes uneconomical.

3.6 THE USE OF ARRAYS OF SOURCES

In this discussion of arrays or patterns of sources, the sources themselves are considered omnidirectional. That is, they individually radiate equally in all directions. The effect of the arrays can then be compared to the action of diffraction gratings (or crystal lattices in x-ray diffraction) as means for obtaining high-intensity beams in certain directions. In seismic reflection prospecting, arrays are used to reduce the effects of waves traveling in a near horizontal direction, while at the same time amplifying the effects of near vertically traveling waves. As seen later, not all reflected waves can be considered as traveling vertically, since large horizontal offsets between source (patterns) and receiver (patterns) cause off-vertical rays to be considered. However, for simplicity, we assume that the discrimination required is between

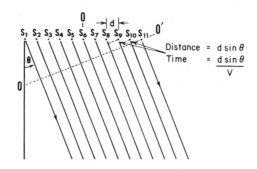

Figure 3.13 Diagram showing an array of equidistant sources $S_1, S_2, S_3, \ldots, S_{11}$ contributing to the illumination in a direction θ to the vertical. In the general case, neither the spacings nor the intensities of the source are equal.

interfering horizontally traveling waves and vertically traveling reflected waves. The effects are obtained by the same physical mechanism, namely, constructive or destructive interference between the component waves produced by the individual sources constituting the array. Figure 3.13 illustrates such an array, and we investigate the strength of the composite signal as it is transmitted at an angle θ to the vertical. For the present, all sources are assumed to have the same strength and to be oscillating in synchronism. While it is possible to consider each source as producing a signal of some arbitrary shape, we constrain them to oscillate at a constant angular frequency ω. So, the problem is resolved into finding a method for summing the contributions from a series of individual sources. As usual, the theory deals only with the results in the far field—distances great compared with the size of arrays or the wavelengths concerned. We can therefore examine the results as though the sources were producing plane waves. The way in which the composite amplitude varies with the angle θ is called the directivity function.

If the individual sources have their own directivity function (as surface sources and buried sources near the surface have), the final directivity function is given by the product of the directivity function of the individual sources (taken as a class) and the array directivity function.

The method of obtaining the array directivity function is illustrated by considering a line of N discrete sources separated by equal distances d and examine the directivity in the plane perpendicular to the earth's surface and containing the line of sources. In Figure 3.13 we see that once the waves from the individual sources reach the line $00'$ (perpendicular to the required direction of propagation) they will have equal time paths to a point at infinity. So the directivity arises because of time differences in reaching $00'$ from all sources. We can see that the difference in distance of successive sources to the line $00'$ is $d \sin \theta$ and that the time difference is $d \sin \theta/V$, where V is the velocity of waves in the material. Since we are dealing with a single-frequency wave of angular frequency ω (period $2\pi/\omega$) the waveform from one source is delayed compared with that from its right-hand neighbor by a fraction of the period equal to $\omega d \sin \theta/2\pi V$. If the component from one source has the form

$$A_0 \cos \omega t$$

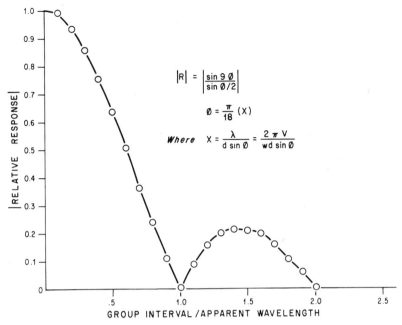

Figure 3.14 Directivity function for an 18-source array.

the next one will have the delayed signal

$$A_0 \cos(\omega t - \phi)$$

where $\phi = \omega d \sin\theta/V$

Summing all the contributions gives the array directivity function as

$$A(\omega, \theta) = A_0 \sum_{K=0}^{N-1} \cos(\omega t - K\phi)$$

Such a directivity function has the same form on either side of the vertical axis and has the formula

$$|A(\omega, \theta)| = A_0 \frac{\sin N\phi/2}{\sin \phi/2}$$

An example for $N = 18$ over an array length of $17d$ is shown in Figure 3.14. Here the directivity function is plotted with the source interval or apparent wavelength along the array as the independent variable. The main features to be noted are:

a. The large maximum for vertical transmission (the apparent wavelength is infinite).
b. A series of zeros and subsidiary maxima corresponding to angles for which $\sin 9\phi$ is zero or unity. In seismic prospecting the initial maximum and the first zero are principally of importance.

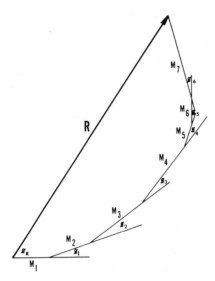

Figure 3.15 The vector addition of contributions from weighted elements of an array for a particular angle, frequency, spacing, and wave velocity. In this m_k are the weights of the sources and $\phi_j = \omega d_j \sin \theta / V$.

If the source strengths are unequal, the contribution of each source will be given by $m_k \cos(\omega t - \phi_k)$, and this can be represented, as in Figure 3.15, as a vector sum:

$$R_N = \sum_{k=0}^{N-1} m_k \cos(\omega t - \phi_k)$$

The same general method applies when the *distance* between the individual sources is no longer constant.

   If it can be assumed that a vibrator is consistently of the same strength no matter how the earth material changes its characteristics, these nonuniform sampling methods may have some advantages. One claimed advantage makes use of different sampling with receivers and sources in order to achieve freedom from aliasing (i.e., inadequate sampling for small wavelengths on the surface) and to gain otherwise unobtainable pattern characteristics (Morrison, 1973). However, in the field, such consistency of vibrator action is hard to obtain, and the advantages of these claims are hard to substantiate. Similar problems with suggested weighted patterns of receivers are considered later.

   The question of aliasing is nevertheless an important one, and considerable care needs to be taken that the waves be rejected (generally surface waves and refracted waves) are sampled by sources and/or receivers so that more than two, preferably four or more, geophones or sources occupy the shortest wavelength expected. Rayleigh waves in the weathered layer may have phase velocities of 300 m/sec and, if the highest frequency of interest is 60 Hz, the wavelength will be 5 m and sources or receivers should occupy positions only 1.25 m apart along the pattern length. It is seen that the directivity of the pattern itself is such that a null occurs where the pattern length is equal to one or more complete apparent wavelengths. If the additional cosine directivity function for vertical vibrators on the surface is disregarded, it will be necessary to choose a pattern length sufficiently long to discriminate significantly against the *longest* wavelength in the interference to be rejected. This is of course the highest phase velocity divided by the lowest frequency. It should not, however, be mandatory that the source pattern alone reject this lowest-frequency interference, and economics usually dictate that some optimum rejection system be found by employing arrays of both sources and detectors.

The discussion of overall rejection of unwanted interference is resumed after a consideration of detectors.

3.7 TYPES OF RECEIVERS

Although geophones of other types have been designed to be used on land, by far the greatest number in field use are of the moving-coil variety (see Figure 3.16). A radial magnetic field, between a center pole piece and a permeable magnetic case, passes through a cylindrical coil suspended within the field. Any motion of the coil in the magnetic field then generates a voltage in the coil, which is proportional to the velocity of the coil with respect to the field. The mass of the coil is supported by some type of spring, and the coil itself is usually constrained to move only vertically by the action of diaphragms. The latter are made from thin metal and are used for electrical connections to the coil also. They exert negligible restoring force on the vertical movement of the coil. By selection of the masses of the coil form and wire and the sizes of the supporting spring, different resonant frequencies can be obtained. Usually, the coil and form are designed in such a way that, when connected across the proper load resistor, the geophone is nearly critically damped. Typical resonant frequencies in use now are 4.5 and 7.5 Hz. The geophones are equipped with spikes to ensure proper planting in the earth—good coupling—and are usually employed in multiple groups, or strings. Thus a typical unit may be a string of 15 geophones separated by 7 m (20 ft) to give a string of length 105 m (345 ft). The geophones are connected in series or parallel to match the input impedance of the amplifiers.

To a first approximation, the geophone moves with the surface of the earth. If a seismic wave is perpendicularly incident on the surface, the surface motion will be

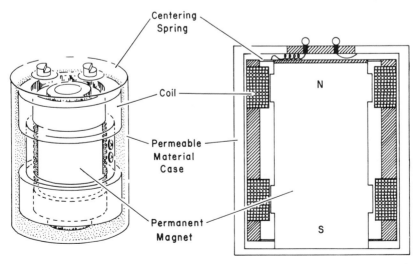

Figure 3.16 Schematic of a moving-coil geophone element (not to scale, electrical connections are diagrammatic).

double that of the wave amplitude because of the necessity of including the effect of the reflected wave.

For this type of geophone, the output voltage is proportional to the rate of change in the flux through the coil, or roughly proportional, over part of the frequency range, to the velocity of the earth's surface. A mathematical analysis of the motion of the geophone on the surface of an elastic earth has been given by Wolf (1944). The geophone, however, is a mechanical-electrical transducer with the mass, spring coil system linked to the electrical parameters of the output circuit through the magnetic field linkage with the coil. As a consequence, the geophone system, considered by itself, has a resonant frequency and damping, and the output is not strictly proportional to geophone velocity except over a limited bandwidth. The frequencies for which proportionality exists are those above the resonant frequency. An additional complication due to phase distortion occurs, since at the resonant frequency (with 70% critical damping) the phase is $\pi/4$ radians different from the phase at frequencies well above resonance. For frequencies well below the resonant frequency, there is a total phase shift of $\pi/2$ radians.

Like many other factors affecting applied seismology, the choice of geophones is a compromise. Low-resonant-frequency geophones are not as rugged as higher-resonant-frequency units and are usually heavier and larger. For most modern interpretations of seismic results, it is desirable that equipment effects, particularly distortion of the waveform, be minimized, and this means that the geophones should preferably have a resonant frequency below the frequencies of interest in the reflection band. To complicate matters still further, the surface waves are usually strongest at the lower end of the frequency spectrum and are usually not required. To accept them in any geophone means that the geophones and their associated amplifiers must have a wide dynamic range over which their linearity (proportionality of output to input) is good. With modern amplifiers and digital recording systems, the tendency is to accept all types of waves without filtering by geophones and then to remove the unwanted events in postrecording processing. This is possible only if the system records all amplitudes linearly. A detailed account of seismic instruments has been given by Anstey, Evenden, and Stone in the book, *Seismic Instruments*.

The sensitivities of geophones (Wolf, 1942) must be such that a string of geophones produces a signal from earth motion that is substantially above the intrinsic electrical noise level of the amplifiers used. This establishes a minimum electrical output voltage level for the minimum seismic signal motion desired. In practice, the limiting sensitivity is not that of the Brownian movement of the mass. Reflection signal levels at the lower limit become obscured by ground noise motion caused by wind and other types of microseismic noises. Motions of 10^{-8} cm can be detected, and voltages supplied to amplifier inputs are in the microvolt to millivolt range.

In the best geophones, sensitivity to motions other than those along the major axis of alignment has been reduced to a few percent. The high-frequency limit of moving-coil geophones is usually set by parasitic oscillations, while the low-frequency limit is a mechanical one of total allowable motion before harmonic distortion becomes unacceptable. In most field geophones, the allowable motion (limited by mechanical stops) is of the order of 2 mm peak to peak, although stronger-motion moving-coil geophones are available for special purposes. Geophones based on other electromechanical principles have been made, such as those employing magnetostrictive or piezoelectric principles, but they are not much used for land work.

In marine seismic prospecting, requirements are entirely different and have been met by making up streamers of geophone elements. These elements are contained in a fluid-filled (kerosene or silicone liquid) flexible tube. They consist of piezoelectric (ceramic) devices in which the output voltage is proportional to the stress (hydrostatic excess pressure) caused by the seismic wave. It has been found convenient to use many elements in parallel in order to achieve a reasonable impedance match to the cable-input transformer system and, at the same time, some noise cancelation. The seismic signal has to pass through the plastic tubing and through the liquid filling the tube in order to act on the piezoelectric elements. For this reason, materials are chosen so that the acoustic impedance of both the material of the tubing and the fluid is close to the acoustic impedance of sea water. For other purposes, special rubbers and silicone liquids have been manufactured, which have very desirable acoustic impedances, but for commercial seismic streamer cables plastics and kerosene are usually employed. A strain-member steel cable runs through the entire cable to reduce the possibility of breakage due to the usual frictional and unusual (snagging) tensions developed when the streamer is pulled through the water.

The piezoelectric units are usually responsive to bending produced by acoustic pressure. Similar bending results from acceleration of the cable through the water and, in order to avoid noise due to this acceleration, modern units installed in marine cables are acceleration-canceling units. Two bender or diaphragm units are placed back to back so that they both respond positively to acoustic pressure, but one has a response to acceleration negative to that of the other. When connected in series, the acceleration-induced noise voltages cancel, while the acoustic pressure voltages add. Because these ceramic (piezoelectric) elements are responsive to acoustically produced excess pressure, the streamer cannot be located close to the surface of the water.

For low-frequency operation, it is usual to drag the cable through the water at a depth of approximately 13 m. Since the output characteristics of the cable are so much affected by its depth, special depth controllers (such as the pressure-controlled Condep® devices) are placed on the cable at intervals of a few hundred meters to maintain the cable at a constant preset depth. Because of the reflection of the seismic waves at the water surface, a pressure geophone at depth has a variable frequency characteristic due to the effective presence of a negative image at an equal distance above the water surface. Assuming that the sea surface is smooth, the received signal is

$$A\{\cos \omega t - \cos [\omega(t + \delta)]\} = -2A \sin \left[\omega\left(t - \frac{\delta}{2}\right)\right] \sin \frac{\delta\omega}{2}$$

for a constant frequency ω. δ is a delay due to a travel time of $2D/V$, D is the depth of the cable in the water, and V is the velocity of sound in water (1500 m/sec approximately). The signal suffers a phase change dependent on the frequency (a received waveform is distorted). The amplitude also is a function of frequency. For very low frequencies, the output is *proportional* to the frequency. In deep water, it is common to set the depth of operation of the cable near 12 to 13 m.

The change in output, for both amplitude and phase, with frequency is only one of the changes in characteristics of the received signal caused by source-receiver characteristics. The cumulative effect is a filter, and its effect must be removed before the seismic trace can in any way be regarded as a near approximation of the signal caused by the layering characteristics of the earth.

To distinguish between sources (which are almost always used at full output) and receivers, it is possible to use weighted receiver patterns. Since the marine environment is so consistent, such patterns can be used with greater fidelity to their calculated characteristics than in land operations.

3.8 MOTION OF A VERTICAL GEOPHONE AS A RESPONSE TO COMPRESSIONAL AND SHEAR WAVES INCIDENT ON THE SURFACE

For angles of incidence other than perpendicular, the action of a vertical geophone on the surface of the earth or in a marine environment is not as simple as that discussed above. Compressional waves, with which we are mainly concerned, give rise to both compressional and shear waves on reflection. Figure 3.17 shows the geometrical considerations for both incident compressional and shear (SV) waves (after Ewing, Jardetsky and Press, 1957). A full discussion is given in this reference (p. 24), and only the main results are given here, namely:

If the velocities of the compressional and shear waves are given by α and β, respectively, and c is the phase velocity of a wave along the surface, for compressional waves,

$$\sqrt{\frac{c^2}{\alpha^2} - 1} = \tan e$$

and, for shear waves,

$$\sqrt{\frac{c^2}{\beta^2} - 1} = \tan f$$

If Poisson's ratio $\sigma = \frac{1}{4}$, then $\alpha^2 = 3\beta^2$ and $\cos^2 e = 3 \cos^2 f$, and under these conditions

$$\frac{A_2}{A_1} = \frac{4 \tan e \tan f - (1 + 3 \tan^2 e)^2}{4 \tan e \tan f + (1 + 3 \tan^2 e)^2}$$

$$\frac{B_2}{A_1} = \frac{-4 \tan e(1 + 3 \tan^2 e)}{4 \tan e \tan f + (1 + 3 \tan^2 e)^2}$$

It can be seen that the shear wave amplitude B_2 is zero only for cases of grazing incidence of the P wave and for perpendicular incidence. For all other angles, part of the P-wave energy is converted to SV-wave energy. A useful concept is the angle of emergence $\bar{e}$ which is defined to be the angle given by

$$\bar{e} = \tan^{-1} \frac{A_v}{A_h}$$

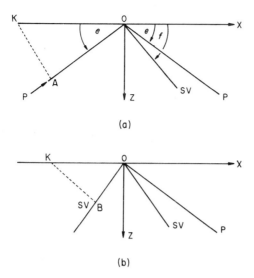

Figure 3.17 Reflection of P and SV waves at the free surface of an elastic solid. (From Ewing et al., *Elastic Waves in Layered Media.* Copyright © 1957, McGraw Hill, Inc. Used with the permission of McGraw Hill Book Company.)

A_v being the vertical motion and A_h the horizontal motion of the point O. It can be shown that

$$\tan \bar{e} = \frac{1 + e \tan^2 e}{2 \tan f} = \frac{\tan^2 f - 1}{2 \tan f} = -\cot 2f$$

Thus, measured at the surface by the vertical and horizontal motions A_v and A_h, the compressional wave appears to arrive at an angle different from the true angle. To see how much the effect can be for an extreme reflection seismology case, we take a P wave arriving below the base of the weathered layer at an angle of incidence of 45°. Then if $\alpha = 2000$ ft/sec, $\beta = 1154$ ft/sec, and $\sigma = 6000$ ft/sec for the sub-weathering, it can be shown that

$$e = 76.37°$$
$$\bar{e} = 74.37° \qquad f = 82.185°$$

and the difference is only 2°.

These relations may be used for rare cases in which the angle of emergence at the surface is not nearly vertical.

It should be noted that SH waves are always reflected as SH waves alone, with a reflection coefficient of unity.

3.9 THE USE OF ARRAYS OF GEOPHONES

When surface sources are used, the large amplitudes of waves propagated along the surface or in the near surface region (waves we call interference to differentiate them from random noise signals) make it almost mandatory that arrays of detectors as well as arrays of sources be used. The amplitude of these interference signals must be reduced before the total received signal is amplified and recorded on magnetic tape. In the older analog tape recording (FM or AM) the range of signals that could be recorded linearly (i.e., retrievable with the same ratio of amplitudes and the same frequency spectrum as the original signal) was about $100:1$ (40 dB). In present-day digital recording this range has been extended considerably, but the advisability of reducing the interference/reflection signal ratio still remains. Because of the cost of the individual sources, source patterns are usually occupied sequentially and the effect of the source array realized only in later summation. It is therefore the task of the detector, or geophone, pattern to reduce the surface wave/reflection signal ratio before energy due to a single source is recorded. Geophones are much cheaper than sources and can be prewired in arrays or strings.

Once these strings of geophones are made up, however, they must be usable at any point in the reflection field setup, since it would be much too difficult to find a string for a particular location among hundreds available. We noted previously that the amplitude of surface waves decreases at least as $r^{-1/2}$, where r is the distance from the source. The greatest part of this decrease occurs in the first few hundred meters from the source. Our interest in this springs from a need, when designing an array of geophones, to have either equal contributions from each geophone or a knowledge of its likely contribution to the overall array response. Usually arrays are made up using equal contributions from all geophones and, if this is done, the array so manufactured must be used at distances no less than several hundreds of meters from the source so that the wave amplitude over the entire string can be sensibly constant. For conventional P-wave recording in the 5- to 50-Hz frequency range, a minimum offset of about 300 m (1000 ft) is usual.

In this discussion of receiver arrays, the presence of a weathered layer is ignored, as in fact it was for sources. Later, this omission is discussed in terms of the equivalent input signal for the exploration system.

The physical principles involved in geophone array design are the same as for sources. The receivers respond to the same plane wave signal but receive it with a different phase, depending on the angle of incidence. The vector addition of signals was illustrated in Figure 3.15, and we note that unequally weighted contributions can be obtained. Equally weighted geophones in any array thus have the same directivity function as the same array of sources, if the question of the directivity function of the individual elements is disregarded. Thus Figure 3.14 describes the response of an 18-element receiver array.

Vertical geophones have a response that is approximately $\cos \theta$, where θ is again the angle of incidence of the seismic wave. Horizontal geophones used for SH recording, however, record the full amplitude of the wave independent of its angle of incidence. In the case of SH strings therefore only the array directivity is available for discrimination on a direction basis.

In the discussion that follows, we pursue the question, from a theoretical point of view, whether or not arrays of receivers can be designed that reject surface waves

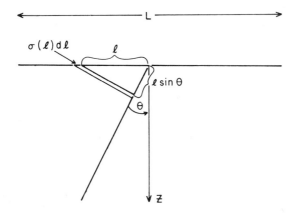

Figure 3.18 The concept of a continuous linear geophone system.

even better than the constant-amplitude arrays customarily used. Since we make use of Fourier transforms—which are developed in Chapter 4—the reader unacquainted with this technique is advised either to consult that chapter or refer to standard textbooks such as R. Brace, *The Fourier Transform and Its Applications.*

It is instructive to consider the concept (shown in Figure 3.18) of a continuous linear geophone system in which the response function is represented by

$$A(\omega) = \int_{-L/2}^{L/2} \sigma(l)e^{i\phi} \, dl$$

where $e^{i\phi}$ = delay factor

$$\phi = \omega l \sin \theta / V$$

$\sigma(l)$ = weight of an output element at the point l

Putting $c = (\sin \theta / V)(1/2\pi)$,

$$A(\omega) = \int_{-L/2}^{L/2} \sigma(l)e^{2\pi i \omega l c} \, dl$$

This is a Fourier integral, and known procedures exist to determine $A(\omega)$ if $\sigma(l)$ is known.

The $A(\omega)$ so obtained gives the output of the continuous system (for a given angle θ) as the angular frequency (ω) is varied.

However, in designing field methods or in using arrays, it is often more convenient to think of rejecting signals of a given frequency, which are approaching the surface at a known set of angles. It can easily be seen that this amounts to the ability to reject certain apparent wavelengths (along the surface) or to reject the corresponding apparent wave numbers k_{app}. The subscript is omitted, but understood, in:

$$\phi = \frac{\omega l \sin \theta}{v} = 2\pi k l$$

Hence

$$A(k) = \int_{-L/2}^{L/2} \sigma(l)e^{2\pi ikl}\, dl$$

and the Fourier inverse is

$$\sigma(l) = Q \int_{-\infty}^{\infty} A(k)e^{-2\pi ikl}\, dk$$

where Q = constant

The fact that the response of all the arrays so far discussed varies with frequency or wavelength, even within the acceptance band, is a source of concern. It is much better to have an array that does not disturb the form of the incoming pulse.

An interesting example of the Fourier transform procedure has been suggested by Goupillaud (1955). When infinite Fourier transforms are used, it is known that the Fourier transform of a bandpass in frequency (or k) is equivalent to a $(\sin kx)/kx$

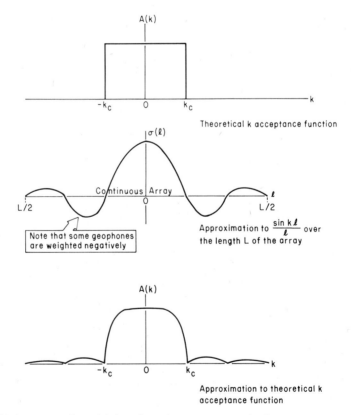

Figure 3.19 An attempt, by weighting of geophone outputs, to develop an acceptance function that is constant over a given range of wave numbers and zero outside this range. Note the negative weighting of some geophones.

pulse in time ($x = t$) or length ($x = l$). This suggests that for a continuous geophone system to accept all k between $-k_c$ and $+k_c$ (cutoff wave numbers), the weighting function for the system should be of the form $(\sin ql)/ql$. In any finite system, it is (as indicated in Figure 3.19) not possible to use the form $(\sin ql)/ql$ without truncation; hence the bandpass in k is not rectangular but approaches it, depending on the ratio of the total length L to the cutoff wavelength $\lambda_c = 1/k_c$. It is noted here that such a pattern or array involves some negative weighting. A further approximation is necessary because of the necessity for using discrete numbers of geophones in the array.

Weighting schemes other than the constant or the $(\sin kx)/kx$ have been suggested (Holtzman, 1963), but all of them rely on the constancy of sampling achieved by the individual geophones. In field practice, the economics of cables and recording has, up to the present, prevented the recording and combining at a later date of individual geophone samples. Many recorders sum the individual source contributions in the field, so these too are unavailable for later weighting. Earlier experiments with permanently weighted geophones in an array showed clearly the lack of consistency of sampling, and such schemes of weighting as one half-cycle of a sine wave proved to be no better in rejecting unwanted interference than a linear pattern only one-half the length.

3.10 NOISE, NOISE CANCELATION, MULTIPLE GEOPHONES AND SOURCES, AND TWO-DIMENSIONAL ARRAYS

In this entire chapter, so far, the materials through which waves are to be propagated and the conditions of observation have been assumed to be perfect. Under these conditions, the paths taken from the source to the receiver are those predicted geometrically, and the plane of propagation of the various waves generated and received is a vertical plane through the source and receiver. It was therefore legitimate to discuss only linear arrays which could be oriented on the surface to lie in this vertical plane. Furthermore, the conditions of observation were perfect, so that the geophones picked up only signals (wanted or unwanted) that were generated by the sources provided. The unwanted signal is classed as interference and is differentiated from noise which comes from extraneous sources beyond the control of the experimenter. Discontinuities in the materials (cracks, sudden lithological changes), particularly in the near surface layers, are often sufficiently intense and random in nature that the seismic waves from the artificial sources used can become incoherent and can approach the geophones of the linear arrays in directions other than along the length of the array. Even if the interference were coherent, the array would not have been designed to reject it. Further, if it is incoherent over the array, the design of the array will have little to do with its rejection.

Noise, however, can be coherent over the dimensions of the array, and some cancelation can be achieved if the wave number (in the direction of the array) is within the reject band, but this is not certain and in general is fortuitous. Noise is generated by local and distant earthquakes, by artificial sources such as traffic, railroads, and aircraft, by the wind swaying trees, bushes, and grasses, and by the fluctuating pressures in the air caused by wind variation due to topography (Frantti, 1963; Junger, 1964; Olhovich, 1964; Gupta, 1965).

Additional problems therefore have to be solved in some areas. These involve:

1. Incoherent noise.
2. Coherent energy coming from outside the source-receiver vertical plane.

The first of these has the characteristic that, once the geophone or source samples lie outside a given distance, the noise samples are uncorrelated, that is, samples taken at different locations or at different times at the same location tend to add randomly and the average amplitude increases only as the square root of the number of samples taken. Signals that are coherent, however, add linearly with respect to the number of samples. Thus the technique for handling random noise is to employ many geophones at separations greater than the coherence radius and to add them together. The coherence radius can be regarded as the average distance between points in the area for which the correlation coefficient between noise samples is .5. It does no good to increase the number of geophones indefinitely, regardless of separation, because then the samples are correlated and tend to add like the signal does. The statistical characteristics of the noise (spatially and temporally) must therefore be known.

The technique of reducing noise must further be linked to signal strength, because an increase in the signal strength of N is equivalent to adding together N^2 samples as far as the signal/noise ratio improvement is concerned. However, if the noise is spatially random but results from incoherent scattering of the signal (in other words, it is interference), no good will come from increasing the source strength.

The second problem, that of coherent noise coming from points not in the source-receiver vertical plane, must be solved by employing samples in the plane of the surface rather than in a single line in the plane. If there is a preferred direction from which the noise is coming, the two-dimensional pattern can be optimized to give the highest signal/noise ratio. The design of weighted two-dimensional arrays has been treated by Parr and Mayne (1955), although not from the point of view of two-dimensional Fourier analysis, which appears to be simpler now that computer

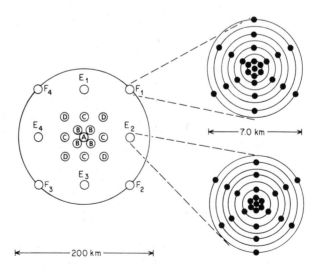

Figure 3.20 A complex compound array used in earthquake seismology at the Long Aperture Seismic Array, Montana. [After Capon et al. (1968), reprinted with permission from *Geophysics*.]

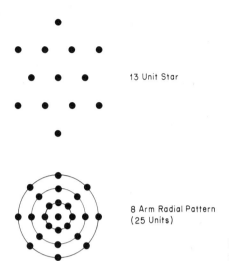

13 Unit Star

8 Arm Radial Pattern
(25 Units)

Figure 3.21 Different types of two-dimensional arrays used in exploration seismology.

methods are available. In exploration practice, these arrays are usually radial or star patterns of seven units upward.

Much larger compound fixed patterns (Figure 3.20) (Capon et al. 1968) have been used in earthquake seismology in order to achieve higher angular resolution and noise cancelation. Modern digital processing techniques allow beam steering and optimization of the noise cancelation performance (Capon, Greenfield, and Kolker, 1967; Burg, 1964).

In modern exploration practice, two-dimensional patterns are rarely used because of the noise cancelation afforded by the common depth point (CDP) method to be described later. However, in particularly stubborn areas, they are still used. Rectangular patterns of several hundred meters on a side are more usual than star or radial patterns, and both are only used as a last resort since economic considerations weigh heavily against their use. Two such patterns are shown in Figures 3.21.

3.11 SUMMARY

Explosive sources of seismic energy have now largely been replaced by repeatable, consistent sources which operate either on the surface of the earth or in the water. Different types are required for the two different environments. The newer sources are always of less strength than the explosive sources but, because of their mobility and ease of use, can be employed in multiple units either simultaneously or sequentially. The results of these separate unit sources can then be summed to give a result equivalent to that obtained from a single stronger source. In the presence of random noise, the use of N separate sequential sources results in an increase in the signal/random noise ratio of $\sqrt{N}$. In a like manner, signals from detectors having different noise environments also add like $\sqrt{N}$ as far as the noise is concerned.

Patterns of sources and detectors (geophones) can be used to give directivity to source-created energy or equivalent directivity in the reception of the receivers. Such pattern directivities are multiplicative, as is the directivity of the individual source (detector), with that of the pattern itself.

The calculated responses of patterns cannot in practice be fully realized, largely because of the inadequacy of the model used in the calculation. The earth parameters are not consistent enough to allow realization of the calculated weighting factors.

APPENDIX 3A: SIMPLISTIC ANALYSIS OF THE AIR CUSHION VIBRATOR

For this analysis, the system (Figure 3A.1) consists of a mass B, the platform, forming a tight-fitting but frictionless piston in a cylinder of area A and height h_0. The mass of the piston is held up by the dc pressure P_0 of the air in the chamber so that

$$P_0 A = Bg \tag{3A.1}$$

We neglect the escape of air at the base of the skirt, the compensating inflow of air produced by the dc air pump, and the flexing of the skirt. An auxiliary system X causes an alternating flow of air ($V \cos \omega t$) into the chamber which is thermally insulated and operates adiabatically.

First, for equilibrium at a perpendicular temperature, Boyle's and Charles' laws are followed:

$$P_0 A h_0 = \frac{m}{M} R T_0 \tag{3A.2}$$

where P_0, h_0, and T_0 = initial pressure, initial height, and temperature
R = Rydberg's constant
M = molecular weight of gas (g)
m = actual weight of gas in the chamber

Substituting (3A.1) into (3A.2),

$$h_0 = \frac{m}{MBg} R T_0 \tag{3A.3}$$

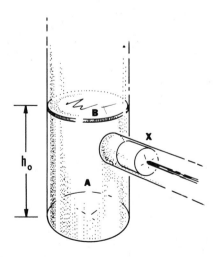

Figure 3A.1 A simplistic model of an air cushion vibrator.

Under nonequilibrium conditions,

$$B\ddot{h} - \delta PA = 0 \qquad (3A.4)$$

where δP = excess dynamic pressure,
$\ddot{h}$ = (upward) acceleration of the mass

It is assumed that, in the following, the dimensions of the chamber are very much smaller than the wavelengths of sound at the frequencies used. So we do not analyze the sound wave propagation in the chamber but the quasi-equilibrium pressures and volumes at the frequencies involved.

For adiabatic operation, no heat is lost from the enclosure. It is assumed that the piston X decreases by $X \cos \omega t$ the volume of air in the chamber, and that the volume is adjusted as necessary by the piston B rising a distance h.

The new volume is

$$V = V_0 - X \cos \omega t + Ah$$

$$\delta V = V - V_0 = -X \cos \omega t + Ah$$

Then, since $P_0 V_0^\gamma = PV^\gamma$ (adiabatic operation),

$$(P_0 + \delta P)(V_0 - X \cos \omega t + Ah)^\gamma = P_0 V_0^\gamma$$

$$\left(1 + \frac{\delta P}{P_0}\right)\left[1 - \left(\frac{X \cos \omega t - Ah}{V_0}\right)\right]^\gamma = 1$$

$$\frac{\delta P}{P_0} = \frac{\gamma(X \cos \omega t - Ah)}{V_0} \qquad (3A.5)$$

Combining (3A.4) and (3A.5),

$$B\ddot{h} + \frac{\gamma P_0 A^2}{V_0} h = \frac{\gamma P_0 AX}{V_0} \cos \omega t$$

This is recognized as the equation of a tuned circuit having a resonance at

$$\omega_R = 2\pi f_R = \sqrt{\frac{\gamma P_0 A^2}{V_0 B}}$$

To see if this leads to a resonant frequency of the correct order of magnitude for the air cushion vibrator under construction

$$
\begin{aligned}
B &= \text{mass of platform} = 17{,}000 \text{ lb} = 7.7 \times 10^6 \text{ g} \\
A &= 10 \text{ ft} \times 20 \text{ ft} = 1.9 \times 10^5 \text{ cm}^2 \\
V_0 &= 10 \text{ ft} \times 20 \text{ ft} \times 2 \text{ ft} = 1.13 \times 10^7 \text{ cm}^3 \\
P_0 &= 15.25 \text{ lb/in}^2. = 1.05 \times 10^6 \text{ g sec}^{-2}/\text{cm} \\
\gamma &\text{ (for diatomic gases)} = 1.4
\end{aligned}
$$

From these,

$$\omega_R = 24.7$$

$$f_R = 3.93 \text{ Hz}$$

This is very close to the resonant frequency (3.5 Hz) actually found.

APPENDIX 3B: THE VIBROSEIS® SYSTEM

3B.1   INTRODUCTION—GENERAL PRINCIPLES

The Vibroseis® system of exploration was devised in order to remove a dependence on explosive sources with their relatively unpredictable characteristics. The source now becomes a prolonged signal—of several seconds length—and causes a stress on the surface or in the earth or water, which follows some preselected form. The increased length of the signal serves two purposes:

1. It makes the input signal easier to control by analog circuitry.
2. The total power emitted by the source is increased in proportion to the length of time the signal is emitted.

Facsimiles of the source output are reflected at every interface in the reflection path, and there is some modification of the amplitude spectrum by attenuation in the earth. However, the receivers on the surface receive a total signal which is the sum of all the reflected signals—different in amplitude and delayed in time. It is not recognizable as a reflection record at this stage, because of the complexity of interference of the overlapping signals of strength a_i and delay time τ_i

$$F_2(t) = \sum_{i=1}^{N} a_i F_1(t - \tau_i) \tag{3B.1}$$

The second part of the Vibroseis® system involves correlation of the received trace F_2 with the original signal F_1, the two having been recorded on the same medium so that a time reference for them has been established. The correlation can be regarded as a mechanism for seeking out and marking the individual signals $F_{1(i)}$ in the received signal. The correlation process is simply a convolution with the signal turned around in time:

$$\phi_{12}(t) = F_2(t) * F_1(-t) = \sum_{i=1}^{N} a_i F_1(t - \tau_i) * F_1(-t)$$

$$= \sum_{i=1}^{N} a_i \phi_{11}(\tau_i) \tag{3B.2}$$

where $\phi_{11}(\tau)$ = an autocorrelation pulse occurring with its maximum at the delay τ

To convert to the frequency domain, we write

$$G_1(\omega) = \frac{1}{\sqrt{2\pi}} \int_{-\infty}^{\infty} F_1(t)e^{-i\omega t}\, dt$$

$$F_1(t) = \frac{1}{\sqrt{2\pi}} \int_{-\infty}^{\infty} G_1(\omega)e^{i\omega t}\, d\omega$$

$(3B.3)$

the standard Fourier integral pair. In the frequency domain, the correlation process is simply

$$\phi_{12}(\tau) = \frac{1}{\sqrt{2\pi}} \int_{-\infty}^{\infty} G_1^*(\omega)G_2(\omega)e^{i\omega t}\, d\omega = \frac{1}{\sqrt{2\pi}} \int_{-\infty}^{\infty} \gamma_{12}(\omega)e^{i\omega t}\, d\omega \qquad (3B.4)$$

where the asterisk denotes the complex conjugate of the spectrum $G_1(\omega)$. It is equivalent in the time domain of turning the trace around in time. $\phi_{12}(\tau)$ is the cross-correlation function in delay time, and $\gamma_{12}(\omega)$ is the spectrum of the cross-correlation function. We note that this shows that the spectra of the time functions to be cross-correlated are multiplied together and the phase of the pulse $G_1(\omega)$ is subtracted from the phase of $G_2(\omega)$. Thus the correlation of a time function with itself results in a zero-phase (symmetrical) pulse having the square of the amplitude of the original time function.

We can see now that, if $G_2(\omega)$ is replaced by the sum of the individual $G_1(\omega)$ pulses, with different delays,

$$G_2(\omega) = \sum_{j=1}^{N} a_j G_1(\omega)e^{-i\omega\tau_j} \qquad (3B.5)$$

$$\gamma_{12}(\omega) = \sum_{j=1}^{N} a_j G_1(\omega)G_1^*(\omega)e^{-i\omega\tau_j} \qquad (3B.6)$$

the cross-correlation with the control trace causes each of the long reflection signals to be compressed to zero-phase pulses at the proper arrival times τ_j. The amplitudes of the reflected signals are preserved after correlation.

Thus we have disclosed the general principles behind the Vibroseis® method, originally invented by Crawford, Doty, and Lee (1960). These principles are simple enough—it is the implementation both in the field and in the processing lab that requires further explanation. Much of the mystique of the Vibroseis® system was associated with the early need for analog recording practices, including methods of preserving the time relation between the reference signal and the geophone group signals. The control signal had to be laid down on one track of a magnetic recording sheet and then picked up and radioed to the vibrators as the geophone signals were being laid down on the same recording sheet—often with the use of narrow-head recording to facilitate later compositing. Further, methods had to be devised to perform analog correlations between the reference signal and the field traces (Anstey, 1964). All these ingenious pieces of equipment and the field vibrators had to work together precisely and reliably, and their completion represented a tour de force in experimental geophysics.

The onset of digital processing and analog-digital conversion (or straight digital recording) made many of these ingenious methods obsolete—and rendered the Vibroseis® method more accessible to the reliability and precision of digital correlation, filtering, and stacking.

3B.2   SELECTION OF INPUT SIGNAL

It has just been established that the shape of the autocorrelation pulses that make up a Vibroseis® record (sometimes called a correlogram) is dependent *only* on the amplitude spectrum of the signal input to the earth. Since the phase remains unspecified in this input signal, this indicates that any signal having a given amplitude spectrum suffices, and in fact this is true to a limited extent. Later we talk about the use of this concept (pseudorandom sweeps) in designing a system that allows more than one source to operate and produce independent data, although received by the same recording setup. For the moment, however, we confine ourselves to a straightforward single system. What determines the choice of an input time signal?

First we require that the output signal, after correlation, be as clean as possible, in two senses:

a. The near-in shape [$\phi_{11}(t)$ for $t < T_0$, where T_0 is about one period of the mid-frequency needed] must be as clean as possible.
b. The pulse must die away as rapidly as possible, without ghosts or local increases in amplitude when $t > T_0$.

Figure 3B.1 shows the situation. Here we are looking at one side of an auto-correlation function—the other side is a mirror image about zero time—whose

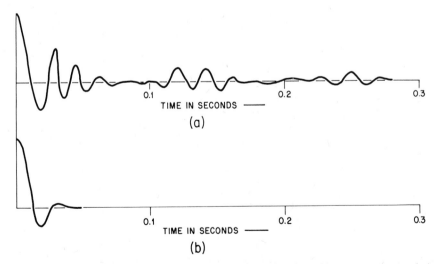

Figure 3B.1   In these two illustrations of autocorrelation functions (*a*) represents an undesirable type which as a very ringy near-in shape and shows local increases in amplitude at 0.13 and 0.25 sec, and (*b*) represents a more desirable, fast-decaying pulse with a near-zero background for $t > 0.05$ sec.

frequency spectrum is unspecified. A fast-decaying near-in shape is obviously desirable, because this indicates a reflection and a very "ringy" shape can easily hide several reflections and obviously reduces the resolution obtainable with the Vibroseis® method. The second requirement is related to the same problem: that a given autocorrelation wave shape occur at each reflection time. We know in practice that reflections generally decrease in amplitude with time, so that it may be necessary, during the processing step, to increase the gain of the system to make later reflections visible. But, if this is done, all the ghosts that occur on a single autocorrelation pulse will be amplified and will appear to be legitimate, strong reflections. However they portray only a much shallower structure and therefore constitute noise.

These requirements become stringent for the input wave shape—the control signal—and have been shown to allow only signals that

a. Are wide-band.
b. Have no sudden changes in amplitude.
c. Have *instantaneous* frequencies that do not repeat.
d. Have *instantaneous* frequencies that are smoothly varying.
e. Have, predominantly, a flat-amplitude spectrum.

To some extent, these qualifications overlap. For example, a constant-amplitude time signal that dwells temporarily (or even changes the rate of change in instantaneous frequency) causes the amplitude spectrum not to be flat.

Some qualifications conflict. For example, the desire for a wide-band, flat-amplitude spectrum within certain frequency limits causes a sudden amplitude change at the ends of the spectrum, and this results in ringing at these two frequencies and, as a consequence, a poor autocorrelation pulse both near in and at later times, since these particular frequencies do not die out quickly. All these phenomena are well known in electrical engineering practice and have been discussed at length by Edelmann (1966) and by Krey (1969).

As a consequence of these requirements, a standard linear swept frequency signal has been adopted for the control signal for use in normal situations. It has the form

$$F(t) = A(t) \sin 2\pi \left(at + \frac{bt^2}{2} \right) \qquad (0 < t < T) \qquad (3B.7)$$

This signal (looking at the sine function first) consists of a sinusoidal function of gradually (and uniformly) increasing frequency—starting at $\omega_0 = 2\pi a(f_0 = a)$ and ending at $\omega_f = 2\pi(a + bT)$. Thus a sweep going from 10 to 100 Hz in 10 sec has values of $a = 10$ and $b = 9.0$.

The purpose of the function $A(t)$ is to have a multiplying factor in the sinusoidal function so that the sweep does not start or stop abruptly in amplitude. The taper at each end is usually linear over a few cycles at each end of the sweep but is sometimes sinusoidal or a more complex function (see Figure 3B.2).

Suggestions have been made that filtered random noise can be used as a control signal for the Vibroseis® method. Random noise is chosen because, with a suitable distribution for the amplitudes of the noise trace, the spectrum can be made flat over a very wide band of frequencies. This is obviously a good starting point in view of requirement *e*. However, the original bandwidth is controlled by the sample time, and

(a) Linear Sweep-Linear Taper

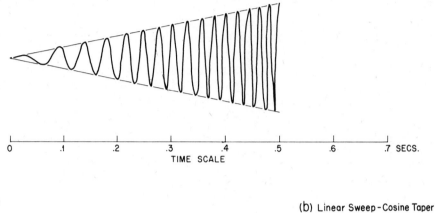

(b) Linear Sweep-Cosine Taper

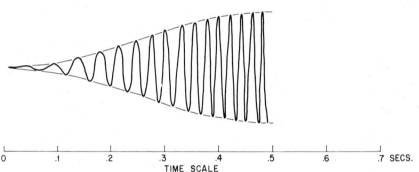

Figure 3B.2 Two forms of taper for sinusoidal sweep signals. The linear taper (*a*) is usually considered adequate and is the type featured in automatic sweep generators now widely used. Type (*b*) gives a quieter autocorrelation signal at times >0.05 sec.

the spectrum can be flat for bandwidths much greater than required—or greater than *can be accepted* by the mechanical vibrator system. So, the random noise trace is filtered with a zero-phase pulse designed to leave the spectrum amplitude alone inside the required band and cut it off (with a taper, if necessary or desired) outside this band. It is particularly necessary to remove low frequencies outside the design range of the vibrator, as seen later.

We can call this filtered version a pseudorandom sweep. It is now found that the filtered version no longer has a constant-amplitude envelope but has random variation.

For practical reasons, this type of sweep is not desired, since the vibrators are usually required to operate at maximum amplitude all the time, so that the signal entering the earth will have maximum energy.

For theoretical (or processing and interpretation) reasons, moreover, this sweep is undesirable, because there are local repetitions of instantaneous frequency, and these cause the overall form of the autocorrelation pulse to deteriorate because of the formation of ghosts—where two different parts of the sweep correlate to a greater extent than is desirable.

3B.3 PRACTICAL METHODS OF INTRODUCING THE INPUT SIGNAL INTO THE EARTH—VIBRATOR DESIGN PRINCIPLES

Having selected the signal needed, there remains the problem of introducing such a primary signal into the ground. The conventional method (see Crawford, Doty, and Lee, 1960) has been to vibrate a large plate (called the baseplate) which is kept in contact with the ground during the time of generation of the signal. If the vibratory signal desired has a range of force of $2A$ pounds, the baseplate must be held down on the ground with a force of at least A pounds, since otherwise it would leave the ground and hammering would result. The weight of the vehicle is used as the hold-down weight, and the vehicle is isolated from the vibration by air bags or similar "soft" springs. With the weight of the truck on it, the system corresponds to a very low-frequency-damped resonant system. The period is kept well below the lowest frequency to be generated by the vibrator. Large, heavy vehicles are therefore necessary to have a high enough mass to hold down the vibrator, as well as to carry the weight of the diesel power supplies, vibrator units, and auxiliary equipment. With the continual demand for increasing the low-frequency characteristics of vibrators, a compromise is becoming necessary between the weight of vehicles, their maneuverability, and road weight limits.

The vibrator mechanism is attached to the lift system through the isolating air bags, and it can be raised or lowered onto the ground by hydraulic cylinders capable of lifting the entire truck off the ground so that its weight is fully on the baseplate.

Plate 1   A standard Conoco Model 8 vertical vibrator. Peak force 19,000 lb; low-frequency limit 4.7 Hz.

Plate 2   Low-frequency research vibrator (Conoco). Peak force 36,000 lb; low-frequency limit 2.3 Hz.

Actually, it is desirable to leave a small proportion of the weight on the front wheels to stabilize the truck. The lift system is therefore positioned slightly aft of the center of gravity of the truck and mounted components. Raising and lowering have to be done in a few seconds in order to allow quick moving. In the raised position, the vibrator mechanism is well above the ground, to achieve maximum clearance in rough terrain.

All these features are evident in Plates 1 and 2 which illustrate, respectively, a commercially manufactured field vibrator (Conoco Model 8) and a large low-frequency vertical vibrator assembled for research purposes. The comparative specifications of these vibrators, as well as those of a shear wave vibrator, are given in Table 3B.1.

Shear wave vibrators are built using the same general engineering principles, to be described, but have a different method of introducing motion into the earth. For this reason, the baseplate area has been given as "Not Applicable," although it is later seen that it has an effect.

The usual starting points in the design specifications (Brown and Moxley, 1964) are the maximum peak force and the low-frequency cutoff values. At high frequencies, the system is limited in force because of the pressure available from the power supply. For example, the usual dc hydraulic fluid pressure is 3000 lb/in.^{2}, and there is usually a drop of 500 lb/in.^{2} across the servovalve, leaving 2500 lb/in.^{2} to be applied to the piston area. If, for example, this were 5.0 in.^{2}, the total peak force (one direction) available would be 12,500 lb. The low-frequency limit is usually the total stroke length available. For an actuator whose displacement is given by

$$y = A \cos \omega t \qquad y_{\max} = \pm A \tag{3B.8}$$

THE COMPARATIVE SPECIFICATIONS OF MODEL 8, CONOCO LFV, AND CONOCO SHEAR NO. 1 VIBRATORS

	Model 8	Conoco LFV	Conoco shear no. 1
Vibrator mounting	6 × 4 or 6 × 6	Truck 6 × 4	6 × 6
Piston area (in.^{2})	6.33	2 × 6.11	5.2
Peak force (lb)	19,000	36,600	15,600
Usable stroke	3.75 in.	9.0	4.0
Actuator weight (lb)	4,500	2 × 7,600	2,200
Displacement limit (Hz)	4.7	2.3	5.8
Required average flow at displacement limit (gpm)	58	126	63
Pump model	Kline 1240	Vickers PVB 90	PVB 90
Pump output (gpm @ rpm)	58 @ 1500	93 @ 1,800	93 @ 1,800
Pump horsepower	110	182	182
Vehicle wheelbase (in.)	199	198	212
Front axle weight (lb)	16,560	18,000	10,750
Rear axle weight (lb)	22,500	32,250	24,250
Total weight (lb)	37,110	50,250	35,000
Weight on ground (from baseplate)	36,080	48,750	17,803
Baseplate area (in.^{2})	3,700	3,888	3,500 (approx.) N.A.

the velocity is

$$\frac{dy}{dt} = \omega A \sin \omega t \qquad \frac{dy}{dt}_{\text{max}} = \pm \omega A$$

$$= \pm 2\pi f A \qquad\qquad (3\text{B}.9)$$

the acceleration is

$$\frac{d^2 y}{dt^2} = -\omega^2 A \cos \omega t \qquad\qquad (3\text{B}.10)$$

and the maximum is

$$\pm \omega^2 A = \pm 4\pi^2 f^2 A$$

Since the force available is M times acceleration,

$$\text{Maximum force} = 4M\pi^2 f^2 A \qquad\qquad (3\text{B}.11)$$

In the midband of frequencies, the maximum velocity of the actuator is limited either by the servovalve flow or by the pump flow. It is not entirely necessary to

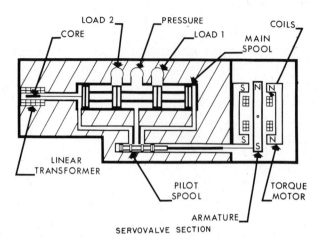

SERVOVALVE SECTION

Figure 3B.3 Schematic of a two-stage hydraulic servovalve with torque motor control and a linear transformer for feedback signals.

provide for a pump flow that is sufficient at all times of the swept frequency signal, since the demand at high frequencies can be supplied by an oil accumulator which is replenished during the idle periods between sweeps.

While fluids are normally considered incompressible, the high dynamic response demanded of the system necessitates consideration of the compressibility of the fluid, which acts as a spring between the hydraulic servovalve and the piston actuator. (Of course, the compliance of any coupling tubing also has to be taken into account.)

The entire system is controlled by the two-stage hydraulic servovalve, controlled by an electric torque motor (see Figure 3B.3). Five watts of electric power results in a deflection of ± 0.015 in., or a midposition force of ± 13 lb. This controls a pilot spool which in turn ports high-pressure fluid to the main spool. The motion of the main spool is monitored by a linear transformer which supplies a signal to be used for feedback control of the torque motor.

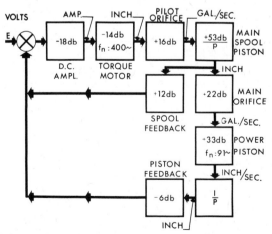

Figure 3B.4 A typical servosystem block diagram showing gains at each stage and feedback controls.

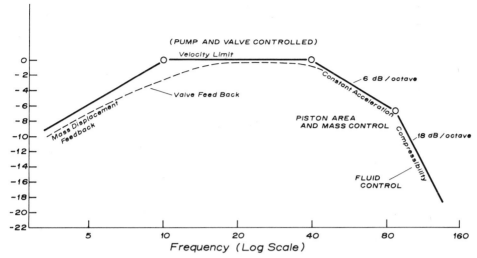

Figure 3B.5 Salient features of vibrator design.

An additional feedback control is used for the main piston, in order to make it constant *displacement* below a given frequency—at a point where there will be no danger of the driven mass exceeding the stroke limit. Bumpers are usually placed inside the framework to prevent damage should the stroke be temporarily exceeded. Over the entire midrange, the servovibrator tends to act as a constant-*velocity* system and, beyond the high-frequency break, as a constant-*acceleration* system. Fluid compression sets in soon and produces an additional 12 dB per octave drop in output (see Figure 3B.4).

Figure 3B.5 is a diagram of the salient features of the restrictions on vibrator output. The solid lines show the restrictions of the primary equipment, while the dotted lines show how these are affected by closing the servoloops.

While the design of the hydraulic servovibrator itself is complicated enough, there are still other considerations. Earlier experiments showed that the earth acts mainly as a spring, although there is a mass of earth to be driven. This mass is usually about that of the baseplate but varies with baseplate area and soil characteristics. Since seismic surveys of necessity cross different soils with different spring constants, the inconsistency of phase is intolerable in a system whose purpose is to measure times accurately. Between the early high-frequency reflections and deeper lower-frequency signals, a change in time interval of as much as 0.035 sec has been known to exist, and this is largely due to phase inconsistencies between the baseplate motion and the control signal.

Of course, baseplate motion can be monitored, recorded, and used as the reference signal with which the remote earth motions are correlated, but an important consideration is the possibility of using only one reference signal—that originally sent out by the sweep generators. Such a unique reference signal makes data processing much simpler. It is for this reason that a phase compensation system was developed.

A phase compensation system is shown, in block diagram form, in Figure 3B.6. The modern field concept is to have a digitally controlled sweep generator in each vibrator truck, as well as in the recording truck. A trigger set off in the recording truck sends a

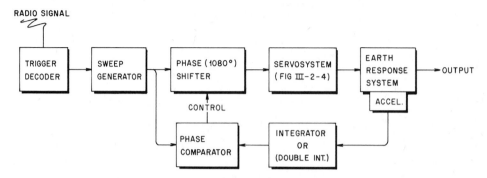

Figure 3B.6 Block diagram of phase compensation system.

coded radio signal to the vibrators, where it is decoded, and the individual sweep generators are all triggered simultaneously. This operational principle prevents a noisy radio signal from feeding into the vibrator(s). These sweep generators usually have controls to allow a change in initial frequency, final frequency, and the linear tapers at the two ends of the sweep.

This control signal is now split between the phase compensator and the phase shifter, the latter having the capability of shifting by a maximum of 1080°. The phase-shifted signal then goes to the servosystem shown in Figure 3B.4, and its output controls the motion of the baseplate against the earth. The earth acts both as an added baseplate mass as well as a spring. Finally, an accelerometer, designed for strong motion, integrated to give baseplate velocity (or sometimes doubly integrated to give baseplate position), delivers a signal to one side of the phase comparator which compares it with the reference signal and issues a (rectified) control signal to change the operating point of the phase shifter in such a direction as to reduce the discrepancy. Note that it does not work by applying a voltage feedback of the same frequency as the signal it is controlling. Instead it controls the phase added to the signal by a rectified signal from the comparator. When this rectified signal acts on the phase-shifting network, the *difference* in phase between the *control signal* and the *output velocity* is kept constant. A system of this type locks in more readily at high frequencies than at low frequencies, so the tendency is to use downsweeps when phase compensators are used, even though this is dangerous from the point of view of harmonic distortion, which is considered later. Note that no attempt is made to keep the amplitude spectrum constant from vibrator position to vibrator position. This type of consistency of output must be achieved through later digital processing.

All the above considerations apply equally to horizontal (shear wave) vibrators. The accelerometers are mounted on the baseplate to measure horizontal acceleration. With the large force available, a good deal of consideration has to be given to the problem of coupling the horizontal vibrator to the ground. Unless the coupling is continually good, some frequencies are inadequately conveyed to the ground. Several different systems have been tried, but the most satisfactory found so far consists of large pyramids welded, inverted, to the light baseplate so that, as the weight of the vibrator truck is placed on the baseplate, the pyramids are pushed into the ground. Thus a sideways force is counteracted by the truck's weight and, even if there is a tendency for the soil to consolidate, the combination of pyramid and truck weight

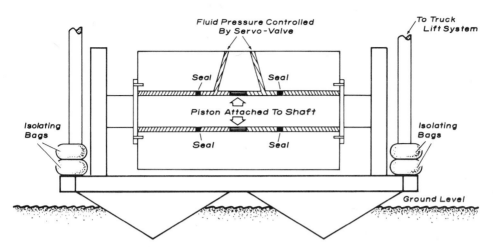

Figure 3B.7 Side schematic of a shear vibrator baseplate and actuator.

keeps the sides of the pyramids in contact with the soil (see Figure 3B.7). The damage to the surface, particularly in intensive patterns, may be severe but can usually be repaired.

Light baseplates are of course desirable, since the baseplate mass has to be accelerated by the hydraulic forces without producing any radiated power. However, considerations of strength usually limit the reduction in baseplate mass to about 2000 lb (~ 1000 kg).

3B.4 USE OF MORE THAN ONE VIBRATOR

Field experiments have shown that the law of linear superposition holds very closely in the case of multiple vibrators operating simultaneously. In other words, the use of N vibrators, all synchronized with the control signal, produces a signal in the earth, vertically under the group, which is N times that of a single vibrator. At other points in the earth, the N vibrators produce a signal which is the vector sum of each vibrator signal, thereby taking into account different travel times.

The mutual interaction effect, whereby the action of one vibrator affects the transfer impedance between the other vibrators and the ground, must exist but also must be very small.

For all practical purposes therefore, multiple vibrators can be used and can be disposed on the surface as the geophysicist wishes (see Figure 3B.8). In many cases, the vibrators stay within a few feet of each other, either abreast or in line, for one or more sweeps; then they all move up to give energy at the next location in the pattern required. In open country (pasture, desert, etc.), it may be desirable to use vibrators abreast to give a lateral pattern which will reduce interference coming from the side of the recording line. When the trails are narrow, this is a deterrent, and in-line patterns have to be used. Some companies dispose the available vibrators so that, after moving several times, they generate a tapered pattern.

Since the principle of superposition holds, the positions in a pattern can be occupied sequentially or simultaneously—whichever is expeditious.

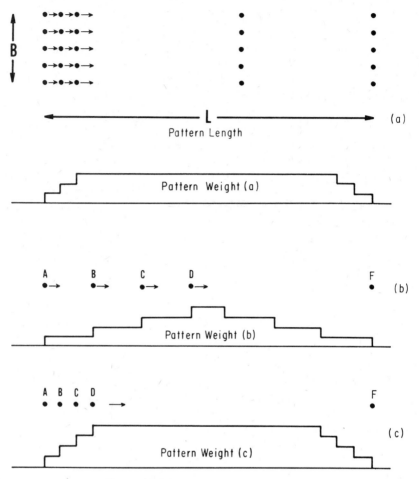

Figure 3B.8   Vibrator deployment schemes.

3B.5   HARMONIC DISTORTION

All vertical vibrators and, to a lesser extent, horizontal vibrators react with the earth in a nonlinear fashion when high-amplitude (usually low-frequency) signals are required. Seriff and Kim (1970) have discussed this phenomenon at length. Essentially, the harmonics of the low-frequency end of the sweep correlate with the proper fundamental frequency in the control signal and produce a ghost. The time of arrival of this ghost can easily be determined. As an example, we take a sweep signal that goes from 10 to 45 Hz in 7 sec. The rate of change in frequency is thus $35/7 = 5$ Hz/sec. Now, if the second harmonic of 10 Hz is to correlate with 20 Hz in the control signal, the delay will be $(20 - 10)/5 = 2$ sec. Thus *all* events on the Vibroseis® correlated record will have a ghost 2 sec away from the true event.

   It is now up to the geophysicist to design the sweep in such a way that the main body of the seismic cross section will be free from ghosts. Theoretically of course this can be done by using upsweeps, because then the ghosts will appear before zero record time. However, the phase compensator circuitry does not lock in easily on an up-sweep, and most companies prefer to use downsweeps and to alter the length of the sweep so that the ghost is beyond reflections of interest. In the previous example, if it had been known that all reflections of interest occurred before 4 sec, the first ghost (that of the first arrivals) could have been made to appear at 4+ sec simply by making the sweep 15 sec long. The rate of change in frequency is now $35/15 = 2.333$ Hz/sec, and the harmonic ghost of 10 Hz occurs at $(20 - 10)/2.333 = 4.286$ sec. With digital sweep generators, any reasonable sweep length is easy to obtain.

B. J. Heath (personal communication) has designed a digital program to examine and plot the spectral components within a short, sliding window so that a graphical picture of baseplate distortion is relatively easy to obtain. Figure 3B.9 is an example. It is seen that a small amount of third harmonic as well as a second harmonic are present. The symbols allow an estimate of the intensity to be made in 6-dB steps down from the maximum amplitude.

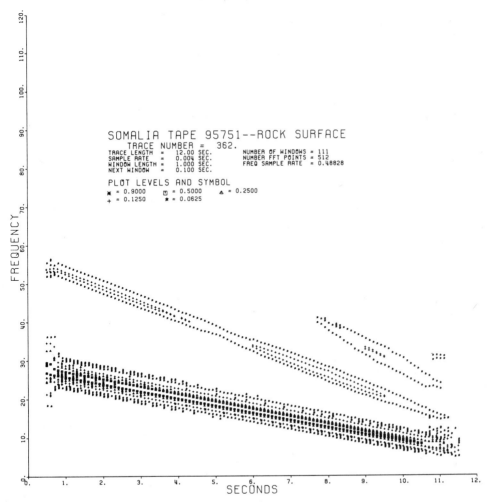

Figure 3B.9 A computer analysis of the motion of the vibrator baseplate. Note that some double- and triple-frequency motion is present because of the nonlinear nature of the load.

3B.6   SPECIAL CONSIDERATION FOR VIBROSEIS® FIELD METHODS

The Vibroseis® method, as normally used, is a surface source technique and therefore generates large-amplitude surface and near-surface waves. In contrast to impulsive sources, the long swept frequency signal usually ensures that the geophone array receives surface waves at the same time it receives reflections.

One primary need, in the case of Vibroseis®, is therefore a recording system that is linear over a very wide amplitude range, although attempts are made in the field to minimize the amplitudes of the interfering waves. This is the primary function of the pattern of geophones, since the source pattern is often occupied sequentially.

The geophysicist must decide if it is possible to sacrifice some amplitude of shallow reflections with respect to deep reflections, since any linear pattern of geophones rejects to some extent, energy from shallow reflections if it is made to reject near-surface interference. Figure 3B.10 makes this clear.

If we take, as an example, a frequency range from 20 to 60 Hz, a presumed pattern length of 400 ft, and the velocities and geometry of the previous figure, Figure 3B.11 shows the positions on the acceptance curve for the receiver pattern occupied by the deep reflector, the shallow reflector, and the refractor. It turns out in this case that the

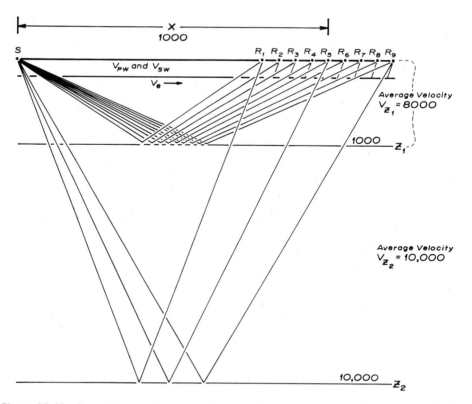

Figure 3B.10   Geometrical considerations for designing the geophone (or source) array. Phase velocities for the various waves are: (a) refractive, V_e; (b) Rayleigh waves, $V_R \approx 0.9\ V_s$; (c) shallow reflection, $V_{Z_1}/(X/2\sqrt{X^2/4 + Z_1^2})$; (d) deep reflection, $V_{Z_2}/(X/2\sqrt{X^2/4 + Z_2^2})$.

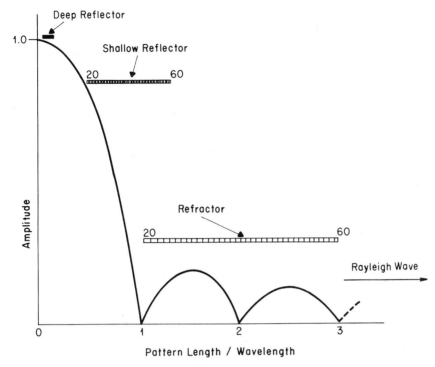

Figure 3B.11 Relative positions on the array acceptance curve for deep reflector, shallow reflector, and refracted events. The Rayleigh wave is off the diagram to the right ($X/\lambda = 10$ to 30).

Rayleigh wave occurs near $X/\lambda = 10$ to 30, so that there is very high rejection. The refractor coverage is within the second and third lobes of the acceptance curve and, since these are 180° out of phase, causes distortion of the refraction pulse. Note now that, with these parameters, the shallow reflector occurs over the steep part of the first and second lobes (including the first null). It is therefore reduced in size, predominantly because of high-frequency rejection. It is obvious that other compromises must be possible and each area has to be treated based on its own characteristics.

There are, however, two further points to be made about this method of rejection by patterns. The simple explanation above rests on the assumption of a continuous detector or a very high sampling rate. While the sampling rate affects the reflections very little, it can be important for waves, such as refraction and Rayleigh waves, that have small wavelengths. The cancelation shown by the curve is valid only if even the smallest wavelength is sampled with no less than two (preferably more) samples.

In the case of a high-frequency (60-Hz) Rayleigh wave, the wavelength is only 13.3 ft, so the geophones have to be no more than 6 ft apart. This requires a large number of geophones per pattern (approximately 70).

The second point to be made is that all the geophones in the pattern must accept the wave with equal amplitude. If the patterns are started too close to the source, the near geophones will be subjected to much higher interference amplitudes than the far ones, and the theoretical pattern effect will not be obtained. This is one reason why Vibroseis® field work is often done with relatively large (1000 ft) minimum offset.

With a geophone pattern of this type, the dynamic range of amplitudes dealt with by the recorder is reduced, and the interference is no longer intolerably larger than the reflection amplitudes during any given sweep injection. Of course, a similar pattern can be used sequentially for the sources or, in some cases, a different pattern can be devised which places zeros (nulls) where the geophone pattern has maximum acceptance of the interference. The proper choices are determined by experiment. The final output at the geophone pattern, after compositing of the results due to all the sweeps, is the product of the acceptance functions of the two separate acceptance functions and results in a high signal/interference ratio for the reflection work.

The second reason for a relatively large initial offset is to prevent too great a change in reflection amplitude with time. As the offset is increased, the rate of change in travel time (hence spherical divergence) decreases. Inasmuch as the primary pulse for Vibroseis® work is an autocorrelation function controlled by the reference signal, it is important that the side lobes of this autocorrelation pulse be maintained at a low level compared with other primary peaks occurring at the same time. The rate of decay of the autocorrelation function can theoretically be made 100:1 or better in 0.1 sec, but in practice the rate of change in gain needed for proper visibility of reflections should be kept small. In shot reflection work, gain changes from 0.1 to several seconds can be of the order of 10^4, but in Vibroseis® work, it should not be necessary to increase the gain over the same interval by more than a factor of 20.

3B.7   INCREASE IN TOTAL ENERGY BY INCREASING TIME OF INJECTION

If Vibroseis® reflection results fail to show an adequate signal/noise ratio, particularly for deeper reflections, the most effective way to increase the available energy is obviously to increase the number of vibrators. However, there is an economic limit and usually a logistic limit to this procedure. Since we are trying to improve the signal and reduce the random noise, another possibility is to increase the time of injection of energy. The sweeps can be made longer and/or more of them used for each reflection source position. The signal increases linearly with the time of operation of the vibrators, whereas random noise amplitudes increase as the square root of the recording time.

As a consequence, a net gain equal to the square root of the time increase is realized. An extreme example of this has been described by Fowler and Waters (1975) to provide sufficient energy for investigation of deep crustal reflections. In this case, 240 separate sweeps were composited, giving a theoretical increase in the signal/noise ratio over one sweep of 15.5. In most circumstances, this would be an uneconomical procedure, but smaller uses of the time increase are often possible in normal reflection practice.

3B.8   DESIGN CRITERIA FOR FIELD WORK AND INITIAL FIELD EXPERIMENTATION

The versatility of the Vibroseis® system in obtaining reflections of high or low frequency, at great and shallow depths and with a diversity of ground surface conditions, means that it is possible to tailor the operating conditions to suit the job being undertaken. Obviously, the exactness of fit depends on how much is known about the work to be done, chiefly about the surface and subsurface conditions in the area.

The depth of the primary reflection objective and some velocities are possibly the most important factors. Essentially they determine, to some extent, the frequency range of the sweep, the maximum and minimum offsets, and the pattern sizes that can be allowed. The experimental work needed immediately is a determination of the surface wave and refractor velocities, since they determine pattern lengths. This is usually done first with a single spread of bunched geophones, using source positions at multiples of the spread length (Figure 3B.12a). Usually, this work should be done with a very wide-frequency band, since later filtering can be used to determine the frequency characteristics of the various events recorded. It is rarely the case that a reflection is seen clearly on this interference record. However, the refraction velocity for the top of the unweathered layer and any subsequent higher-velocity refractors can be easily measured, and the time offset cross section shows clearly those zones that are free from high-amplitude surface waves. The latter are dispersive, and the range of velocities can also be measured.

It is desirable of course to place the geophone spread at an offset distance where the interference amplitude is low for the zone of reflection times that are the most important. On these initial interference records, the proper placement may not be clear, but this decision is made easier by repeating the interference spread using different lengths of geophone patterns. Figure 3B.12b is an example in the same area as Figure 3B.12a and the improvement is clearly seen. It must be remembered that an equal amount of improvement can be expected on the final reflection records as a result of the effect of the source patterns.

In some particularly difficult areas, usually associated with a hard rock surface, rough topography, or surface cracking, elimination of in-line interference may not be sufficient to show the reflections clearly. The interference spreads should then be repeated with geophone "nests" which have lateral coverage and, in some cases, lateral source coverage is necessary also in order to reduce scattering from inhomogeneities in the near-surface formations.

All these increments in field effort necessarily reduce the amount of effective work that can be done per day and therefore have to be used with discretion. There is nevertheless no substitute for an adequate signal/interference ratio in the final reflection record. The examples given are illustrative only, and each area legitimately demands its own interference investigation.

The field problems (Erickson Miller, and Waters 1968; Cherry and Waters 1968) for shear wave reflection generation are in many respects similar to those for compressional waves. An important difference, however, is that shear wave velocity in the weathered layer is much more variable and the contrast between the weathering and subweathering velocities is almost always greater. There are at least three consequences. First, Love waves, guided by the weathered layer, are stronger relative to shear reflections than pseudo-Rayleigh waves relative to compressional wave reflections. They are therefore harder to eliminate. Second, the reverberation in the weathered layer makes the downward traveling pulse more ringy, at a frequency determined by the weathering velocity and thickness. Third, the very low weathering velocity and its variability make the weathering corrections more severe and more difficult to extract from the data. By the same token, the variability of weathering velocity causes scattering from velocity inhomogeneities to be more prevalent.

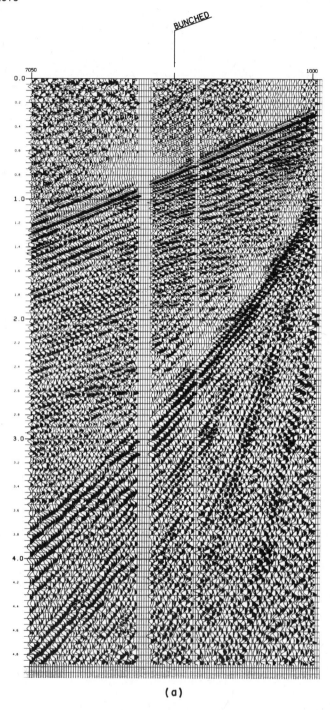

(a)

Figure 3B.12 (*a*) An interference record using bunched geophones (no array).

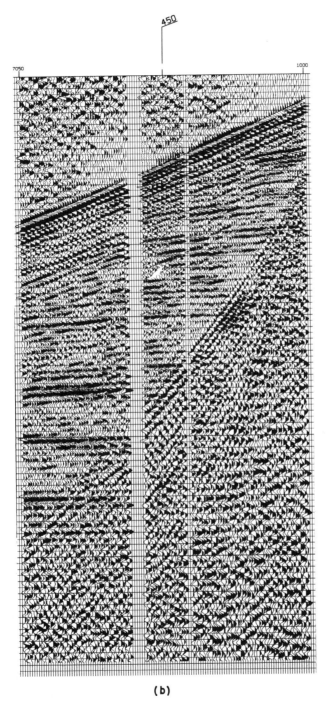

Figure 3B.12 (*b*) The same interference when an array of geophones is used. Array length was 137 m (450 ft).

3B.9   PSEUDORANDOM SWEEPS AND THEIR USES

An allusion was previously made to the possibility of using swept frequency control signals that are not linear increases in frequency with time. The total bandwidth of seismic signals is not very great for most areas. In some cases, by design or otherwise, two field crews, or a composite field crew, may need to have two sets of vibrators working at the same time and yet have the results distinguishable from one another. This problem has been incompletely investigated. Random noise, filtered to the seismic bandwidth, has the major disadvantage that the vibrators are not used at maximum efficiency. This part of the problem can be solved by a method due to Goupillaud (1976) in which a standard linear sweep is decomposed into individual, numbered, positive and negative half-cycles. These numbered half-cycles are then reassembled under the control of a random sequence generator, with the provision that positive and negative half-cycles alternate, It is of course possible to do this (for a sweep signal that may consist of 500 or more half-cycles) in an almost infinite number of ways; some will satisfy other considerations, and others will not. It is necessary to filter the reassembled sweep with a zero-phase band-pass filter in order to eliminate the slope

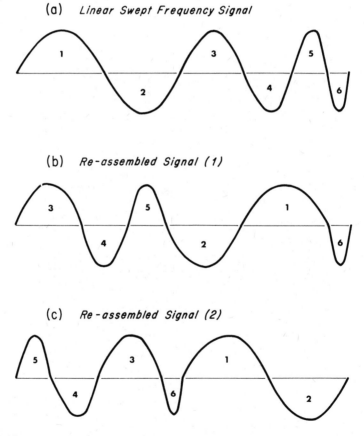

Figure 3B.13 Pseudorandom sweeps (*b*) and (*c*) are derived from the linear sweep (*a*). In the field, the number of half-cycles can be two orders of magnitude greater than the six shown here.

changes that inevitably occur when half-cycles are assembled in a random manner. A (trivial) set of examples is given in Figure 3B.13.

In this case, however, every sweep has to be correlated with the geophone output before compositing, and this adds to the difficulty with present field recording systems. The advent of new, extremely fast modules which can be interfaced with standard computers promises to make new recording systems possible, which correlate in the field and only store the correlated or composited data. Single-trace, real-time correlators already exist and have been used in the field for special applications.

As mentioned earlier, this investigation has hardly begun and will probably be a useful adjunct to the standard Vibroseis® system only when three-dimensional data collection becomes more prevalent.

REFERENCES

Anstey, N. A. (1964), "Correlation Techniques—A Review," *Geophysical Prospecting*, Vol. 12, No. 4, p. 355.

Anstey, N. A., Evenden, B. S., and Stone, D. R. (1970), *Seismic Instruments*, Vols. I and II. Gebrüder Borntraeger, Berlin, Germany.

Blake, F. A., Jr. (1952), "Spherical Wave Propagation in Solid Media," *Journal of the Acoustical Society of America*, Vol. 24, pp. 211–215.

Bracewell, R. (1965), *The Fourier Transform and Its Applications*, McGraw-Hill, New York.

Brown, G. L., and Moxley, S. D. (1964), *IEEE International Convention Record*, Vol. 12, Part 8.

Burg, J. P. (1964), "Three Dimensional Filtering with an Array of Seismometers," *Geophysics*, Vol. 29, pp. 693–713.

Capon, J., Greenfield, R. J., Kolker, R. J. (1967), "Multidimensional Maximum-Likelihood Processing of a Large Aperture Seismic Array," *Proceedings of the IEEE*, Vol. 55, pp. 192–211.

Capon, J., Greenfield, R. J., Kolker, R. J., and Lacross, R, T. (1968), "Short Period Signal Processing Results for the Large Aperture Seismic Array," *Geophysics*, Vol. 33, p. 452.

Cherry, J. T., and Waters, K. H. (1968), "Shear-Wave Recording Using Continuous Signal Methods, Part 1—Early Development," *Geophysics*, Vol. 33, No. 2, pp. 229–239.

Cole, R. H. (1948), *Underwater Explosions*, Princeton University Press, Princeton, N.J.

Crawford, J. M., Doty, W. E. N. D., and Lee, M. R. (1960), "Continuous Signal Seismograph," *Geophysics*, Vol. 25, No. 1, pp. 95–105.

Edelmann, H. (1966), "New Filtering Methods with 'Vibroseis,'" *Geophysical Prospecting*, Vol. 14, No. 49, pp. 455–469.

Erickson, E. L., Miller, D. E., and Waters, K. H. (1968), "Shear Wave Recording Using Continuous Signal Methods, Part 2—Later Experimentation," *Geophysics*, Vol. 33, No. 2, pp. 240–254.

Ewing, W. M., Jardetsky, W. S., and Press, F. (1957), *Elastic Waves in Layered Media*, McGraw-Hill, New York.

Fowler, J. C., and Waters, K. H. (1975), "Deep Crustal Reflection Recording Using 'Vibroseis' Methods—A Feasibility Study," *Geophysics*, Vol. 40, No. 3, pp. 399–410.

Frantti, G. E. (1963), "The Nature of High Frequency Earth Noise Spectra," *Geophysics*, Vol. 28, p. 547.

Goupillaud, P. L. (1954, 1955, 1956), personal communications.

Gupta, I. J. (1965), "Standing Wave Phenomena in Short Period Seismic Noise," *Geophysics*, Vol. 30, p. 1179.

Holtzman, M. (1963), "Chebyshev Optimized Geophone Arrays," *Geophysics*, Vol. 28, p. 145.

Junger, A. (1964), "Signal to Noise Ratio and Record Quality," *Geophysics*, Vol. 29, p. 922.

Kisslinger, C. (1963), "The Generation of the Primary Seismic Signal by a Contained Explosion," Vesiac State of the Art Report, University of Michigan, AD 403708.

Krey, T. (1969), "Remarks on the Signal to Noise Ratio in the Vibroseis System," *Geophysical Prospecting*, Vol. 17, No. 3, pp. 206–218.

Lampson, C. W. (1946), "Final Report on Effects of Underground Explosions," Office of Scientific Research and Development, Report No. 6645.

Merzner, P. (1965), "Abstracts of E.A.E.G. Meeting," *Geophysical Prospecting*, Vol. 13, No. 1, p. 140.

Miller, G. F., and Pursey, H. (1954), "The Field and Radiation Impedance of Mechanical Radiators on the Free Surface of a Semi-Infinite Isotropic Solid," *Proceedings of the Royal Society, A*, Vol. 223, p. 521.

Miller, G. F., and Pursey, H. (1956), "On the Partition of Energy between Elastic Waves in a Semi-Infinite Solid," *Proceedings of the Royal Society, A*, Vol. 225, p. 55.

Molotova, L. V. (1963), "Velocity Ratio of Longitudinal and Transverse Wave in Terrigenous Rocks," *Bulletin (Izvestiya) of the Academy of Science of the U.S.S.R.*, Geophys. Series, No. 12, p. 1769.

Muir, F., and Morrison, J. P. (1973), "Anti-Aliasing of Spatial Frequencies by Geophone and Source Placement," U.S. Patent No. 3,719,924.

O'Brien, P. N. S. (1957), "The Relationship between Seismic Amplitude and Weight of Charge," *Geophysical Prospecting*, Vol. 5, pp. 349–352.

O'Brien, P. N. S. (1960), "Seismic Energy from Explosions," *Geophysical Journal*, Vol. 3, pp. 29–44.

Okado, H. (1964), "Analyses of Seismic Waves Generated by Small Explosions," *Journal of the Faculty of Science, Hokkaido University, Japan*, Series 7, Vol. 2, No. 2, p. 197.

Parlin, B. R. (1958), "A Review of Similitude Theory in Ground Shock Problems," RM-2173, The Rand Corporation, Washington, D.C. and Santa Monica, CA.

Parlin, B. R. (1962), "Elastic Wave Calculations for the Cowboy Program," RM-3105-ACE, The Rand Corporation, Washington, D.C. and Santa Monica, CA.

Parr, J. O., and Mayne, W. H. (1955), "A New Method of Pattern Shooting," *Geophysics*, Vol. 20, p. 539.

Poulter, T. C. (1950), "The Poulter Seismic Method of Geophysical Exploration," *Geophysics*, Vol. 15, p. 181.

Seriff, A. J., and Kim, W. H. (1970), "The Effect of Harmonic Distortion in the Use of Vibrators," *Geophysics*, Vol. 35, No. 3, p. 234.

Sharpe, J. A. (1942), "The Production of Elastic Waves by Explosion Pressures (I)." *Geophysics*, Vol. 7, p. 144.

Sharpe, J. A. (1942), "The Production of Elastic Waves by Explosion Pressures (II)," *Geophysics*, Vol. 7, p. 311.

Smith, S. G. (1975), "Measurement of Airgun Waveforms," *Geophysical Journal of the Royal Astronomical Society*, Vol. 42, pp. 273–280.

Wolf, A. (1942), "The Limiting Sensitivity of Seismic Detectors," *Geophysics*, Vol. 7, p. 115.

Wolf, A. (1944), "The Equation of Motion of a Geophone on the Surface of an Elastic Body," *Geophysics*, Vol. 9, pp. 29–35.

FOUR

Description of Wave Trains
and the Characteristics of
the Reflection Process

4.1 TIME AND FREQUENCY DOMAIN CONCEPTS

The concept of time of arrival of an event and delay time after the instant of initiation are familiar ones related to our everyday experience that one thing happens after another. It is orthodox to plot these arrivals as a graph of their amplitude, particle velocity, pressure, and so on, as a function of time in which the time line is horizontal (the independent variable) and the event parameter is the dependent variable. Thus Figure 4.1, the time domain description of a series of reflected seismic wave amplitudes arriving at a geophone, is a conventional, graphic means of illustration.

From the point of view of any extended analysis (apart from a simple time difference measurement made between two selected points), a time domain description is sometimes inconvenient. An electrical or communications engineer, for example, needs to know how best to design amplifiers, filters, and other processing equipment. Geophysicists are interested in the effects on interpretation of geology due to filtering of the signals—by passage through the layered earth and by instrumentation used both in creating and receiving the seismic waves. Fortunately, as is well known in electrical engineering, there is available an alternative means of describing a time series. This description is known as the frequency domain method. It consists of building up a series of sinusoidal signals of known amplitudes and phases, the frequencies being related to a fundamental frequency in a harmonic manner; that is, the frequencies used for the sinusoids are all integral multiples of the fundamental frequency.

The first step is to decide the length of time (a finite time, for the present) over which the description is needed. If this is T, the fundamental frequency f_0 is established as $1/T$. In the case of seismic signals, for example, we can be satisfied to take a seismic trace (time series) 6.0 sec long simply because any reflections that take longer than 6 sec to arrive come from depths greater than current drilling depths. Then, the fundamental frequency would be 1/6 hertz, and the Fourier theorem says that it is possible to describe any single-valued time series (having only a finite number of discontinuities) within the interval T by adding up a series of sinusoidal signals having frequencies $f_0, 2f_0, 3f_0, \ldots, Nf_0$, and a constant. The degree of approximation to the time function increases as N increases. The amplitude and phase associated with each of these

101

Signal
Amplitude

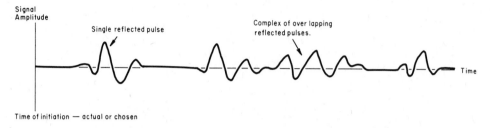

Figure 4.1   Time domain illustration of a series of reflected pulses arriving at a receiver.

sinusoidal signals have to be determined by Fourier analysis, which is described shortly.

Inversely, if somehow the amplitudes and phases of the constituent sinusoids are already known, for example, by calculation, it is obviously possible to add them together to obtain the (unknown) time series. This is referred to as Fourier synthesis.

Alteration of the amplitudes and phases of the constituent frequency sinusoids (on Fourier synthesis) produces a different time series, and this can be regarded as a fundamental definition of "filtering." Thus, to filter a time series, it is necessary to use Fourier analysis to gain knowledge of the amplitude and phase of the harmonic frequencies, alter these in a manner prescribed by the filter action, and then use Fourier synthesis to go back to the filtered trace.

In mathematical terms,

$$S(t) = a_0 + \sum_1^N a_j \cos 2\pi j f_0 t + \sum_1^N b_j \sin 2\pi j f_0 t \tag{4.1}$$

is a statement of the Fourier synthesis procedure, where it is assumed that the coefficients $a_0, a_1, a_2, \ldots, a_N$ and $b_1, b_2, \ldots, b_N$, as well as f_0, are known.

However, if $F(t)$ and f_0 are known, the coefficients $a_0, a_1, a_2, \ldots, a_N$ and $b_1, b_2, \ldots, b_N$ can be found from the formulas

$$a_0 = \frac{1}{T} \int_0^T F(t)\, dt$$

$$a_j = \frac{1}{T} \int_0^T F(t) \cos 2\pi j f_0 t\, dt \tag{4.2}$$

$$b_j = \frac{1}{T} \int_0^T F(t) \sin 2\pi j f_0 t\, dt$$

and it is noted that these processes really describe the calculation of a correlation coefficient between the known $F(t)$ and each of the cosine and sine functions corresponding to the individual harmonics.

Each harmonic is independent of another harmonic, and this is the power of the frequency domain description, since during a filtering operation the various frequencies can be dealt with separately.

If we form the complex number C_j:

$$C_j = a_j + ib_j \quad \text{(where } i = \sqrt{-1}\text{)} \tag{4.3}$$

the two relations above become

$$a_j + ib_j = C_j = \frac{1}{T} \int_0^T F(t)e^{ijf_0t}\,dt \tag{4.4}$$

Since

$$\cos 2\pi j f_0(t + T) = \cos 2\pi j f_0 t$$

and

$$\sin 2\pi j f_0(t + T) = \sin 2\pi j f_0 t \quad \text{(for all values of } j\text{)}$$

we can see that Fourier analysis and synthesis really describe a signal that is periodic and repeats itself for every T interval.

One way of looking at this is that the trace being examined (or synthesized) is in the form of a circle of circumference T, and that as t increases the process goes continually around the circle. There is a reference point on the circle where the beginning of the time trace joins on to the end. This may result in a discontinuity, but the value of the trace at this point is $\frac{1}{2}[F(0) + F(T)]$. See Figure 4.2.

To describe the signal $S(t)$ in the frequency domain then involves two different graphs, amplitude being plotted against discrete values of frequency (separated by

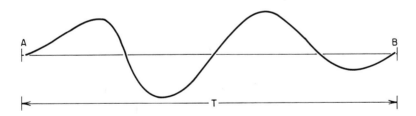

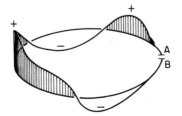

Figure 4.2 Illustration that the seismic trace AB can be joined, head to tail, to give a circular trace of circumference T. Filtering operations can cause some energy to cross the "join" AB.

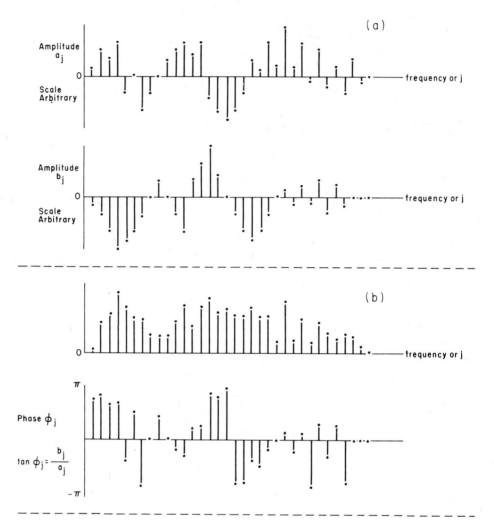

Figure 4.3 Representation of a time function over the interval, $0 < t < 1/f_0$. (a) Representation as real and imaginary components for each frequency. (b) Representation as amplitude and phase spectra.

f_0). One graph is for a_j and the other for b_j (sometimes called the real and imaginary parts, respectively). (See Figure 4.3a.)

We can also define quantities, called the amplitude

$$A_j = \sqrt{a_j^2 + b_j^2} \tag{4.5}$$

and the phase

$$\phi_j = \tan^{-1} \frac{b_j}{a_j} \tag{4.6}$$

and by plotting these, the amplitude and phase spectra, we have an equivalent representation (Figure 4.3b). In both these illustrations, $N = 34$ and the time interval

T must be divided into $2N$ intervals. The $2N + 1$ points to be obtained are equidistant samples throughout the time series. All this is in agreement with the fact that N cosine coefficients, N sine coefficients, and one constant are to be determined.

It is obvious that dividing the time interval T into $2N$ intervals can easily lead to inadequate sampling, since much more sudden (high-frequency) variations may be present in the trace than can be adequately sampled by a given sampling rate.

4.2   SAMPLING CONSIDERATIONS

We wish to consider how often equidistant samples must be taken in order to describe adequately a given time function. We have already seen that it is possible to describe such a function in the frequency domain as a series of sinusoidal functions of different amplitudes and phases and that these amplitudes and phases are independent of each other as one goes from one frequency to the next. It seems intuitively obvious that the gradual variations in a function can be described by widely spaced samples, whereas very rapid variations require very high-frequency sampling.

Figure 4.4 shows the same frequency sinusoidal wave sampled with a sampling interval much less than one-half of the period, again with the sampling interval

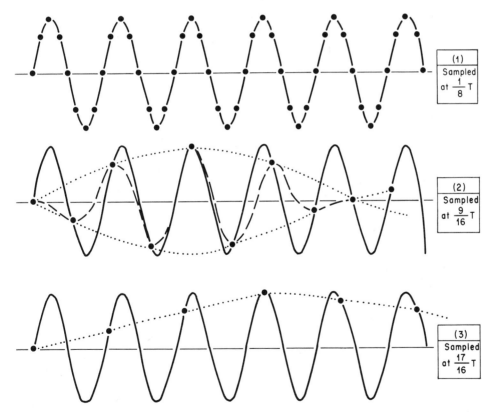

Figure 4.4   How insufficient sampling of a signal (less than two per cycle) causes the introduction of new, or aliased, frequencies which were not present in the original, continuous, signal.

greater than one-half the period, and finally where the sampling is greater than the period of the sinusoid. It is seen that artificial low frequencies are created by inadequate sampling. This process is known as *aliasing*. This is dangerous in the computation of Fourier spectra, hence a cardinal rule in sampling is that there must be at least two sample points per cycle for the highest frequencies present in the signal to be analyzed —not the highest frequencies *of interest* but the highest frequencies actually present. This must be remembered in the sampling of rapidly varying logs. In seismic recording practice, electrical antialiasing filters are provided before the sampling (or digitization) step to ensure that frequencies above an established maximum are severely attenuated by analog filtering. As an example, suppose that it has been established that digital samples are needed every 0.004 sec; this establishes that the highest frequency that can be adequately sampled is 125 Hz (period 0.008 sec). Analog filters are then included in the circuit which rejects frequencies above (say) $62\frac{1}{2}$ Hz very severely. The frequency that corresponds to the two samples per cycle is called the Nyquist frequency or the folding frequency.

This phenomenon is discussed at greater length when the general subject of data processing is taken up. For the present, we use it only to see its effect on Fourier analysis and synthesis.

In going back to the Fourier series analysis of (4.2), it is seen that we have selected N separate frequencies, separated by f_0. Actually, $2N + 1$ separate amplitudes have to be computed, and this means that $2N + 1$ pieces of independent data have to be supplied. It is for this reason that the period over which the Fourier analysis is to take place is divided into $2N$ intervals, giving rise to $2N + 1$ sample points.

Since only a finite number of points has been specified, it is not possible to compute uniquely the integrals in (4.3), and they must now be replaced by summation formulas:

$$a_j = \frac{1}{N} \sum_{j=0}^{2N} S_j \cos \frac{2\pi j}{2N} = \frac{1}{N} \sum_{j=0}^{2N} S_j \cos \frac{\pi j}{N}$$

$$b_j = \frac{1}{N} \sum_{j=0}^{2N} S_j \sin \frac{2\pi j}{2N} = \frac{1}{N} \sum_{j=0}^{2N} S_j \sin \frac{\pi j}{N}$$

(4.7)

These formulas hold for all j except 0 and N, where they become

$$a_0 = \frac{1}{2N} \sum_{j=0}^{2N} S_j$$

$$a_N = \frac{1}{2N} \sum_{j=0}^{2N} S_j \cos \frac{\pi j}{N} \qquad b_N = \frac{1}{2N} \sum_{j=0}^{2N} S_j \sin \frac{\pi j}{N}$$

(4.8)

4.3   FOURIER INTEGRALS—CONTINUOUS-FREQUENCY DISTRIBUTIONS

While the Fourier series represents the only practical way to analyze a time function, it has been seen that, by restricting the length of the function to be represented, a tacit assumption is made that the function is periodic, with the chosen time interval as the fundamental period. In practice, if one requires the transformation of a pulse in the time domain into its corresponding frequency domain spectra, there is little problem, since points having zero amplitudes can be added before and/or after the pulse duration, so that the interval between frequency components f_0 can be made as small as required—so also can the highest frequency be made as high as required by taking the sampling frequency high enough.

There is, however, a formal extension of the Fourier series representation to the continuous Fourier integral representation, in which the time domain and the frequency domain functions are piecewise continuous functions of the time or frequency. Often, these integral forms are used, since the mathematical manipulations are easier to perform. Stratton (1941) gives a readable account of this extension. In the complex notation normally used, this transform pair is given as

$$f(t) = \frac{1}{\sqrt{2\pi}} \int_{-\infty}^{\infty} g(\omega)e^{+i\omega t} d\omega$$

$$g(\omega) = \frac{1}{\sqrt{2\pi}} \int_{-\infty}^{\infty} f(t)e^{-i\omega t} dt$$

(4.9)

4.4   THE CONCEPT OF AN IMPULSE—OR THE DELTA FUNCTION

It is useful, on many occasions, to determine first what information can be gathered from the seismic method if there are no constraints on the frequencies that can be transmitted into the earth and received by the geophones. Then the limitations on such knowledge due to frequency restrictions, of whatever form, can be determined as desired.

For this purpose, the concept of an impulse is useful. It is a curious function, since it has no value of amplitude except at one point in time. It can be approached through several more familiar concepts.

1. It can be regarded as the limit of a rectangular pulse in time, in which the function has a constant amplitude over a range of time, if the product of the amplitude and the time duration are kept constant while the time duration shrinks to an infinitesimal value. See Figure 4.5a.

In mathematical terms (for an impulse at time $t = \tau$),

$$\int_{-\infty}^{\infty} \delta(t - \tau) \, dt = 1$$

$$\delta(t - \tau) \quad = 0 \qquad (t \neq \tau)$$

$$\delta(t - \tau) \quad = \infty \qquad (t = \tau)$$

(4.10)

2. It can be regarded as the limit of the function

$$f(t) = \frac{e^{-(t-\tau)^2/h^2}}{h\sqrt{2\pi}} \qquad \text{(the error function)}$$

(4.11)

as $h \to 0$. See Figure 4.5b.

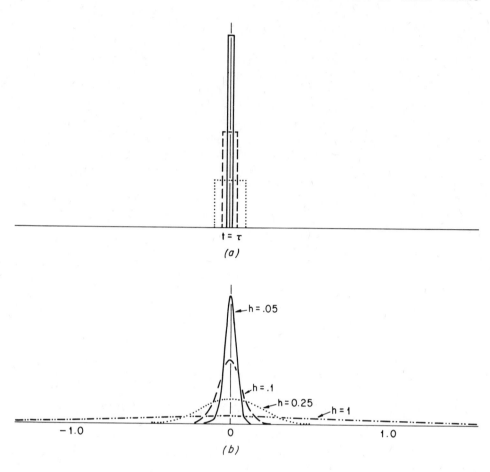

Figure 4.5 Approximations to the delta function. (*a*) Successive approximations of rectangular pulses having a constant area but successively decreasing width. (*b*) Successive approximations based on the error function as the parameter *h* approaches zero.

3. It can be looked at as the pulse that has infinitesimally small but *constant amplitudes of all frequencies*. The location of the impulse is controlled by the slope of the phase function.

Under these conditions

$$g(\omega) = \frac{1}{\sqrt{2\pi}} \int_{-\infty}^{\infty} f(t)e^{-i\omega t}\, dt \tag{4.12}$$

and

$$f(t) = \frac{1}{\sqrt{2\pi}} \int_{-\omega_0}^{\omega_0} g(\omega)e^{i\omega t}\, d\omega \tag{4.13}$$

Since $g(\omega) = A = $ constant and $f(t)$ has to be real,

$$f(t) = \frac{A}{\sqrt{2\pi}} \, \mathrm{Re} \int_{-\omega_0}^{\omega_0} e^{i\omega t} \, d\omega = \frac{A}{\sqrt{2\pi}} \, \mathrm{Re} \left[\frac{e^{i\omega t}}{it} \right]_{-\omega_0}^{\omega_0}$$

$$= A \frac{2}{\sqrt{\pi}} \frac{\sin \omega_0 t}{t} \tag{4.14}$$

This is a $(\sin x)/x$ type function (Figure 4.6) which, as ω_0 approaches infinity, gradually approaches an impulse at zero time and is near zero elsewhere. Thus one can regard the delta function $\delta(t)$ as being one that has equal amounts of all frequencies present. This is a very useful property, since the delta function can be used to find out (for example) the characteristics of the reflection process of the earth—the impulse response—if the input pulse contains all frequencies in equal proportions. Deviations from this assumption can later be corrected by filtering, that is, the adjustment of the amplitudes and phases of all frequencies to conform to what is known about the input pulse.

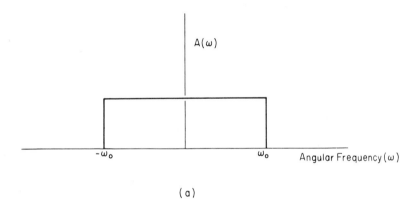

(a)

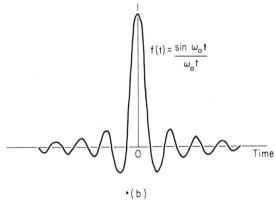

(b)

Figure 4.6 An approximation to the delta function, which shows a pulse having a constant-amplitude spectrum up to the cutoff frequency ω_0.

The duality in description of a time function (i.e., in the time domain or in the frequency domain) appears throughout this book. Sometimes it is convenient to use one representation and at other times the other, but we switch between the two using Fourier analysis or synthesis without further question. As an example, we can take the general problem of filtering and introduce the concept of convolution—a very important one in the science of seismic data processing. The response of any linear system (electrical filter or mechanical mass and spring system) or the response of the earth to a delta function (sometimes called a spike) is called an impulse response. We now filter the impulse response of a system to see how it would behave if a different input pulse were applied.

As we have seen, there are two ways to do this. One, which has been described in outline, is to Fourier-analyze the impulse response, modify the coefficients of the Fourier series to reflect the fact that the delta function has been replaced by a pulse whose Fourier analysis is also known, and then perform a Fourier synthesis on the result. Symbolically, we can describe the entire process by

$$F^{-1}[P(\omega)G(\omega)] \tag{4.15}$$

where

$$G(\omega) = F[I(t)] \tag{4.16}$$

and

$$I(t) = \text{impulse response}$$
$$P(\omega) = \text{known spectral composition of the desired pulse}$$

Note that the two spectra of the impulse response and the desired pulse are multiplied together.

There is another way in the time domain. Let the desired pulse $P'(t)$ be sampled at the same sampling interval as the impulse response. Physically, then, the filtering is done (as shown in Figure 4.7) by replacing each delta function by the properly scaled pulse and adding up the result of the overlapping pulses at each sample time. The values of the resultant signal samples $S(t)$ are then

$$S_1 = a_1 p_1$$
$$S_2 = a_1 p_2 + a_2 p_1 \qquad\qquad S_i = \sum_{j=1}^{i} a_j p_{i-j+1} \tag{4.17}$$
$$S_3 = a_1 p_3 + a_2 p_2 + a_3 p_1$$

This process is known as convolution, and symbolically we write

$$S(t) = I(t) * P'(t) \tag{4.18}$$

These two methods of filtering are equivalent. Convolution in the time domain is equivalent to multiplication in the frequency domain, and convolution in the frequency domain is equivalent to multiplication in the time domain.

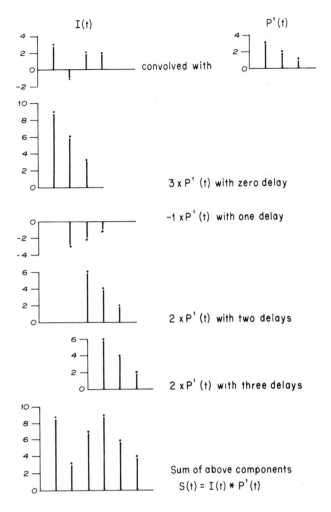

Figure 4.7   The procedure in a simple convolution.

4.5   SYNTHETIC SEISMIC RECORDS FROM WELL LOG CHARACTERISTICS

Until recent years, the knowledge that sound waves could be generated, reflected from geological discontinuities or strata, and recorded at the surface by suitable instruments was all that was required to sustain a very active seismic exploration program. The reflections (although their characteristics had not been related in any way to the lithological sequence) were near enough alike that they could be correlated by eye, and the relative time of occurrence at sequential subsurface points could be determined. After suitable corrections, which are detailed later, a structural map could often be made. This was the era of the search for *structural* hydrocarbon traps, and this era has by no means ended. In more intensively investigated areas, however, this approach is rapidly becoming less fruitful.

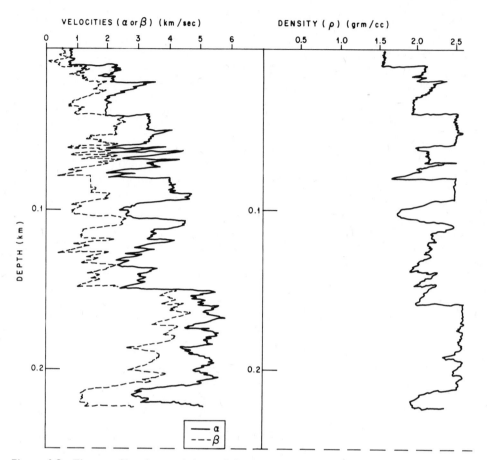

Figure 4.8 The type of log characteristics needed as input for reflectivity determination. At the present time only the P-wave interval velocity α and the density ρ can be obtained in continuous form.

It was nevertheless inevitable that some notice would be given to the form, or character, of seismic recordings, particularly after it became possible to make continuous measurements of seismic compressional wave velocity and the density of formations with instruments lowered into boreholes. Well logging can now provide these continuously measured parameters in a form similar to those shown in Figure 4.8. These logs have a very ragged, or high-frequency, character, and it is evident that the problem of sampling them is a difficult one, if one is to avoid aliasing. For the moment, it is supposed that any digitization is done with sufficiently close sampling that aliasing will cause no problem.

When it was recognized that reflection character could be used to give additional information about the earth, an entirely new field—data processing—was developed to maintain consistency in the treatment of field data and to organize this treatment in such a way that some of the distortions caused by known processes could be removed.

It is our purpose to examine the ways in which seismic reflection sequences are formed and to learn ways of matching their characteristics with the known parameters of the earth's layering.

Although other approximations are possible, the simplest representation of the system of rocks beneath the earth's surface is a series of plane, parallel layers, each having its own characteristic—but constant within the layer—parameters of velocity and density. Both compressional and shear velocities should be used to characterize the rock, but at the present time, shear velocities are not measured accurately. Since we are talking about compressional waves, the lack of shear parameter information is not important, but had the shear velocities been available the treatment of shear reflection output would have paralleled this development.

The type of approximation we are talking about is shown in Figure 4.9. Berryman, Goupillaud, and Waters (1958) have considered a more complex case in which the velocity and/or density can be a linear function of depth within each layer. Our treatment assumes that they are constants.

There are two different ways of tackling the problem of the reflection response. One of these, in the frequency domain, has been described by Berryman, Goupillaud, and Waters (1958). The expressions for up and down propagating waves, of constant frequency, in each layer are written down, and then the necessary boundary conditions are fitted so that all the amplitudes are obtained relative to one another. Such amplitudes are functions of frequency, and it is necessary to employ Fourier synthesis later on the amplitudes in the uppermost layer to obtain the reflection output in the time domain. In Appendix 4A, the details of the method are given, and complex velocities are used to take attenuation into account.

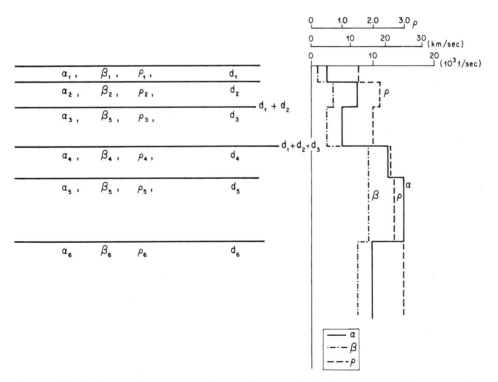

Figure 4.9   The layered system assumed for investigations of reflection response. α, P-wave velocity; β, shear wave velocity; ρ, density.

The other method, in the time domain, is much more revealing as to the effects of multiple reflection, the effects of transmission losses, and the relation of the reflection character to the form of the velocity and density logs. The examination in the time domain is facilitated by arranging the layer thicknesses so that they are integral multiples of the sample interval. If more detail is required, the fundamental sample interval is made smaller.

The geological sequence of rocks can then be drawn in which the vertical dimension is travel time and all layers are of constant "time thickness." The requirement for quantizing the time is a computational feature and not a physical requirement. It is paralleled by the need (in the frequency domain) to compute the complex wave amplitudes for a large but finite number of frequencies when the frequency domain method is used. This too is an approximation demanded by computational needs.

To complete the initial picture, we consider only waves propagating vertically, perpendicular to the layering, although for clarity the various ray paths used are drawn, as in Figure 4.10, at nonnormal incidence. Since the rays are really perpen-

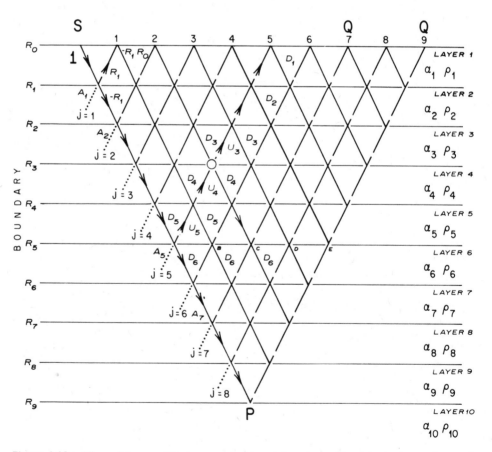

Figure 4.10 All possible ray paths in a constant-travel-time, layered model are shown. Any path starting from S that continually travels with a right component of motion and returns to the surface is a possible path. The output at points $1, 2, \ldots, n$ on the surface is the sum of all constant-time $(2j\,\Delta t)$ paths that emerge at the jth position.

dicular to the boundaries, the diagrams do not show any difference in angle for re-flected and refracted rays.

A pioneering attempt to derive synthetic seismograms from a sequence of velocities as determined by well logging was made by Peterson (1955). If such a log is continuous, we can consider two points infinitesimally separated by δt, for which the velocities are V and $V + \delta V$. When density differences are neglected, the reflection coefficient

$$\delta R = \frac{V + \delta V - V}{V + \delta V + V} = \frac{\delta V}{2V} = \frac{1}{2}\delta(\ln V) \tag{4.19}$$

neglecting second-order quantities.

Peterson neglected multiple reflections and transmission losses and derived a primary reflection trace simply by converting the velocity log into a reflection co-efficient log using a nonlinear logarithmic transformation, following this by a simple filtering scheme which essentially placed a particular-shaped pulse (which he could designate) at each reflection coefficient at the correct amplitude. This was linear filter-ing accomplished electrically. In spite of its success in pointing out a basic relationship, the idea was not physically sound. Nevertheless it was a useful approximation from which further developments have come. The model described here is physically correct and takes into account the facts that:

1. If an interface between two rocks causes a reflection for a wave traveling in one direction, it must give rise to a reflection for a wave traveling in the opposite direction. The reflection coefficients are opposite in sign but of the same magnitude.

2. In crossing a boundary, a transmitted wave has a amplitude different from that of an incident wave. If the down-going reflection coefficient is R, the amplitude of the downward transmitted wave will be $1 - R$. As a corollary, the upward re-flection coefficient will be $-R$, and the amplitude ratio of the transmitted to the incident upgoing wave will be $1 - (-R) = 1 + R$.

The transmission (both ways) through a boundary thus gives rise to multiplication of the original amplitude by $1 - R^2$, which is always less than unity.

This model has been studied in a more formal, mathematical manner by Wuenschel (1960), and his paper forms a useful adjunct to the description given here. However, it is not necessary in order to have an exact physical picture of the process within the confines of the model.

In our system, the input waveform seems to be neglected because we use only ampli-tudes. However, there is an implicit assumption that the input is really a delta function. Since the reflection wave shape is the same as the incident wave shape when reflecting from a single boundary, it is sufficient to record amplitudes as a function of a series of integers, since the times of arrival are always even integral multiples of the one-way travel time through the layer Δt.

After the log has been properly sampled over the interval logged, additional problems arise because of the lack of information in the near surface and the need to ensure that the travel times to certain depths agree with any measured times available. These measured times may be from actual check shots made at the time of the well survey, or they may be based on statistically derived velocities from reflection data. An estimate must be made of the thickness and velocity applicable to the weathered layer. Then a smooth velocity-depth function is derived which:

1. Agrees with the probable subweathering velocity.
2. Blends smoothly into the known interval velocities near the top of the logged interval.
3. Integrates to give the correct arrival time of a wave starting from the base of the weathering and ending at a known checkpoint.

There are no standard, prescribed ways of doing this, but a velocity that is a quadratic function of depth is usually assumed, and the coefficients adjusted to fit the known information.

With this accomplished, it is then possible to try to adjust to the known travel times. This involves multiplying the velocity by the time interval and adding these values until the check shot depth is bracketed. An interpolation is made to fit the exact depth, and the calculated time is compared with the measured time. An adjustment can then be made to all velocities so that there is exact time agreement.

The same process is continued from one check shot to the next deeper one, until all the check shot times are satisfied. It is necessary to be very critical of the check shot times themselves, editing out those that yield unlikely interval velocity averages within the check shot interval. When all this is finished, the velocities are tabulated for each layer and form part of the basis for the synthetic record calculation. Density values are usually entered as read, except that care must still be taken to avoid aliasing; and they must be given at the proper time values instead of the depth values as originally read.

The computational procedure is:

1. The sequential velocity and density values α_i and ρ_i are stored in a computer.
2. The sequential reflection coefficients $R_1, R_2, \ldots, R_N$ are calculated from the velocities and densities:

$$R_i = \frac{\alpha_{i+1}\rho_{i+1} - \alpha_i\rho_i}{\alpha_{i+1}\rho_{i+1} + \alpha_i\rho_i} \tag{4.20}$$

The reflection coefficient R_0 for the air-earth interface has to be assigned. While this value should be very close to unity for perfect earth materials, the attenuation for elastic waves in the weathering can be approximately taken into account by reducing the value of R_0. If this is not done, experience shows that the synthetic record takes on a ringy character, that is, it contains long trains of waves of near-constant frequency, because of reverberations in the weathered layer.

3. Referring to Figure 4.10, we note that the most general situation is that found (say) at point O, where there are both upcoming and downcoming incident waves and upgoing and downgoing reflected or refracted waves. All amplitude symbols used bear the superscript (5), since the energy from point O, going directly to the surface, arrives at $5(2\,\Delta t)$. The subscript refers to the layer in which the wave is traveling.

The contribution to $U_3^{(5)}$ due to $D_3^{(5)}$ is $R_3 D_3^{(5)}$, and to $U_3^{(5)}$ due to $U_4^{(5)}$ is $(1 - R_3)U_4^{(5)}$. Therefore $U_3^{(5)} = R_3 D_3^{(5)} + (1 + R_3)U_4^{(5)}$. Similarly, $D_3^{(5)} = (1 - R_3)D_3^{(5)} - R_3 U_4^{(5)}$. More generally,

$$U_N^{(J)} = R_N D_N^{(J)} + (1 + R_N)U_{N+1}^{(J)}$$

$$D_{N+1}^{(J)} = (1 - R_N)D_N^{(J)} - R_N U_{N+1}^{(J)} \tag{4.21}$$

$$D_N^{(J+1)} = D_N^{(J)}$$

These are the recursion formulas to be used.

4. To start off, the condition at point A_1 is considered, and here $D_1^{(1)} = 1$ (the initial pulse amplitude), and $U_2^{(1)} = 0$ (which would otherwise correspond to a wave generated before zero time), so that

$$U_1^{(1)} = R_1 1 + 0 = R_1$$

$$D_2^{(1)} = (1 - R_1)1 - 0 = (1 - R_1)$$

Any value of U_1 calculated is stored in a location that is representative of its arrival time (in terms of the interval $2\,\Delta t$). The one just calculated arrives at $2\,\Delta t$. Strictly speaking, it should be necessary to multiply U_1 by $1 + R_0$ to arrive at the particle velocity observed at the surface, but this is a constant factor (near 2) and for most purposes can be omitted. It is necessary, however, to calculate $D_1^{(1)} = -R_1 R_0$.

5. The process now continues by moving to A_2 and making iterative calculations for all points on the right uptrending diagonal path. It is necessary to store, at each calculation, $U_N^{(J)}$ and $D_{N+1}^{(J)}$. However, a single common storage location can be used for $U_N^{(J)}$, since it is used only once, but all sequential values of $D_{N+1}^{(J)}$ must be stored for later use. For example, in following up the diagonal line from A_5, $D_6^{(5)}$ through $D_1^{(5)}$ must be generated and stored but, at point 0, $U_3^{(5)}$ is generated and stored for the next step where it is used, but it is never used again. Finally, $U_1^{(5)}$ is calculated for time $5(2\,\Delta t)$, and this value is permanently stored as part of the output.

6. The output then continues to be calculated by incrementing the time and continuing the same basic procedure. By this method, any wave that starts at S and finishes at a surface point Q_j gives rise to an output Q_j which includes all possible reflections and takes into account any change in amplitude due to transmission through the boundaries. The output consists of a sequence of numbers which represent the amplitudes of the signal arriving at integral multiples of $2\,\Delta t$. We can represent the signal by

$$\{a_1, a_2, a_3, \ldots, a_j, \ldots, a_N\}$$

where $2\,\Delta t$ delay between each pulse is implied. If the input to the geological section had been a delta function, the output from the layered section would be a sequence of delta functions, as described above, and it is called the impulse response.

The final step of going from the unreal concept of transmission of an infinite range of frequencies to the real world of a restricted bandwidth now remains. In this chapter, we do not try to deal with attenuation in the earth in which the pulses that have traveled the furthest are restricted the most in bandwidth (high-frequency loss). Rather, at this stage, a simple filtering process or convolution is used.

4.6 CHARACTERISTICS OF THE REFLECTION PROCESS

Having described a means by which well-logged parameters can be converted into synthetic, or expected, seismograms, we can now discuss some experience obtained from wholsesale usage of this process and a comparison with actual seismograms taken in the immediate vicinity of the logged well.

First, however, it should be noted that it is possible to produce synthetic seismograms, without multiples, by essentially the same process as that used by Peterson (1955). The digital computer program is arranged in such a way that, if the option "without multiples" is desired, the work of computing the impulse response of the layered system is bypassed and the desired result achieved simply by filtering the sequence of reflection coefficients

$$\{R_1, R_2, R_3, R_4, \ldots, R_j, \ldots, R_N\} \tag{4.22}$$

by convolution with the desired pulse.

If transmission coefficients, but not intrabed multiples, are desired, the appropriate sequence to be convolved with the filter pulse will be

$$\left\{R_1, R_2(1 - R_1^2), R_3(1 - R_1^2)(1 - R_2^2), \ldots, R_j \prod_{k=1}^{j-1} (1 - R_2^2), \ldots\right\} \tag{4.23}$$

where $\prod_{k=1}^{j-1} (1 - R_k)$ stands for a continued product of all the individual transmission factors $1 - R_k^2$ for the allowable values of k.

Some discussion of the desired pulse is necessary, since it changes, depending on the type of source used for the field records with which the synthetic record will be compared. There are two basic types:

1. The impulsive type.
2. Vibroseis® pulses.

These are fundamentally different, inasmuch as the impulse type gives rise to an indication of motion only after the cause, that is, after an explosion or similar event has taken place. This basic waveform is difficult, if not impossible, to determine except by direct measurement. It must take into account the response characteristics of the recording instruments, as well as the source characteristics. The latter are particularly difficult to determine on land, since they can vary with the source placement, the receiver location, and the characteristics of the near-surface layers. Measurements of marine sources have been made in very deep water, where the hydrophones cannot record anything but the direct pulse. Such pulses are shown in Figure 4.11.

In the case of Vibroseis®, it has already been stated that the source generates a surface motion which is constrained as nearly as possible to be a swept frequency signal. In theory, every reflecting boundary then returns to the surface a replica of this

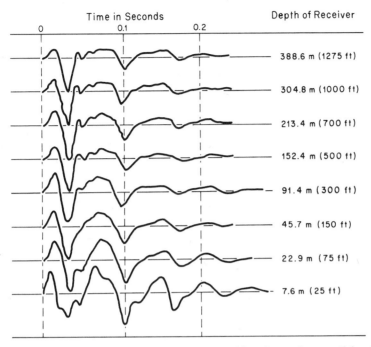

Figure 4.11   Deep-water tests of a pulse for a combination of four Aquapulse guns 12.2 m (40 ft) deep.

signal, multiplied by the reflection coefficient; attenuated by transmission coefficients, and delayed in time by the two-way travel time from the source to the reflector and back to the receiver.

The Vibroseis® method then consists of cross-correlating the received signal with the input signal, with the result that each reflecting horizon is represented by a band-limited, zero-phase (symmetrical) signal in which the peak occurs at the proper travel time. Thus the Vibroseis® seismogram apparently displays earth movement before, as well as after, the effective cause. There are of course effects of the near-surface layering (such as multiple reflections in the weathered zone at the source and the receiver) that have to be taken into account in establishing the effective pulse to be used but, in comparing field records with synthetic records, many of these effects have (to the amount known) been included. A Vibroseis® basic pulse is shown in Figure 4.12, which is the autocorrelation function of a signal having a constant-amplitude spectrum between 5 and 20 Hz. It is of course possible to alter this basic waveform when the Vibroseis® system is used, by changing the form of the swept frequency control signal. The autocorrelation pulse shape is a function of the ampli-tude spectrum only, and theoretically this allows an infinite number of ways of con-structing a time series, since the phase spectrum has not been specified. It should be noted, however, that the pulse given in Figure 4.12 does not take into account the characteristics of the weathered layer. It is presumed that this is adequately dealt with in calculating the reflection coefficients and therefore included in the impulse response of the earth.

A display of synthetic traces calculated for a shallow well (using velocity only for calculating the reflection coefficients) is given in Figure 4.13. Synthetic records made

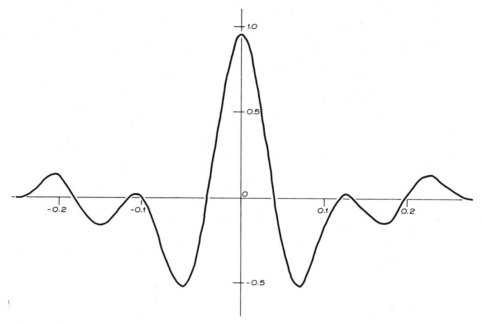

Figure 4.12 Autocorrelation pulse for a rectangular amplitude spectrum in the range 5 to 20 Hz.

both with and without multiples are included. It serves to illustrate some of the characteristics of the reflection process. This well penetrated the Permian and part of the Pennsylvanian section of Oklahoma, and the section generally consisted of thin sands and shales with occasional thin limestones. Reflection coefficients were therefore quite large and close together. The only thick, high-velocity layers were concentrated in the first 1000 ft and, for the lower-frequency synthetic records (the spectrum is given above the trace), this is where the low-frequency seismic reflection energy is concentrated. The general impression is that there is a correspondence between the filtered velocity (or acoustic impedance) log and the filtered reflection coefficients. A closer examination reveals that there are also time shifts, so that the peaks of the reflection coefficients occur at times when the velocity is changing most rapidly.

Reflection records without multiples, then, follow closely reflection coefficient traces filtered with the appropriate filter. The only difference is that the rate of decay is faster than for the reflection coefficients alone, and this is due to transmission losses.

There is a surprise awaiting us when we compare the synthetic record with multiples, because the rate of decay is reversed; that is, a synthetic with multiples decays less per unit time than one without multiples. Both have the transmission losses included in the calculation. This apparent paradox is explained by consideration of the change in the downward traveling pulse as it travels deeper into the geological section. The form of this downward traveling pulse can be examined relatively easily, using the same synthetic record program.

Referring to Figure 4.10 again and looking at a particular line of constant j (say, $j = 5$), we can see that $D_6^{(5)}$ consists only of energy that has traveled directly downward from the surface. $D_5^{(5)}$, however, as illustrated in Figure 4.14a, has only energy that, *at some point in its path*, has traveled up one layer and then downward again. In other

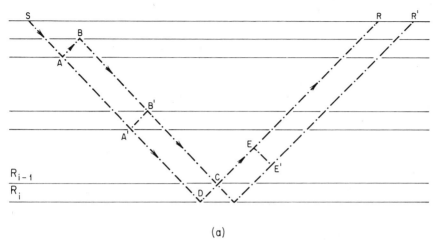

(a)

There is a single primary reflection SAA'DCR. There are many possible "peg leg" multiples of the first order which are reflected along paths such as SABB'CR or SA'B'CR. These occur at the same time as the primary shown. On the other hand, multiple reflections, including reflection from the R_i boundary, will arrive with one incremental delay behind the primary. Note that these multiples can also occur on the upward path, such as SAA'DEE'R.

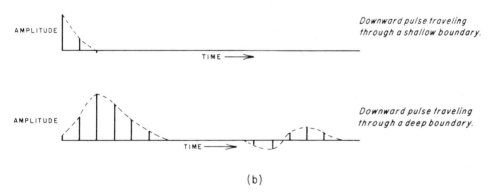

(b)

Figure 4.14 Illustrating the formation of peg-leg multiples and the effect of these reverberations on a downward traveling pulse.

words, $D_5^{(5)}$ contains all the energy that has reverberated *once* in some layer. Similarly, if we look at D_4 ($j = 5$), we will see only energy that has reverberated twice (in any combination of layers above). Of course, if we follow a line of constant j, all these energies are composited together. However, if we deal with points, such as A, B, C, D, and so on, all at the same level (reflector), the downward values $D_6(j = 5, 6, 7, 8, \ldots)$ give the successive amplitudes of the resultant pulse traveling down through reflector R_5. These values are easy to segregate from the mass of values calculated during the synthetic record computation.

Thus, as shown in Figure 4.14b, the downward traveling pulse gradually accumulates trailing energy which not only delays the reflection from a given reflector but also can cause it to lose some high frequencies. The deeper the reflector, the more these "peg-leg" reverberations tend to accumulate after the theoretical time of arrival. This process of accumulating peg-leg energy occurs not only on the downward path to a reflector but also on the upward path from the reflector to the surface.

It is not immediately obvious that the peg-leg reflections should have predominantly the same sign as the initial pulse. However, consideration of several acoustic impedance situations, such as those in Figure 4.15, shows how this occurs in cases of geological interest. In case a where the acoustic impedance log is very erratic—consisting of alternating hard and soft layers and alternating signs for the reflection coefficients—all the first-order multiples are positive, as is the primary. Although the multiples have amplitudes involving the product of small reflection coefficients, with enough layers, the amplitude of the composited multiples can exceed that of the primary. It must be remembered that this process occurs on the upward path of the reflected ray also.

In case b where the velocity increases at each step, it is found that the first-order multiple has a sign opposite that of the primary pulse. Thus, if it becomes large enough through the superposition of events from many layers, the initial reflection traveling upward could have the wrong polarity. Again the upward path is as effective in multiple production as the downward path. Thus the polarity of the reflection is reversed, and the main energy is carried by the (delayed) multiple event.

The case of continually decreasing velocity can be argued in the same manner as that continually increasing velocity and leads to the same conclusion.

O'Doherty and Anstey (1971) have pointed out that there are two types of contributors to the intrabed multiples, ramplike changes in velocity (in which the reflection coefficients must be small because of the limited velocity range) and alternating velocity changes which give rise to alternating reflection coefficient sequences with relatively large reflection coefficients.

Information on the size of intrabed multiples of delays greater than one is given by the autocorrelation of the impulse response. It varies from one velocity log to another.

In general, the amplitude for all reverberations of all orders can best be obtained by the synthetic calculation. The gradual change in form of the effective downward traveling pulse and its transformation on the upward path may at times be of great importance to the ability of the reflection method to define lithological change. A complex pulse from a simple interface gives a complex reflection and, if the shallow section changes appreciably, this deeper reflection, besides being complex, will be inconsistent in character.

One result of great importance, however, has been deduced, and this is that the internal reverberations help to sustain the amplitudes of the reflections compared with primary reflections alone, but at the cost of a small, but measurable, time delay. This effect is essentially one of scattering, dispersing the original very sharp impulse and providing a second—sometimes severe—form of attenuation for higher frequencies. This may be the final constraint on our ability to achieve high resolution with a seismograph when dealing with deep reflectors. It is not an accident then that, as a rule, reflections tend to become lower and lower in frequency as the depth (or time of occurrence on the reflection record) increases.

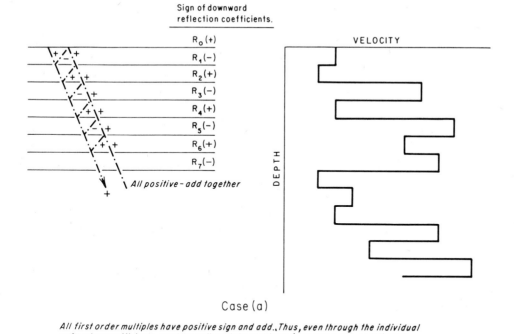

Case (a)

All first order multiples have positive sign and add., Thus, even through the individual reflection coefficients may be small, and the multiples (two reflection coefficients multiplied together) individually much smaller, eventually the summed multiples grow bigger than the transmitted primary.

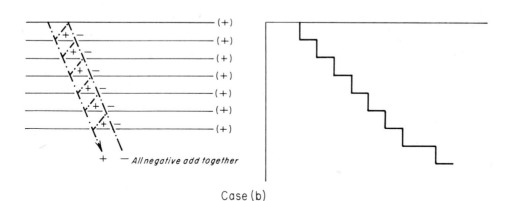

Case (b)

Reflection coefficients all positive. A negative peg-leg multiple builds up behind a diminishing positive primary.

Figure 4.15 The addition properties of reverberations. (*a*) Reflection coefficients of alternating sign. (*b*) Reflection coefficients that are all positive.

4.7 SUMMARY AND CONCLUSIONS

There are two different ways in which continuous seismic record traces can be described. One of these, the time domain method, simply describes the variation in amplitude as a function of time. The other, the frequency domain method, selects a basic time interval to be described and then establishes both the amplitudes and phases of constituent sinusoidal waves that are integral multiples of the fundamental frequency, $f_0 = 1/T$.

A seismic trace can be filtered by Fourier transformation into the frequency domain, changing the amplitudes and phases of the constituent harmonic frequencies and then inverse Fourier-transforming back into a time domain trace.

A duality exists also in filtering methods corresponding to the duality in descriptions. For example, filtering in the time domain is done by the process of convolution, corresponding to complex multiplication in the frequency domain.

Seismic signals can be prepared for digital computer operations by sampling at fixed time intervals. Such time intervals must be carefully selected so that at least two samples exist for the highest frequency present in the record to be sampled. Failure to do this gives errors in the representation of the trace and errors due to aliasing of high frequencies into lower frequencies.

An impulse function has been described, which is convenient for the analysis of any process since only its amplitude needs to be specified and its time of occurrence is assumed to be at specified sampling points only. Its most useful characteristic is that it can be shown to contain equal amounts of all frequencies. Hence, if the impulse response of a circuit or a process is found, the response to other waveforms can be obtained by filtering in the frequency domain or by convolution in the time domain.

If a simplistic model of the earth is used, such as one consisting of a large number of layers having constant travel time thickness, it is possible to use a digital computer to calculate the impulse response of such a system using velocities and densities derived from the appropriate well logs. Contrasted with the simple (but physically impossible) synthetic record constructed by Peterson, these synthetic traces are realistic and include the effects of transmission losses and multiple reflections.

These multiples, principally peg-leg multiples, explain many of the experimentally derived characteristics of seismic records. For example, the scattering contained in the notion of reverberation accounts for anomalous attenuation, some delay of reflections, and a gradual decrease in average reflection frequency with depth. In some cases of geological interest, deeper reflections may be complexly altered in form as a result of the scattering involved in the multiple process.

APPENDIX 4A: SYNTHETIC RECORDS USING COMPLEX VELOCITIES

As mentioned earlier in Chapter 4, the frequency domain provides a convenient method of computing synthetic seismograms, but details were omitted there because the time domain approach gives an easy physical picture of the multiple process. However, when frequency-dependent attenuation must be included, the frequency domain approach is simpler, although it involves the use of complex arithmetic computer subroutines.

This appendix gives a detailed derivation of the frequency domain approach and the use of an iterative formula starting from the bottom layer and working up to the layer just below the surface. This is done for every frequency of interest, and the response of the uppermost layer is related to either the stress or the initial velocity applied to the surface.

Fourier synthesis finally takes all the coefficients (responses) and adds them together to obtain a time domain output.

4A.1   THEORY

The plane wave equation is

$$\frac{\partial^2 \zeta}{\partial t^2} = C^2 \frac{\partial^2 \zeta}{\partial x^2} \tag{4A.1}$$

where the velocity is complex and can be represented by

$$C = V + vi \tag{4A.2}$$

Assume a solution of (4A.1) of the form

$$\zeta = \zeta_0 \varepsilon^{i\omega[t - (x/c)]} \qquad \text{(for waves advancing in the positive } x \text{ direction)}$$

$$= \zeta \varepsilon^{i\omega\{t - [V - vi/(V^2 + v^2)]x\}}$$

where $v^2 \ll V^2$

$$\zeta = \zeta_0 \varepsilon^{i\omega[t - (x/V)]} \varepsilon^{(-\omega vx/V^2)}$$

$$= \zeta_0 \varepsilon^{-\alpha x} \varepsilon^{i\omega[t - (x/V)]} \qquad \alpha = \frac{\omega v}{V^2} \tag{4A.3}$$

This is a wave that attenuates with distance as the first power of the frequency—the solid friction model—and

$$\frac{1}{Q} = \frac{2v}{V}$$

We now take two consecutive layers in the earth and in each we have upgoing A^- and downgoing A^+ waves. Since these waves attenuate in going through the layer, we write the continuity equation with the convention that the amplitudes A_m^- and A_m^+ are measured at the extreme uppermost point of the layer.

As Figure 4A.1 shows, the downgoing waves attenuate to the right and the upgoing waves to the left.

Continuity of displacement at the mth boundary gives

$$A_m^+ \varepsilon^{-\alpha_m x_m - (i\omega x_m/V_m)} + A_m^- \varepsilon^{\alpha x + (i\omega x_m/Vm)} = A_{m+1}^+ + A_{m+1}^- \tag{4A.4}$$

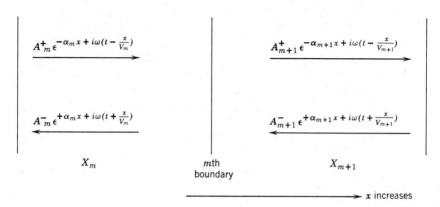

Figure 4A.1 Up- and downgoing waves in two consecutive layers.

Continuity of stress at the mth boundary gives

$$C_m^2 \rho_m \left(\frac{\partial \zeta_m}{\partial x}\right)_{x=X_m} = C_{m+1}^2 \rho_{m+1} \left(\frac{\partial \zeta_{m+1}}{\partial x}\right)_{x=0} \tag{4A.5}$$

Note that ζ_m is the particle displacement due to the combination of upgoing and downgoing waves.

$$C_m^2 \rho_m \left[A_m^+ \left(-\alpha_m - \frac{i\omega}{V_m}\right) \varepsilon^{-\alpha_m X_m - (i\omega X_m / V_m)} + A_m^- \left(\alpha_m + \frac{i\omega}{V_m}\right) \varepsilon^{\alpha_m X_m + (i\omega X_m / V_m)} \right]$$

$$= C_{m+1}^2 \rho_{m+1} \left[A_{m+1}^+ \left(-\alpha_{m+1} - \frac{i\omega}{V_{m+1}}\right) + A_{m+1}^- \left(\alpha_{m+1} + \frac{i\omega}{V_{m+1}}\right) \right] \tag{4A.6}$$

For brevity we write

$$q = \varepsilon^{-\alpha_m X_m - (i\omega X_m / V_m)}$$

and

$$K_m = \frac{\alpha_{m+1} + (i\omega/V_{m+1})}{\alpha_m + (i\omega/V_m)} \frac{C_{m+1}^2 \rho_{m+1}}{C_m^2 \rho_m}$$

Equations (4A.4) and (4A.6) reduce to

$$A_m^+ q + \frac{A_m^-}{q} = A_{m+1}^+ + A_{m+1}^- \tag{4A.7}$$

$$-A_m^+ q + \frac{A_m^-}{q} = K_m(-A_{m+1}^+ + A_{m+1}^-) \tag{4A.8}$$

Divide (4A.7) by (4A.8) and let

$$R_m = \frac{A_m^-}{A_m^+} \qquad R_{m+1} = \frac{A_{m+1}^-}{A_{m+1}^+}$$

Then

$$R_m = q^2 \left(\frac{R_{m+1} + r_m}{1 + R_{m+1} r_m} \right) \qquad (4A.9)$$

where

$$r_m = \frac{1 - K_m}{1 + K_m}$$

We now make a further simplification of K_m, neglecting v^2 with respect to V^2 and substituting $\alpha = \omega v / V^2$:

$$K_m = \frac{C_{m+1}^2 \rho_{m+1}}{C_m^2 \rho_m} \frac{\alpha_{m+1} + (i\omega/V_{m+1})}{\alpha_m + (i\omega/V_m)}$$

and, after some tedious algebra,

$$K_m = \left[\frac{V_{m+1} + iv_{m+1} - (2v^2/V_{m+1})i}{V_m + iv_m - (2v_m^2/V_{m+1})i} \right] \frac{\rho_{m+1}}{\rho_m}$$

Again neglecting v^2/V compared with v,

$$K_m = \frac{(V_{m+1} + iv_{m+1})\rho_{m+1}}{(V_m + iv_m)\rho_m}$$

so that K_m is the ratio of the acoustic (complex) impedances across the boundary, and

$$r_m = \frac{1 - K_m}{1 + K_m}$$

is the complex reflection coefficient associated with the mth boundary.

Thus we obtain the general iteration formula

$$R_m = \left(\frac{R_{m+1} + r_m}{1 + R_{m+1} r_m} \right) q^2 \qquad (4A.10)$$

There now remains the problem of relating the motion in the layers to the motion of the surface induced by the source (see Figure 4A.2.)

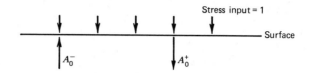

Figure 4A.2 Relation between surface stress input and the up- and downgoing waves in the near-surface layer.

Since the upper boundary is free,

$$-A_0^+ \left(\frac{i\omega}{V_0} \right) + A_0^- \left(\frac{i\omega}{V_0} \right) + 1 = 0 \qquad (4A.11)$$

if the stress on the boundary due to the source is 1 and

$$\frac{A_0^-}{A_0^+} = R_0 \qquad \text{(determined from the iteration procedure)} \qquad (4A.12)$$

$$A_0^- = \frac{V_0 R_0}{i\omega} \left(\frac{1}{R_0 - 1} \right)$$

Now, for Vibroseis ®-type sources, we compare (correlate) the input velocity with the output velocity, hence we need to convert from a stress input to a velocity input constant with frequency.

The velocity just below the surface due to a stress of unity is given by

$$\left(\frac{\partial \zeta}{\partial t} \right)_{x=0} = i\omega A_0^- + i\omega A_0^+$$

$$A_0^+ = \frac{A_0^-}{R_0}$$

$$\therefore i\omega(A_0^+ + A_0^-) = V_0 R_0 \left(\frac{1}{R_0 - 1} \right) \left(1 + \frac{1}{R_0} \right)$$

$$\text{Velocity input} = V_0 \left(\frac{R_0 + 1}{R_0 - 1} \right)$$

(4A.13)

Thus the upward displacement A_0^- due to a *unit velocity* input is [combining (4A.12) and (4A.13)]

$$A_0^- = \frac{V_0 R_0}{i\omega} \left(\frac{1}{R_0 - 1} \right) \left(\frac{R_0 - 1}{R_0 + 1} \right) \frac{1}{V_0} = \frac{R_0}{i\omega} \left(\frac{1}{R_0 + 1} \right)$$

Finally, the upward velocity $\dot{A}_0^-$ is $i\omega A_0^-$:

$$\dot{A}_0^- = \frac{R_0}{R_0 + 1} \qquad (4A.14)$$

4A.2 SUMMARY

1. We have described a method of making synthetic records with multiples and attenuation, using a complex velocity $C = V + iv$.
2. V is the velocity measured on the continuous-velocity logs, and v must be estimated in some manner.
3. The density ρ is the conventional density as measured on density logs or computed from porosity and particle density.
4. α_m is an attentuation constant for the mth layer and is given by

$$\alpha_m = \frac{\omega v_m}{V_m^2} \qquad (4A.3)$$

5. It is related to the Q by the formula

$$\frac{1}{Q} = \frac{2v}{V}$$

6. R_m is the ratio of the amplitude of the up-traveling wave to that of the down-traveling wave, at the upper boundary of a layer:

$$R_m = \frac{A_m^-}{A_m^+}$$

7. r is the complex reflection coefficient of the mth boundary (at the base of the mth layer):

$$r_m = \frac{1 - K_m}{1 + K_m}$$

8. K_m is the ratio of the complex acoustic impedance of a layer $m + 1$ to that of a layer m

$$K_m = \left(\frac{V_{m+1} + iv_{m+1}}{V_m + iv_m} \right) \frac{\rho_{m+1}}{\rho_m}$$

9. A general iteration formula (for layers of unequal thickness) is

$$R_m = \frac{R_{m+1} + r_m}{1 + R_{m+1} r_m} \varepsilon^{-2[\alpha_m x_m + (x_m/V_m)i\omega]}$$

10. If the layers are of equal time thickness, $\tau = x_m/V_m$,

$$R_m = \varepsilon^{-2\tau i\omega} \varepsilon^{(-2vm/V_m)\omega\tau} \frac{(R_{m+1} + r_m)}{(1 + R_{m+1} r_m)}$$

where the factor $\varepsilon^{-2i\omega\tau}$ is a complex constant for a given frequency.

11. When the layer stops, that is, when the last layer is assumed infinite (no reflection comes from deeper than a given depth), $A_{m+1}^{-} = 0$ and R_m can be calculated.
12. The iteration continues until R_0 is obtained.
13. The final output is $R_0/(R_0 + 1)$.
14. Since all these quantities are complex and dependent on frequency, the values of $R_0/(R_0 + 1)$ are computed for all relevant frequencies and a Fourier synthesis is performed to obtain the time domain reflection output. The usual precautions with Fourier syntheses are needed to ensure no overlap of head and tail and to give the requisite signal length.

4A.3 EXAMPLES

An example is given in Figure 4A.3, in which a synthetic record with attenuation is compared with synthetic records made in the time domain without attentuation but otherwise the same.

REFERENCES

Berryman, L. H., Goupillaud, P. L., and Waters, K. H. (1958), "Reflections from Multiple Transition Layers—Theoretical Results," *Geophysics*, Vol. 23, pp. 223–243.

O'Doherty, R. F., and Anstey, N. A. (1971), "Reflections on Amplitudes," *Geophysical Prospecting*, Vol. 39, No. 3, pp. 278–291.

Peterson, R. A., Fillipone, W. R., and Coker, F. B. (1955), "The Synthesis of Seismograms from Well Log Data," *Geophysics*, Vol. 20, pp. 516–538.

Sengbush, R. L., Lawrence, P. L., and McDonal, F. J. (1961), "Interpretation of Synthetic Seismograms," *Geophysics*, Vol. 26, p. 138.

Sherwood, J. W. C. (1962), "The Seismoline, An Analog Computer of Theoretical Seismograms," *Geophysics*, Vol. 27, p. 19.

Stratton, J. A. (1941), *Electromagnetic Theory*, 1st ed., McGraw-Hill Book, New York, p. 284.

Trorey, A. W. (1962), "Theoretical Seismograms with Frequency and Depth Dependent Absorption," *Geophysics*, Vol. 27, No. 6, pp. 766–785.

Woods, J. P. (1956), "The Composition of Reflections," *Geophysics*, Vol. 21, p. 261.

Wuenschel, P. C. (1960), "Seismogram Synthesis Including Multiples and Transmission Coefficients," *Geophysics*, Vol. 25, pp. 106–129.

FIVE

Seismic Data Gathering Methods

5.1   INTRODUCTION

Since from experience it is known that oil fields occur in random sizes and at random locations in basins about which little is known, it appears reasonable to seek information on the subsurface at points uniformly distributed on the surface. The earliest attempts to gather geological information with the use of a reflection seismograph were made in this manner, through a technique known as isolated correlation. The technique presupposed that the character of the seismic trace was caused only by a succession of reflections—as described in the discussion of synthetic records. Of course, with a single shot point and receiver, resulting in a single seismic trace, there was no way to check on the occurrence of other types of events on this trace. Therefore a single shot point was recorded at a small number of single geophones separated by a few tens of meters with minimum offset from the shot point of no more than 35 m (100 ft).

The several traces then allowed assessment by eye of the occurrence of some types of unusual events. For example, the Rayleigh wave and the refracted (head) wave could be distinguished by the rate of change in time of occurrence as the distance of the geophone from the shot point increased—a finite rate for near-surface disturbances and an extremely small rate for reflections (or indistinguishable multiple reflections). Nevertheless this type of information was gathered at section corners or at other quasi-equidistant points so that a grid coverage of the area was achieved. The recording apparatus was simple and easy to lay out, so the rate of coverage of an area was limited by the density of coverage, the difficulty of drilling, or sometimes by the rate at which adequate land surveying and permitting the landowners could be achieved. The ground roll could be reduced by proper placement of the charge in the hole and by increasing the low cutoff frequency of the geophones and amplifiers— to 25 Hz at times. This was the era of seismography as an art. Figure 5.1 illustrates the method.

The idea of uniform area coverage was good, but there were too many deleterious factors to deal with to allow it to remain the sole means of obtaining both reconnaissance and detailed information. Changes in seismic character with shot depth and medium, rapid changes in subsurface geology, and the presence of noise all made correlation by eye between successive recording locations too difficult and ambiguous.

Most of the difficulty was, correctly or incorrectly, assigned to the problem of rapid geological variation, and the next attempts, reversed profiling or continuous profiling, used the available geophones and amplifiers—which tended to increase

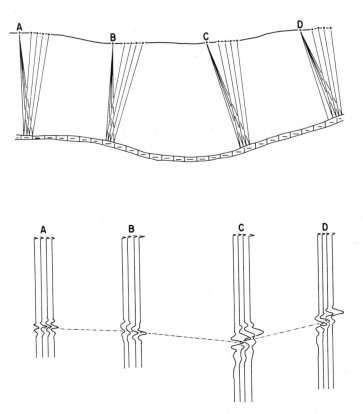

Figure 5.1 Schematic of isolated correlation method and interpretation in a simplistic case.

in number to 20, 24, or even 40 channels—but still usually only one geophone per channel to cover spread distances between holes of about 400 m (1320 ft). The method emphasized, as seen in Figure 5.2a, the provision of seismic traces equally spaced on the surface, the idea being that it is easier to make 12 small correlation decisions than possibly 1 larger one. With this concentration of holes between which cables and geophones had to be laid, the idea of equal coverage by area was abandoned.

The continuous-profiling method has been well described in the literature, including the methods of correcting near and far traces (from a given shot point) so that a time tie for each selected reflection could be made on offset traces shot from opposite directions. Corrected times, or depths if a velocity survey were available, were computed individually for the near and far traces, and time sections consisting of bars joining corresponding reflection points were made. This method, with minor modifications of overall field layouts, continued until almost 1962. However, the advent of surface sources had given considerable emphasis to the need for groups of geophones connected to give an average ground displacement over a substantial surface distance—sometimes a few hundreds of meters. As described previously, these arrays of both sources and receivers caused a considerable differential gain in reflection amplitude compared with that of surface waves. In addition, the number of geophones averaged, aided by accumulated knowledge of the necessity for firm planting or spiking of the geophones into the ground, caused a considerable drop in the

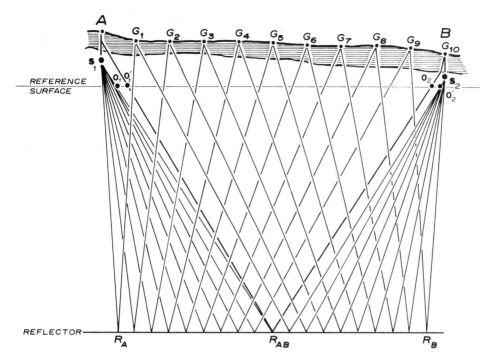

Figure 5.2a Illustration of the continuous, reversed profile reflection method, using subweathering sources. It is necessary to correct the arrival times to a reference surface before time ties can be effected.

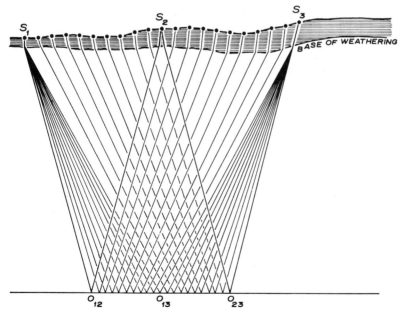

Figure 5.2b Illustration of the time-tying method when offset geophone spreads are used, as in the Vibroseis® method. The surface-to-surface time $S_1 O_{12} S_2$ is correlated between traces to the time $S_1 O_{13} S_3$, which is equal to the reverse time $S_3 O_{13} S_1$ on the next but one record. Further correlations then carry the reflection to $S_3 O_{23} S_2$.

random noise level caused by wind, traffic, and other kinds of microseismic noise. Offset continuous profiles (Figure 5.2b) were commonly used.

Other kinds of seismic spreads were used in areas where reflection continuity was poor, such as isolated dip spreads, both along a single line and with components in two directions at right angles. In these cases, phantom horizons had to be drawn on the cross sections. These phantom horizons started at an arbitrary point and were drawn so that they were parallel to dips nearby (in time and space). The lack of reflection continuity, although it could obviously occur in clastic sections, or in highly faulted areas, was not as prevalent as at first supposed—as later reflection seismograph methods testified.

5.2   MODERN LINEAR SEISMIC REFLECTION METHODS

Dependence on large arrays of source points and receivers can be carried only to the point where these multiple elements become large enough that an appreciable amount of the subsurface is used as a reflector. Beyond this point, the averaging of subsurface energy tends to modify the structure and impairs the resolution. However, these uses had pointed out the vital point that noise can still be a controlling factor in interpreting subsurface geology from geophysical data.

One method of avoiding this problem is to use reflection traces, employing different offsets, all of which use the same reflecting point on the subsurface. Of course, the geometrical effect of increased arrival time of reflections as the offset is varied has to be corrected by using an adequate velocity function to remove the normal move-out (NMO).

As early as 1938, Green had suggested the use of multiple offsets centered about a common depth point as a means of eliminating the effect of dip on velocity determinations. It was Mayne (1956), however, who first proposed the modern CDP method. This technique found acceptance largely because of the increased need to find oil fields in difficult areas and also because surface sources could be utilized easily. The economics of seismic surveying had reached a point where the increased effort and cost were affordable. An additional consideration was that recording on magnetic tape had reached a point that the transcriptions from one tape to another— necessary for analog correction and reshuffling of seismic traces—could be performed without greatly increasing the signal/noise ratio.

Nevertheless, these analog procedures for correction and reordering data were cumbersome and subject to human error during the tape handling process. As a result, it was only when digital processing of seismic data established itself that the CDP method took hold. It was the fortunate combination of a viable idea with a technological development that resulted in one method of vastly improving data quality.

As far as seismic cross sections are concerned, the CDP method is now practically standard. It is therefore necessary to discuss the data gathering method in detail, including the type of information that can be generated. However, data processing techniques are deferred to a later chapter.

Figure 5.3 illustrates the different ray paths that, after appropriate time corrections, are stacked together to form one composite trace. It is noted here that these traces all have different sources and receivers as well as different paths through the upper formations. Thus it would be expected that random events would (on stacking

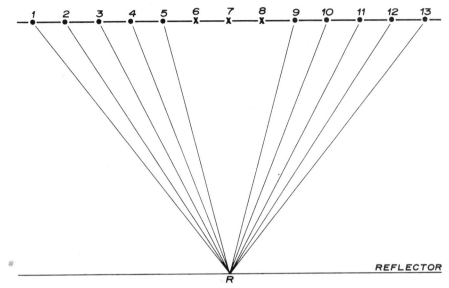

Figure 5.3 Different ray paths in a simplistic model, which are added together after a geometrical time correction, to give a fivefold stacked trace (applicable to position 7 on the surface).

several traces together) add up only as the square root of the number of traces, whereas the reflection amplitude, common to all traces at the same time, would add up linearly as the number of traces. Thus an increase in the signal/noise ratio as the square root of the number of traces is to be expected. Moreover, in any geological section in which the velocity increases with depth, long-period multiples do not have their geometrical time increases corrected by the standard NMO correction for a primary reflection at the same time and, given the proper disposition of stacked traces, tend to cancel (see Figure 5.4). This cancelation is, however, a sensitive process, needing seismic traces which have a constant time difference for the multiple—and such traces are not spaced uniformly along the surface. A brute force stack of all traces probably does not cancel multiple reflections as well as the addition of certain selected traces. Moreover, since this effect is exactly the same as the use of arrays on the surface, the sequence of multiple reflections over the traces used (after NMO correction) must cover a time interval of at least one period of the predominant frequency in the reflected energy. Sometimes, with small increases in velocity with depth, this requirement means the use of extremely long offsets—with the consequent danger of error in NMO removal due to lack of adequate knowledge of the reflection ray paths. Weathering and elevation corrections are also extremely important (Mayne, 1967; Marr and Zagst, 1967).

The layout in the field to accomplish the CDP procedure in the most efficient way is simple, flexible, and ingenious. Source points and receiver points, which are at least partly shared, are laid out equidistantly along the line of profile (see Figure 5.5). The cable is divided into sections which can be plugged together. Sections can be removed from the beginning of the spread and plugged in at the end, thus providing a system that proceeds efficiently and with a minimum of labor. The recording truck is equipped so that two sets of geophone inputs can be handled, with a switch to select the appropriate traces from those available as the shot points (or other source positions) are

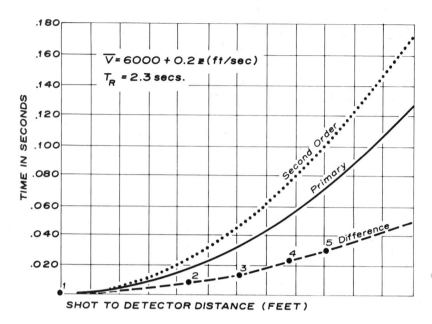

Figure 5.4   NMO for primary and multiple reflections (and their difference) as a function of the offset distance. Positions 1, 2, 3, 4, and 5 have equal differences and traces can be stacked to give a cancelation of waves of period 0.035 sec. [After Mayne (1962). Reprinted with permission from *Geophysics*.]

occupied. Thus the recording vehicle moves only after one complete spread is used up. The number of traces has now increased considerably over the original 20 to 24; 48 to 96 trace units are not uncommon, and this makes possible 24-fold, or higher, stacking where such an effort is demanded by poor record quality.

In marine environments of course the situation is even simpler, because the recording boat, carrying the source(s) and towing the cable, needs no such division of the cable. It is arranged instead that the sources operate in synchronism with the motion of the boat and cable, so that new source energy is introduced at the appropriate points, which the receivers *will later occupy* as they are towed. In principle of course there is nothing new. In practice, there are some differences. While a marine cable does have the advantage that the offsets of receivers are always constant and can always be occupied, the effects of cross-currents can be a severe problem, since the cable does not lie along the course pursued by the towing vessel (see Figure 5.6). In this case, while the offsets are known, the reflection points are not common. An even more disastrous situation occurs when the cross-current is not constant in speed and direction over the length of the cable.

While such problems are known (from observation of a buoy attached to the end of the cable) at the time of the exploration, little has been done about them. A suggestion has been made (Waters, 1976) to employ one or more servocontrolled variable lift paravanes to keep the cable on the line of traverse, rather than dead astern. The positioning relative to the water surface is equally important, and this problem has been solved by the use of Condep® controllers. These devices allow the cable to pass through them and then position the cable below the water surface using water

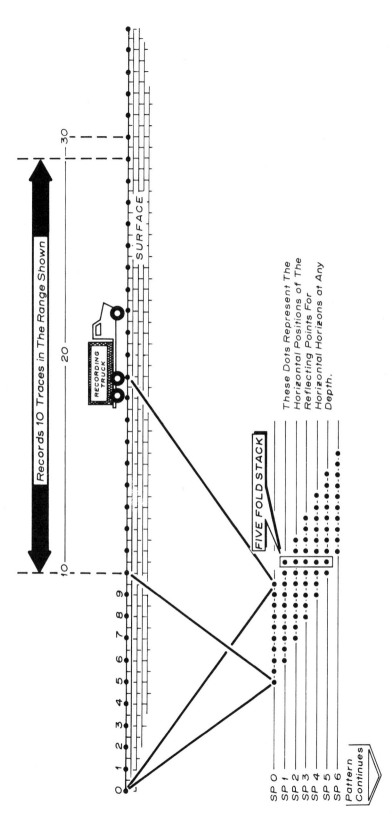

Figure 5.5 The basic arrangement of a typical (fivefold) CDP reflection spread. The recording truck contains a switch which allows selection of the correct sequence of geophones as the shot point is moved up. As convenient, portions of cable and geophones are moved from the left end and connected to the right end of the cable.

137

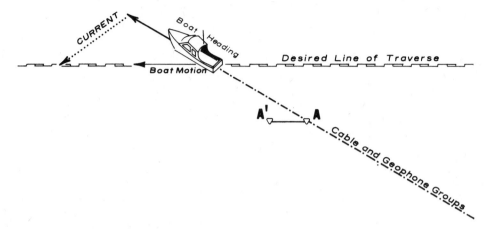

Figure 5.6 Effect of cross-currents on the motion of a boat and recording system. Points on the cable, such as A, move parallel to the desired traverse (to A', for example). Different offset depth points travel along different subsurface lines and do not stack.

pressure to actuate a vane. The vane, as the cable passes through the water, raises or lowers the cable unless it is at the correct (preset) depth. This device, simple in concept and action, has proved very effective and is almost universally used.

5.3 METHODS OF OBTAINING TWO-DIMENSIONAL CDP COVERAGE

It is axiomatic that the subsurface lithology varies in three dimensions, and a conventional CDP linear spread provides a cross section showing only apparent two-dimensional variation. For structural geological measurements where the third dimension (at right angles to the line of cross section) may give rise to only slow variation, the problem is not very important but, for stratigraphic trap problems, where the changes in sedimentation may be rapid (channel sands, offshore sand bars, reefs, etc.), there is a need to search for variation in three dimensions. This has led to the use of methods whereby three-dimensional variation can be examined.

First, it must be realized that a series of parallel CDP profiles can be used if terrain and cost are not deterrents. However, if for reasons of surface conditions the quality of subsurface records is dependent on the location of the sources and receivers, every effort should be made to maintain, as far as possible, communality of source and receiver positions. The next step therefore could be two parallel CDP lines recorded simultaneously, since a row of depth points (as shown in Figure 5.7) can be obtained between the two lines as well as beneath them (Davis, 1975). These center points are recorded from both sets of shot points and, as a consequence, have a double stack compared with those below the surface spreads. This method is difficult to stabilize at sea. On land, CDP lines are usually constrained by available roads and trails, hence the positions of the depth points and the distance apart of the lines of reflecting points are constrained. We return to the original recording philosophy that the points of information about the geology should be equally spaced in two orthogonal directions.

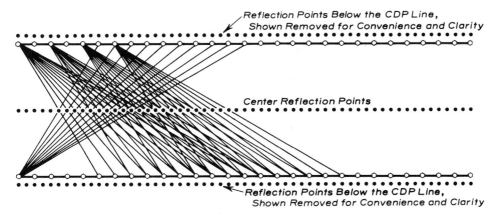

Figure 5.7 The method of formation of CDP reflecting points between two simultaneously recorded lines is illustrated. The number of fold of the stack is arbitrary, depending on the source-receiver relation and the number of traces recorded.

In order to satisfy this need, the idea of recording points (geophones) on lines at right angles to the line of sources was conceived. Actually the recording lines can be at any angle to the line of sources, but when such lines are a mile or two long and the line of sources is a continuous line several miles long, a right angle is most convenient, particularly in sectionized country. Figure 5.8 illustrates this concept and shows how a swath of reflecting points is generated. Note that the number of geophone stations per line and the number of source locations (or intervals) between the geophone

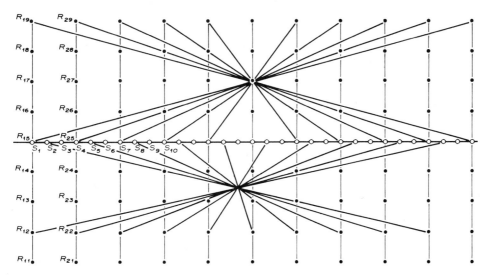

Figure 5.8 The seismic swath arrangement. Sources stay along the central horizontal line. Receivers occupy lines at right angles. If 99 traces can be recorded simultaneously, this will yield a 10-fold stack, or more.

lines are arbitrary. The number of offset distances that can be stacked together is related to several factors:

1. The total number of geophone groups that can be recorded simultaneously.
2. The number of geophone groups per line.
3. The maximum and minimum offset distances that can be used.
4. Receiver separations.
5. Source separations.

Economic and logistic factors contribute to the decision, too. It is intuitively obvious that this method of laying out the sources and receivers—and communication of the collected information to the recording truck—must be more time-consuming and/or require more manpower than a conventional CDP profile; however, a great deal more information is being gathered. Probably the factors that control the use of the method most are economics and the need for advanced communication methods from the geophone groups to the recording vehicle.

Additional field problems are:

1. If linear groups of geophones are laid out at each group location and a linear set of sources is used, they cannot be colinear for all pairs of sources and receivers employed. Consideration should be given to the use of radial patterns of geophones.
2. The vertical and horizontal locations of geophones must be known—just as in linear methods of surveying—but are more time-consuming to obtain. New geodetic surveying methods are needed to do this quickly and accurately.
3. The permit problem may be much more difficult in some areas.

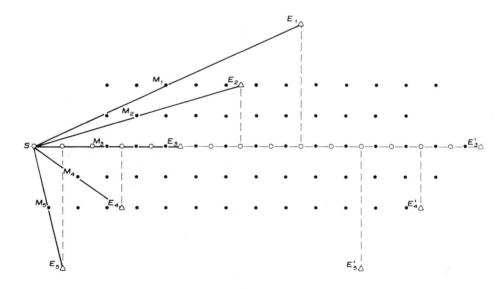

Figure 5.9 The method of Wide Line Profiling®. E_i represents a shot point, S a receiver, and M_i a corresponding reflecting point. (After Michon. Reprinted with permission from the *Oil and Gas Journal*, Nov. 27, 1972.)

The difficulties are not discussed further, but they do represent problems that will have to be solved before CDP information in two dimensions, with the same quality as the linear CDP method, becomes a routine procedure.

The Wide Line Profiling® method of C.G.G. treats the need for three-dimensional information from a slightly different point of view. The main objective is to provide knowledge of the component of dip at right angles to the main line of profile. This is useful in order to differentiate between arrivals recorded from different locations and directions but has not been emphasized as a method of tracing sedimentary trends within a wide swath of data. The field method differs little from that previously described and is shown in Figure 5.9.

All these methods have been used on land, but sometimes their implementation in marine surveys is more difficult. Hedberg (1971) proposed the use of sources, whose positions can be controlled in two dimensions by means of paravanes, to give three-dimensional information in marine surveys. Numerous other suggestions based on the use of paravanes have also been made. Figure 5.10 shows Hedberg's scheme. In theory of course there is no reason why such methods should not be successful. The difficulties occur largely because of currents and tides in the ocean and the characteristics of paravanes—lift versus drag, and so on. Furthermore, if the swath method could be carried out, it would be difficult to navigate such a system in relatively crowded waters, for example, around already existing drilling platforms.

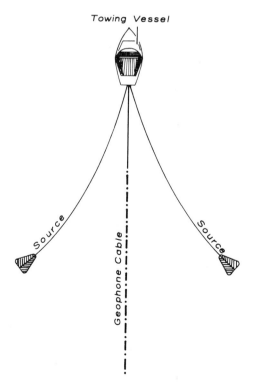

Figure 5.10 Marine three-dimensional method suggested by Hedberg (1971). It has since been suggested that the source paravanes can be steerable and servocontrolled.

5.4   THE MEASUREMENTS OF AUXILIARY SEISMIC INFORMATION

In addition to the generation and reception of seismic reflections, other data are needed to make the necessary corrections for the topography and varying thickness of the low-velocity layer. Although attempts are now being made to make weathering corrections for surface source usage from the reflection data by methods to be described later, these techniques have disadvantages which sometimes can be alleviated by making direct measurements.

In the case of source shots of dynamite in holes, the procedure is very simple, simply consisting of positioning a geophone (cluster) near the hole and recording its output. The uphole time, the depth of the shot, and measurements of the weathering and subweathering velocities then provide all the information necessary. Figure 5.11a shows this diagramatically and also shows a method of obtaining the necessary data. Alternatively (Figure 5.11b), use can be made of the fact that the arrival times for the first arrivals (first breaks) are those of waves whose travel paths to successive geophones are predominantly horizontal through the subweathering region and then

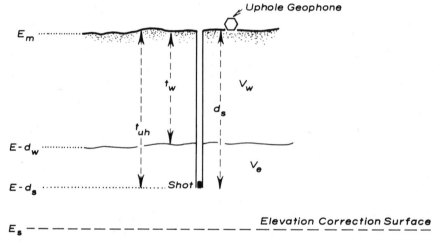

Figure 5.11a   Determination of corrections for a shot below the weathered layer.

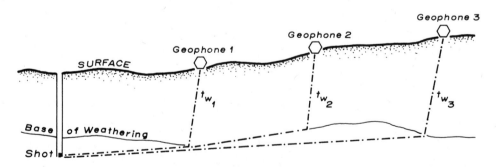

Figure 5.11b   Determination of differential weathering information by refraction arrivals.

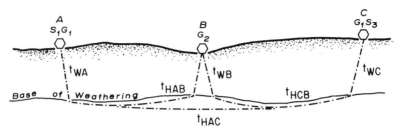

Figure 5.12   The *ABC* refraction weathering method.

almost vertical through the weathered zone. If allowance is made for horizontal travel distance differences, the difference in weathering time can be calculated. By comparison with uphole time information, these times can be made absolute.

For surface sources, however, a different weathering time procedure is necessary (Figure 5.12). A succession of shot points is shot into receivers at the next two successive shot points, and the arrival times are recorded. This procedure is repeated down the line. For any successive three points A, B, and C, the times t_{AC}, t_{AB}, and t_{CB} will (eventually) be known, and the weathering time at B is

$$t_{WB} = \frac{t_{AB} + t_{CB} - t_{AC}}{2}$$

This is a simple procedure and works well when the weathered layer can be regarded as a simple two-layer case. When, however, there is an appreciable gradient in velocity below the base of the weathering, the horizontal travel times are not additive and the method fails.

5.5   SUMMARY AND CONCLUSIONS

Data gathering methods for exploration seismology have changed from the early attempts to use isolated measurement points (as is common for geological well data) to continuous lines, in which the measuring points are much closer together— thereby substituting decisions on correlations with several much smaller changes in record characteristics for a single, larger decision between points widely separated. Some reasons for this shift, which is undesirable from the exploration viewpoint, are:

1. The frequency bandwidth obtained with seismic data is insufficient to give a good chance of making undisputed correlations of reflections (and therefore their associated geological boundaries).
2. A need existed to improve the signal/noise ratio on seismic recordings, so that surface waves, multiple reflections, scattered interference, and random noise did not complicate the process of picking corresponding reflection times. Much of this improvement has been achieved by the stacking together (after normal geometrical corrections) of traces with a common reflection point but different source-to-receiver offsets.

An additional need for gathering three-dimensional information is now felt. This can be accomplished of course by running several closely spaced parallel CDP lines, but a swath method in which the source or receiver positions are common for several parallel lines of reflection points, and which are recorded simultaneously, is preferred. Such swath methods rely on the simultaneous recording of a large number of traces, and the methods of communicating such data from the recording stations to a central recorder—or methods of recording several traces locally and interlacing the data later—are of great importance. It is likely that substantial advances in this art will be achieved soon.

In marine prospecting, single-line CDP methods are easily carried out, but three-dimensional information is difficult to obtain simply.

For shear wave recording, land methods are very similar to those used for recording compressional wave reflections. However, since horizontally polarized shear waves are picked up by horizontal geophones and are generated by horizontal forces or horizontal vibrators, much care must be taken in planting the geophones (and using the vibrators) so that the same polarity combination is always used. In marine environments, the zero transmission of shear waves through water has precluded any progress.

REFERENCES

Davis, J. L. (1975), U.S. Patent No. 3,890,593, June 17, 1975.

Hedberg, R. M. (1971), U.S. Patent No. 3,581,273, May 25, 1971.

Marr, J. D., and Zagst, E. F. (1967), "Exploration Horizons from New Seismic Concepts of CDP and Digital Processing," *Geophysics*, Vol. 32, No. 2, pp. 207–224.

Mayne, W. H. (1962), "Common Reflection Point Horizontal Stacking Techniques," *Geophysics*, Vol. 27, No. 6, Part II, pp. 927–938.

Mayne, W. H. (1967). "Practical Considerations in the Use of Common Reflection Point Techniques," *Geophysics*, Vol. 32, No. 2, pp. 225–229.

Michon, D. (1972), "Wide Line Profiling offers advantages," *Oil and Gas Journal*, Vol. 70, No. 48, pp. 117–120, November 27, 1972.

Waters, K. H. (1976), patent pending.

SIX

Seismic Data Processing

6.1 INTRODUCTION

The data obtained in a typical seismic survey consist of measurements of the motion of the surface at a series of surface positions following the operation of a source of seismic waves. This motion may be particle displacement or velocity, which in its most general form is a vector function of time or, in certain cases, the scalar excess pressure can be measured as a function of time. An important property of sets of time functions is that they are recorded simultaneously or at least have an indication of operation of the source.

For an exploration geophysicist, usually the most relevant data are from reflections from the geological strata and, for this reason, methods are needed for treating the raw data, which deemphasize all but the reflection information. As seismic reflection exploration methods developed, more and more data processing became necessary. The advent of magnetic tape recording in the early 1950s began to demonstrate the possibilities of filtering, stacking, and rerecording. Such methods dealt with continuous recordings of time functions (analog data).

The Geophysical Analysis Group (GAG) project, sponsored by several oil companies at M.I.T. (1953 onward), was the first major indication that digital computing, using sampled values from continuous time recordings, would eventually become a versatile means of seismic data processing. Many of the concepts of filtering using sampled data originated there (Treitel and Robinson, 1967), but a major advance in digital technology—convolution filtering equipment (CFE)—was necessary before a full range of data processing techniques could be economically employed.

It is probably not an overexaggeration to say that the CDP method, involving the correction for geometrical (offset) effects and the addition of many corrected traces selected from a large number of available ones to give a final CDP trace, owed its final overwhelming acceptance to the speed and consistency of the digital computer. In turn, it was the data quality improvement provided by the CDP method that initially led to wholesale employment of digital methods. Once computers were available and the speedup provided by CFE made other processes economical, the digital revolution was on its way.

This chapter deals with the processes, and methods for accomplishing them digitally, that are now in use to improve seismic *reflection* data. In only a few cases of exceptional interest is reference made to analog methods.

6.2   SAMPLING (MULTIPLEX) PROCEDURES

Sampling of multitrace seismic records can be done either directly in the field or by recording continuous time traces on analog magnetic tape and later replaying these records to provide electrical signals which are sampled. In either case, the process normally involves taking a sample sequentially from a number N of traces and then returning to the first trace (well within the sampling interval) and cycling through the sampling process again. This system provides a series of numbers in the sequence

$$\{a_{11}, a_{21}, \ldots, a_{N1}, a_{12}, a_{22}, \ldots, a_{N2}, \ldots, a_{NM}\}$$

where the first index is the trace number and the second index is the sample number (at a predetermined number of milliseconds apart) on a particular trace. The electronic consequences are easy to see (Figure 6.1). The allowed time for the digitization of N traces must be less than the sample interval (typically 0.002 or 0.004 sec) which in turn is prescribed by the frequency spectrum to be retained. In the recording system itself, an antialias filter is provided, which rejects very strongly (-50 to -60 dB) frequencies that, with the selected digital interval, may give rise to aliasing errors. For example, suppose a digital interval of 0.004 sec is chosen; the Nyquist (or folding) frequency will be $1/0.004 \times 2$ or 125 Hz. That is, at 125 Hz, there are two samples per cycle. Since this is a theoretical limit, it is usual to provide an antialiasing filter with a high reject capability above 62.5 Hz (the four-samples-per-cycle frequency). The exact filter cutoff points as well as the design of the filters are manufacturer's decisions.

Thus a decision to digitize at 0.004-sec intervals makes the assumption that energy with frequencies above (say) 62.5 Hz are mainly noise and are of no interest for providing geological information. With digital recording in the field, this important decision must be made before the survey is started. Usually, the next step provided is 0.002 sec, which raises the accepted frequency band to 125 Hz but also doubles the number of digital sample points for a given maximum recording time. The cost of digital processing is not always a linear function of the number of sample points, but

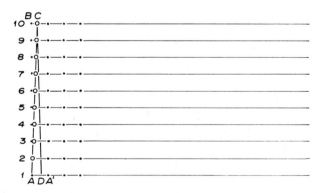

Figure 6.1   Deviation of digitizing from ideal sampling. Digital samples are required at the dots on the time traces 1 through 10. Actual samples are obtained at the positions of the open circles because of finite digitizing time (the path AC). Resetting the digitizer occurs along CD. DA' represents the waiting time until the next timing pulse initiates further digitizing.

the increased cost of processing double the number of points is never insignificant, and this is therefore an important decision.

It has been stated that the time for sampling N traces must be significantly less than the smallest digital interval available. For illustration purposes, we suppose that the sampling time for 40 traces must be 0.001 sec or less. Then each trace must be digitized within 25 μsec, and the digitizer must step over to the next trace voltage to be sampled.

Details of the electronics are not within the scope of this book, but the general methods adopted are explained. Almost universally, digitization, recording, and subsequent digital computation are conducted in *binary* notation. This is a method of representing a number using as a basis the number 2. As in decimal notation, where we consider an integer (say) 461 as being related to powers of 10:

$$4 \times 10^2 + 6 \times 10^1 + 1 \times 10^0$$

so can the same number be expressed as powers of 2

$$1 \times 2^8 + 1 \times 2^7 + 1 \times 2^6 + 0 \times 2^5 + 0 \times 2^4 + 1 \times 2^3 + 1 \times 2^2 + 0 \times 2^1 + 1 \times 2^0$$
$$256 \quad + \quad 128 \quad + \quad 64 \quad + \quad 0 \quad + \quad 0 \quad + \quad 8 \quad + \quad 4 \quad + \quad 0 \quad + \quad 1$$

or, in contracted notation,

$$111001101$$

where the power of 2, starting at 0 on the right, is indicated by the position of the digit in the number. A binary digit is called a bit. Sometimes eight of them together are called a byte.

These numbers can be continued as fractions, since $\frac{1}{2}$ is represented as 2^{-1}, $\frac{1}{4}$ as 2^{-2}, and so on. Thus a binary fraction is written

$$110011$$

In digital engineering practice, the use of binary numbers is convenient because use can then be made of switching circuits in which 1 (say) represents the circuit in one stable state and 0 represents the *only other* stable state. Such *bistable* circuits are the basic components of all computer engineering. For most geophysicists these internal procedures are of little interest, since they are concerned with only the visible recordings of electric signals generated in the reflection process. For them, it is proper to point out that the computed results (still in binary form) are rendered tractable to plotters through the use of digital-to-analog converters.

After this brief diversion we return to the subject of sampling (or analog-to-digital) procedures.

If the total range of the voltage to be measured is considered 1 unit (perhaps 10 V) on the binary scale, this will be 2^0 and 5 V will be $\frac{1}{2}$ unit $= 2^{-1}$. Alternatively, the smallest increment in voltage to be measured can be set equal to 2^{-N} units, and voltages measured in terms of this increment. A voltage given as 1011 is 11 (decimal) increments.

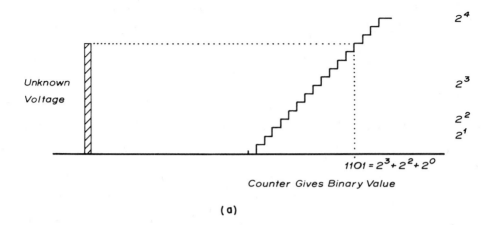

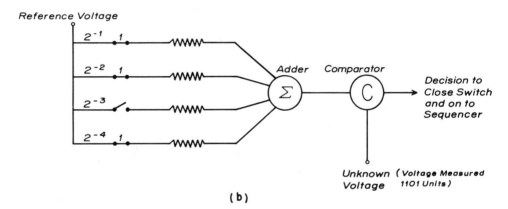

Figure 6.2 Digitization methods. (*a*) Sequential counter method. (*b*) The faster sequential adder method.

The first method (see Figure 6.2*a*) consists of comparing a voltage provided against the required voltage. The comparison is a stair step which increases in equal increments according to the stepping of a binary counter. Thus the reading of the counter when the comparator indicates equality (or change in sign) gives a measure of the unknown voltage. This method is slow, because it has to cover the entire range from zero to the unknown voltage each time. The sign of the voltage is an additional bit which is set separately. When a total of eight bits is provided (one byte), the range can be from (almost) -2^8 to $+2^8$ or, in binary notation, (1)1111111 to (0)1111111, where the digit in parentheses is the sign bit.

The second method (Figure 6.2*b*) is faster. The logic consists of dividing up the total voltage into ranges related by the binary scale and having a system in which such binary-related voltages can be added and compared with the unknown voltage. The most significant bit is tested first and, if it is less than the unknown voltage, a switch is set to keep it permanently connected to the adder. Then the next most significant bit is added, and the process repeated. When all the bits have been added and tested, the bit arrangement of the switches set is a measure of the unknown voltage. Although this method is sequential, it involves only N comparisons compared with an average of 2^{N-1}.

Thus, if N is 8, the ratio of times for same-speed electronic circuits is 8:128 or 1:16. If higher values of N are used, the ratio increases.

For very high speeds, parallel instead of sequential circuitry is available. Under such circumstances, digitizing speeds of 50 to 100 MHz are available.

The discussion above applies only to seismic traces that have a known overall range so that the scale factor (or maximum range of digitization required) can be set. The precision of digitization, that is, the smallest incremental voltage change that can be detected, is set by the number of binary digits employed. The overall dynamic range between the largest number and the increment is also set by the number of bits employed. For example, a voltage range of ± 255 to 1 implies a dynamic range of 48 dB in power. To deal with some of the larger ranges that may be required in seismic reflection work (e.g., when surface wave amplitudes as well as threshold noise amplitudes must be correctly recorded in a linear manner), amplifiers have been built that have a binary ranging feature. These are very wide-band amplifiers arranged so that the gain can be very rapidly changed in steps of 2—as indicated schematically in Figure 6.3. By employing these amplifiers, it is arranged that the final output of the amplifier is a voltage that goes to the digitizer within a known range. The digitizer then produces a binary output to be recorded but, in addition, the scaling factor of the amplifier must also be recorded. In logarithmic terms, the digitizer output is the mantissa and the amplifier scale factor is the exponent. Since both are in binary notation, they can be adjoined in a (larger) binary word such as

Sign

$$(1)\underbrace{0110}_{\text{Exponent}},\underbrace{01111001}_{\text{Mantissa}}$$

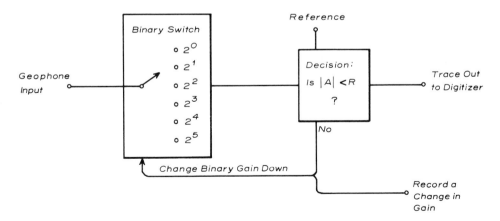

Figure 6.3   Schematic showing simplistic action of the binary gain control. It is normally more important to change the gain downward, to prevent truncation of the trace amplitude, than it is to change it upward. In an operational amplifier, provision is made for both directions.

The exponent can be evaluated and used to shift the mantissa to the left. The above number is equivalent to

$$(1)1111001000000 \quad \text{(in binary notation)}$$
Sign

or

$$-7744$$

The effective linear range of the amplifier (dynamic range) is therefore considerably increased, but the accuracy (in any binary range) is still the same.

6.3 DEMULTIPLEX PROCEDURES

The binary number sequence obtained in the digitization process, namely,

$$\{a_{11}, a_{21}, \ldots, a_{N1}, a_{12}, a_{22}, \ldots, a_{N2}, \ldots, a_{NM}\}$$

has multiplexed the information from N channels into a signal channel. For most digital processes, it is required to demultiplex this single-channel sequence into a different single-channel sequence (on magnetic tape, for example) in which the sequential values of a single trace appear in order, followed by the sequential values of the second trace, and so on, until all the N traces are recorded in order. The new sequence is

$$\{a_{11}, a_{12}, \ldots, a_{1M}, a_{21}, a_{22}, \ldots, a_{2M}, \ldots, a_{N1}, a_{N2}, \ldots, a_{NM}\}$$

This is a digital computer process involving reading numbers from a magnetic tape, reordering them, and then recording a new digital tape. During the reordering process, the format of the numbers may be changed, if desired. That is, numbers having binary gain may be changed to floating-point (computer-compatible) numbers or to straight binary integers or fractions. The main consequence of demultiplex programs is that the sampled values of each single trace are now in order, as are the sequential traces. Headers, that is, information words for the trace, can be added as well as end-of-file marks, and so on. Some degree of standardization of formats has been achieved by the work of the Society of Exploration Geophysicists (Northwood, Weisinger, and Bradley, 1967; Meiners et al., 1972), but there are still various formats, styles of tape recording, numbers of track widths, parity codes for reading accuracy checks, and so on, so demultiplex programs are largely specialized for a computer installation. Usually, demultiplexing is done off-line from the main computer, the resulting tapes being used as input to the main computer, but this again depends on the individual computer system design. In all future discussion of digital data, it is assumed that the demultiplexing has been done and that trace-by-trace sequential sets of values are available. While data from the field recording equipment must be maintained, the header record of the demultiplexed tape is made large enough to contain all the auxiliary information needed to continue the data processing.

6.4　LINE DESCRIPTION AND REORDERING

Auxiliary information is partially provided by the land surveyor who lays out the seismic line or, in the case of marine surveys, by the electronic positioning equipment. On each trace, there has to be information that allows access to the elevation and position of the source and receiver for that trace. Information common to more than one trace can be kept in a central computer file, available for access during the entire data processing run. The horizontal position of the source and receiver determine the offset distance and the coordinates of the reflecting point (for assumed zero-dip reflectors).

After the demultiplex operation the traces are (for CDP operation) in the order in which they were recorded, namely, a particular shot point into several sequential receiving points, followed by the next shot point with receivers having an increase of 1 in their index number. For other field geometries, the sequence is different. However, for stacking operations, it is convenient to reorder the traces so that those that contribute to information at a given reflecting point occur in sequence, followed by those at the next reflecting point, and so on. This reordering process is strictly a sorting process which is handled by the computer and its peripheral gear under program control. The use of a few parameters enables the computer to do consistently a task that was formerly a hazardous tape handling operation in the earlier analog tape days of CDP. In the full schedule of data processing, reordering can be done at any convenient time before the removal of geometrical effects due to offset and stacking. This is the chief reason the traces carry header information with them. No matter what their number in a particular sequence, they can always be operated on, using the information they can supply.

6.5　CORRECTIONS FOR NEAR-SURFACE DELAYS AND GEOMETRICAL EFFECT OF OFFSET

These corrections are often referred to as static and dynamic corrections because one is assumed to be constant for the entire trace, whereas the correction for offset changes with record trace time. Although at first sight these are simple problems, the issue is often clouded by the fact that, from an economic viewpoint, no provision is made for special near-surface measurements or special velocity measurements. As seen later, this forces the system to determine velocities and near-surface corrections from the CDP data. This section deals first with methods of obtaining the velocity and the estimates of low-velocity-layer corrections, and then shows how these values are used to correct the data in a preparation for stacking.

Figure 6.4 shows the geometrical ray paths when a single constant velocity is assumed. From this we can see that the relationship

$$\frac{V^2 T_0^2}{4} = \frac{V^2 (T_0 + \Delta T)^2}{4} - \frac{X^2}{4}$$

holds, and this reduces to

$$\Delta T = \frac{X^2}{V^2 (2T_0 + \Delta T)} \tag{6.1}$$

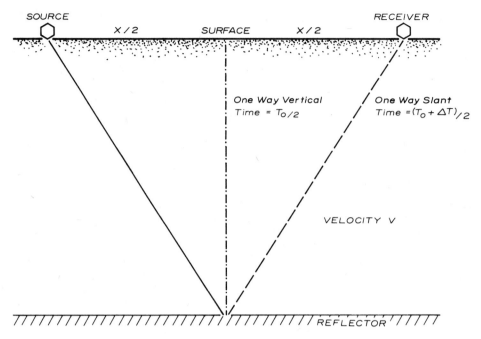

Figure 6.4   Geometrical ray paths for offset and perpendicular reflection paths.

For computer determination of the proper value of ΔT, we can use a successive approximation, so that

$$\Delta T_i = \frac{X^2}{V^2(2T_0 + \Delta T_{i-1})} \tag{6.2}$$

where the subscript denotes the order of the iteration. $\Delta T_0 = 0$ is the first assumption. However, the convergence is faster if one writes

$$\Delta T_i = \frac{2X^2}{V^2(4T_0 + \Delta T_{i-1} + \Delta T_{i-2})} \tag{6.3}$$

since the values determined oscillate in a damped fashion about the true value.

It is usual, in applying these corrections, to prepare a table of ΔT versus T_0 for use in the computer instead of calculating the values needed each time. Figure 6.5 shows a series of primary seismic profiles (sometimes called 100% records since there is no multiple fold stack) which differ from one another only in the constant velocity used to correct the NMO. For each reflection, there is one velocity for which the best ΔT removal has been done. The reflection "pick" moves in a straight line across the record (the straight line may not be a constant time line, since dip may be present). Thus, in this simple way, an average velocity to each reflector can be picked that will best align reflection events for later stacking. Since a change in delay times in the weathered layer may account for some curvature of the reflected event, this method

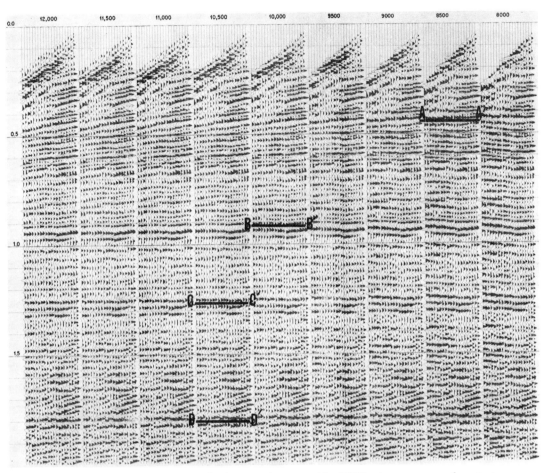

Figure 6.5   A set of single primary reflection records corrected for NMO using a sequence of constant velocities.

has to be regarded as an approximate one only and *serves only for preliminary processing.* Several such determinations are needed on each line in order to reach an acceptable velocity-time function for this preliminary processing.

Whenever possible, a special experimental field method, known as an expanding reflection spread (Musgrave, 1962) (Sattlegger, 1965), should be used to compensate for the effects of dip. It can be argued that the true velocity is not needed, but only a sufficient approximation to the true velocity that the data quality is not affected during the stacking operation to follow. Velocity variations in both the horizontal and vertical directions are important, and the real need is to make a consistently good approximation so that such changes can be given some credence.

Another approach to the measurement of velocity is to plot the square of the arrival time of a reflection against the square of the offset distance. Then, using the formula

$$\frac{V^2 T^2}{4} = Z^2 + \frac{X^2}{4} \qquad\qquad (6.4)$$

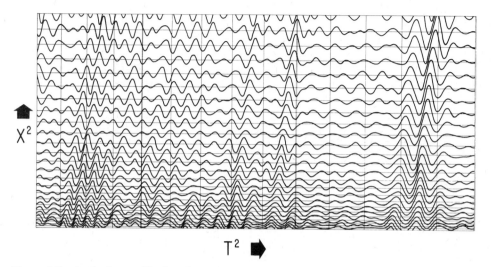

X^2

T^2

Figure 6.6   A seismic record is plotted to nonlinear scales. The offset X and amplitude A are plotted on a scale proportional to X^2. Arrival time is plotted on a scale proportional to T^2.

the slope of the (near) straight line gives V^2. In Figure 6.6, the record is plotted in such a way that the distance down the record is proportional to T^2, and the distance across the traces is proportional to X^2. The best-fit straight lines can be drawn on such a display directly.

Several more sophisticated computer programs have been written to compute (and display) the velocity as a function of time (Taner and Koehler, 1969). The early ones determined a velocity to a constant record time by considering a series of primary records and allowing a computer program to arrange for a comparison of the stacking effectiveness of various constant velocities for each record time. The decision criteria range from simply measuring the maximum sum taken across the traces within a given time interval to more sophisticated ones employing multiple-trace coherence coefficients which are discussed later. As computing speeds have increased and unit costs of computer operations have dropped, these velocity determinations have become more and more routine.

For some purposes, connected with lithological implications of velocities, there are two further needs. First, it is necessary to use a statistical velocity program which measures velocities to a given reflection horizon (rather than a constant time), and second, interval velocities computed from average velocities need to be corrected for path curvature (deviation from the straight-path assumptions made previously). Dix (1955) showed that the true nth-layer interval velocity is related to two sequential average velocities by the formula

$$V_n^2 = \frac{\overline{V}_n^2 t_n - \overline{V}_{n-1}^2 t_{n-1}}{t_n - t_{n-1}} \tag{6.5}$$

where $\overline{V}_{n-1}$ and $\overline{V}_n$ = average velocities from the datum to reflectors above and below the layer t_{n-1} and t_n = the corresponding reflection times.

The reflection arrival times ideally obey the hyperbolic law

$$T^2 - \left(\frac{X}{2V}\right)^2 = T_0^2 \tag{6.6}$$

If, however, the reflecting horizon is itself curved, this curvature will cause an incorrect hyperbola to be calculated, so long as the shot point remains fixed or if the shot points differ but the receiver remains fixed. If the reflector is convex upward, its curvature will be added to that of the wave front returned to the surface and a velocity which is too small will be calculated. The reverse is true for a surface that is concave upward. This fault was common to all early statistical methods and has been fully discussed by Dix (1955).

However, for CDP methods when the receiver and source move equal distances in opposite directions on the surface, and if the normal to the reflecting surface is vertical at the CDP, this effect does not occur and theoretically the correct velocity is calculated no matter what the curvature.

The effect of dip on statistically calculated velocities is not zero for CDP methods, since the CDP migrates up the dip slope as the offset is increased. Levin (1971) showed that this effect is about 1% for dips of 8°, but increases as the angle of the dip squared. The velocity obtained is equal to the true velocity divided by the cosine of the angle of dip. If dip and curvature are both present, each will have an effect on a CDP-derived velocity that is similar to, but not as large as, the effect of curvature described earlier.

It is to be noted that any method of fitting by least squares a hyperbola to reflection times implies a method by which these times can be picked. Further, the accuracy needed is such that the normal digital interval is too coarse a sampling interval, and a method of interpolation is needed. The usual procedure is to pick the differential time between successive traces which correspond to the maximum correlation coefficient between short "windows" of the seismic trace which are centered at the record time under consideration. Other criteria have been used, and for details the reader is referred to articles listed in the references.

6.6 NMO CORRECTIONS AND THEIR EFFECT ON SEISMIC PRIMARY RECORDS

Figure 6.7 shows a schematic set of impulses, such as might be received on a seismic record, with the offsets of the traces varying from 0 to 10,000 ft and with a constant velocity of 10,000 ft/sec. The impulses denote the times of arrival only. The individual reflection layer impulses form a hyperbola. These successive hyperbolas are always closer together on the far offset traces than on the near offsets. When impulses (infinite frequency response) are used, the removal of the correct amount of NMO simply places all the impulses at a constant time for a given reflector regardless of offset distance.

However, when reflections have a finite bandwidth, they overlap one another on the field record, and all the traces are recorded with the same bandwidth. A serious wave shape and spectrum distortion then appears on the long-offset traces as NMO is removed. This is nonlinear distortion, so that filtering again after NMO removal

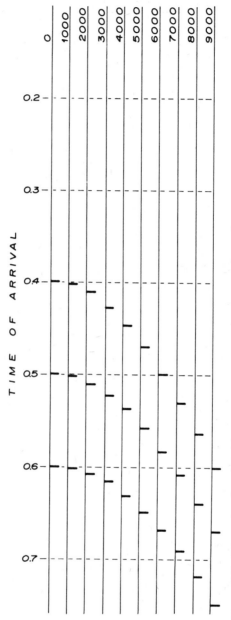

Figure 6.7 Schematic of an impulsive (nonreal) record showing comparative times of arrival. $X = 0(1000)9000$ ft and $V = $ 10,000 ft/sec.

does not give the same result as would have been obtained if all the traces had been recorded at very wide bandwidth and the NMO removed, followed by band-pass filtering. The latter procedure is to be preferred, but it means that a higher sampling rate on recording is necessary, followed by a resampling after NMO removal. Another way of looking at this problem is to determine the NMO for a particular time. If the reflection event consists of only a spike, the proper correction to render the reflection time independent of the offset distance may well be applied. However, the true seismic event, in impulsive-type reflection work, is a pulse possibly 0.1 sec in length, with only its beginning occurring at the proper reflection time. The tail of the reflection event occurs at a later time, but its NMO is removed as though it *were* a later reflection. In fact, all velocities are incorrect, as the main energy occurs with some delay after the true reflection time that applies to it.

It is possible to correct impulsive-type records, by a filtering process known as *deconvolution*, so that, as is the case with Vibroseis® records, the main amplitude occurs at zero time. If this is done, the main amplitude of a reflection will occur at the true reflection time. It is true that there are now positive and negative tails (precursors and afterrunners), but these are minimal and should not cause the same problem as one-sided pulses in the same bandwidth. The reduction of impulsive-type records to zero-phase records is only as good as the information about the primary pulse shape. Deconvolution methods are considered later in this chapter.

The conversion of the offset trace to a pseudozero trace can be done in several different ways, one of the better ones being as follows.

Each digitized amplitude A_i of the offset trace is mapped onto position $j + a$ of the pseudotrace where a is a fraction. In fact, it maps between two digital positions. Its amplitude A_i is then proportioned between positions j and $j + 1$. However, each position j on the pseudotrace receives contributions from more than one i position on the offset trace. It is therefore necessary to sum the contributions at the j positions before the trace is used. The pseudozero offset trace produced should be filtered with a zero-phase finite bandwidth filter. For some purposes, this can be done after stacking, but the option should exist to filter here.

6.7 MUTING AND TRACE ENERGY MEASUREMENT AND STACKING

One further operation is common before the traces are ready to be stacked. It is known as muting and consists of zeroing out each trace to remove the high-energy near-surface arrivals. Usually, the trace is zeroed before a time given by the offset distance divided by the muting velocity (sometimes a constant is added). Muting can be done at any convenient time.

It is convenient also to be able to measure the overall energy of a primary seismic trace. This is done by forming the sum of the squares of the amplitudes of the trace at each of the digitally sampled points:

$$E_j = \sum_{i=1}^{M} A_{ij}^2 \tag{6.7}$$

where there are M values of the amplitude sampled on trace j.

Sometimes, the sum of the absolute values is used instead of the square:

$$E'_j = \sum_{i=1}^{M} |A_{ij}| \tag{6.8}$$

The traces that contribute to a single stacked trace are usually scaled to a constant energy, and then the sample values at a constant time are averaged (where there are N traces to be added):

$$S_K(i) = \frac{1}{N} \sum_{j=1}^{N} \frac{C}{E_j} A_{ij} \tag{6.9}$$

where C = a constant

Optionally, E_j can be replaced by E'_j.

The assumptions for treating the stacking procedure in this manner are that there is a constant signal/noise ratio and that the individual traces have different amplitudes because of other factors, such as geophone placement and gain setting. Of course, the far offset traces have traveled further—more so at the beginning of the trace than at the end—and this should be compensated for. As seen later, when true amplitudes are discussed, the overall assessment of various factors that affect amplitude is difficult.

In areas where some traces may have considerable noise components, the simple scaling to constant energy allows the noisy traces to affect the final trace too much. In this case, each balanced trace can be weighted inversely as the energy on the assumption that most of the energy measured is noise energy. Then,

$$S_K(i) = \frac{1}{N} \sum_{j=1}^{N} \frac{C'}{(E_j)^2} A_{ij} \tag{6.10}$$

where C' = a constant

Optionally, E'_j, as defined before, can replace E_j. This type of method has been called diversity stack. When applied to CDP primary traces, the reduction in the contribution of one or more traces because they are noisy may unbalance the system (as far as canceling unwanted events is concerned), and diversity stack must be employed with caution, bearing in mind the unequal contribution of members of an array of traces.

Finally, the contributions of the individual traces to the final composite can be assessed on the basis of the coherence (similarity or semblance) between the trace under consideration and some standard. Because of the sensitivity of the coherence coefficient to shifts between events on the two traces, this criterion can be used only if the NMO and static shifts have been removed. The correlation coefficient is defined as

$$C_{12} = \frac{\sum_1^M a_{1j}a_{2j}}{(\sum_1^M a_{1j}^2 \sum_1^M a_{2j}^2)^{1/2}} \tag{6.11}$$

If the standard chosen is the average of N component traces, the minimum value of C_{12} will be $1/N$, and its maximum value unity. Then the stacking relation will be

$$S_K(i) = \frac{1}{N} \sum_{j=1}^{N} \frac{C''}{C_{12}} A_{ij} \tag{6.12}$$

where $C'' = $ a constant

The range of determination of C_{12} need not be the full length of the seismic trace(s) but can be made over a window, obtained by multiplying the seismic traces by a window function, the simplest of which is a boxcar:

$$\begin{aligned} W(t) &= 1 \quad (t_1 < t < t_2) \\ &= 0 \quad (t < t_1 \quad \text{or} \quad t > t_2) \end{aligned} \tag{6.13}$$

Other window functions and their advantages are discussed later.

6.8 EXPANSION OR GAIN CONTROL

There is evidently some need for a process that allows change in gain of seismic traces as a function of time. Spherical divergence of the seismic energy from the source gives rise to a change in amplitude with time which must be compensated for. The modern practice of using minimum offsets of several hundred to thousands of meters has reduced the range of data recorded, but compensation is still needed to provide traces in which relative amplitudes of reflected events are preserved. The method of gain compensation is simple. It consists of multiplying the amplitude of the trace, point by point, by the amplitude of the computed gain function:

$$A'(i\,\Delta t) = A(i\,\Delta t)G(i\,\Delta t) \tag{6.14}$$

The simplest of the gain functions (apart from a constant multiplier) is a linear increase in gain with time:

$$G(i\,\Delta t) = i \tag{6.15}$$

which should compensate for spherical divergence of the simplest type (homogeneous, constant-velocity medium without attenuation). However, true attenuation, loss of amplitude by passing through many thin layers (scattering), and instrumental effects of all kinds tend to make the loss of amplitude with time a more complicated function of time. Two additional approaches to compensation for these losses are given here in outline. Individual processing centers implement their own versions.

The first method is to determine the average amplitude over a series of windows—whose shape, length, and location can be set by the geophysicist. For example, in Figure 6.8, three different zones, A, B, and C, have been set up, and the average amplitude for each zone is denoted by $|A_A|$, $|A_B|$, and $|A_C|$. The gain function is then set up to compensate for these average gains at the applicable times, T_A, T_B, and T_C, and to grade between them.

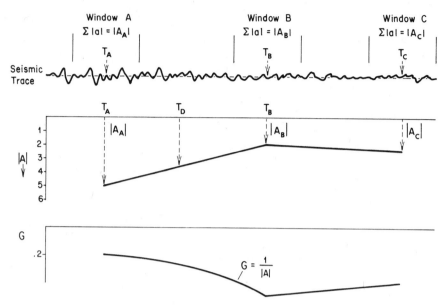

Figure 6.8   The average amplitude for three different zones is determined, and the gain function is then set up so that, after application, the average amplitudes in the three zones will be equal.

For example, in the range $T_A < T_i < T_B$, the gain at T_i is given by

$$G(i\,\Delta t) = G(T_A) + \frac{T_i - T_A}{T_B - T_A}[G(T_B) - G(T_A)] \qquad (6.16)$$

Each interval is treated individually. If the gain function should be smoothly varying, a cubic spline function can be fitted to the gain function values at each of the determination points.

The second method simply tries to fit the gain needs by simulating an automatic gain control in the computer (as shown diagrammatically in Figure 6.9). A simple way of doing this is to start, as above, with the summation of the absolute values of the trace over a given interval of $2l$ values:

$$|\bar{A}_k| = \frac{1}{2l}\sum_{j=k-l}^{j=k+l}|A_j| \qquad (6.17)$$

Because of the use of the multiple points, the change in $|A_k|$ is smoothed. In this case, some high or low values *yet to come* have been used so that a gain change can be anticipated—in fact, an attack time can be set.

The gain, if there is no more action than this, will be

$$G(k) = G(0) + \frac{2ql}{\sum_{j=k-l}^{j=k+l}|A_j|} \qquad (6.18)$$

where q = a constant
   $G(0)$ = initial gain

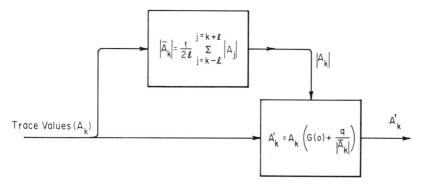

Figure 6.9   The method of obtaining a simple form of automatic gain control is illustrated. By varying the length $2l$ of trace used to determine the gain and its position on the trace, different attack and release times can be obtained.

As k advances, a point is dropped off at the beginning, and an additional point added at the end. A weighting system within the window can be used, if desired. Much more complex schemes will be evident to the reader, but this one has the virtue of simplicity.

6.9   THE NEED FOR OR DESIRABILITY OF GAIN CHANGES

Prior to about 1971, the desire of geophysicists was to obtain seismic cross sections that were optimum from the point of view of continuity and signal/noise ratio. With the use of such sections, continuous reflecting horizons could easily be followed and marked so that structure and gross stratigraphic traps could be seen. Under such conditions, traces were often expanded individually and then adjusted for constant power over the entire section.

The discovery of direct hydrocarbon indicators, particularly of use for gaseous hydrocarbon exploration, has placed much more emphasis on the preservation of true amplitude. Now, in this type of work, amplitude variation with space and time is a useful parameter for making visible the very large reflection coefficient changes that can occur when gas is present in the usually liquid-filled pores of the reservoir rock. Under these conditions, the only gain changes that should be allowed are those (constant from trace to trace) that correct for gradual amplitude changes with time, such as spherical divergence, and those for which there is a physical basis, such as a change in the input power. Figure 1.3 shows how such amplitude changes can be translated into color changes which make them stand out strongly. It is not necessary to use color, but the standout is more visible than in monochromatic representations. The fallibility of such direct hydrocarbon indicators, or "bright spots," is discussed later.

6.10   FILTERING AND DECONVOLUTION IN GENERAL

It is very important to stress the fact that, at this point, full seismic traces are now available as digitized data and that complete signal amplitudes over the important time span are available. Processes can therefore be devised which require future values of the trace, whereas in a *real-time system* the filters and other operators can work only on data that have already been received. There are of course very important differences. Correlation, for example, is possible only after a recording time at least equal to the duration of the input signal. Analog filters, however, deal with electric signals as they arrive. Much of the power of the computer in processing seismic data derives from this property of "memory," whereby the entire seismic signal is available.

In Chapter 4, essentials of the description of a seismic trace either in the form of a complex spectrum or as a sequence of amplitudes separated by a constant digital interval were given. We recapitulate some of the chief results.

First, in the frequency domain, the seismic trace is described in terms of the amplitude and phase of a series of sinusoidal waves separated by a frequency of Δf, where $\Delta f = 1/T$, T being the length of the trace being considered. This description, if continued beyond time T, gives rise to a repeating signal. By "filtering" we mean changing, in a systematic manner, the amplitudes and phases of the constituent sinusoids. Thus one can describe a filtering operation as the complex multiplication of the spectrum of the trace by the spectrum of the filter:

$$T'(i\omega) = T(i\omega)F(i\omega) \qquad (6.19)$$

where $(i\omega)$ means that the functions T', T, and F are complex.

Second, in the time domain, the seismic trace is described in terms of real amplitudes at points separated by a constant time interval ΔT. At each one of these points, we can visualize a delta function (Section 4.4) of the proper amplitude. Filtering is then done by systematic replacement of the delta function by a scaled pulse which has amplitudes at several digital sample times. This process is called convolution.

$$T'(t) = T(t) * F(t) \qquad (6.20)$$

where $T'(t)$, $T(t)$, and $F(t)$ are now the time descriptions of the traces, or pulses, described in the frequency domain by $T'(i\omega)$, $T(i\omega)$, and $F(i\omega)$.

Since the seismic trace is normally given originally in the time domain, filtering by convolution can be done directly and, if T consists of n samples and F consists of m samples, the amount of work done by the computer is $n \times m$ multiplications and additions. However, filtering in the frequency domain involves first two Fourier analyses, and then complex multiplication of all the amplitudes, followed by a Fourier synthesis.

A Fourier synthesis or analysis normally involves N^2 complex multiplications, so that the whole process should take about $2N^2 + N$ complex multiplications. However, the recently discovered fast Fourier transform (FFT) (Brigham, 1974) has reduced this effort considerably. By proper arrangement of the computer workload, a single transform takes about $N \ln_2 N$ multiplications and additions. When N is large, as it is on seismic traces (up to 1500 to 3000 points), the saving of time is considerable. As an example, compare the time taken for analysis of a 6-sec trace digitized at 0.004-sec intervals. There are 1500 sample points which would require about 2,250,000 multiplications and additions in performing the Fourier analysis the old way. However, if the trace is *increased* in length by the addition of zeros, to 2048 points $= 2^{11}$, a FFT analysis will take about $11 \times 2048 = 22,528$ multiplications and additions or about one one-hundredth of the time. While filtering operations are still usually done by convolution, filtering in the frequency domain is now a viable alternative. It should be noted that the filtered trace is longer than the initial trace by the length of the filtering pulse. This indicates that, when filtering in the frequency domain, the original trace length should be long enough that this tail will appear at the end of the trace. This is usually done by adding zero amplitude in sufficient number to the end of the original trace before the Fourier transform is made. In the case of a FFT, it is usual to make its length (number of samples) an integral power of 2, that is, 2^N. Under the circumstances discussed here, it may be necessary to double the trace length.

If the trace is not adjusted to have a sufficient number of zeros on the end, the tail caused by filtering is added to whatever amplitudes exist at the beginning of the trace by the circulating effect described earlier. This effect is particularly troublesome if the complex amplitudes of the frequency spectrum are generated from some formal relationship, and then the time domain trace is required. Examples are the synthetic record calculations in the frequency domain and the generation of surface waves from their dispersive characteristics.

Filters commonly used in the data processing of seismic traces are usually specified to have certain amplitude characteristics and zero phase, in order not to change the time relationships of reflected events. Examples of these simple filters are shown in Figure 6.10. A sharp cutoff of any filter amplitude spectrum causes the time domain pulse to ring, that is, to contain an excessive amount of visible oscillation at the cutoff frequency. It is for this reason that cutoffs are tapered—usually by a straight line but

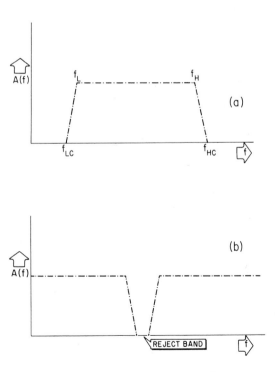

Figure 6.10 Amplitude spectra for a band-pass filter with a tapered cutoff (*a*) and a band-reject filter with tapered cutoffs (*b*).

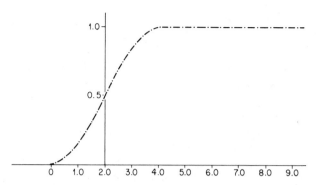

Figure 6.11   The form of the hanning cutoff for a spectrum, which is used to diminish the ringing resulting from the application of a filter with a sharp cutoff. This form or its mirror image, can be used instead of any discontinuity in the amplitude spectrum.

sometimes by a more sophisticated taper such as the hanning cutoff (Blackman and Tukey, 1958).

$$H = \frac{1}{2}\left(1 - \cos \pi \frac{f}{F}\right) \qquad (6.21)$$

where F = range of frequencies over which the cutoff is desired (see Figure 6.11)

There is, however, another form of filter (or antifilter) used to correct the trace for the effect of a filter already applied. The process is known as deconvolution and takes several forms. We deal first with forms that are applicable to impulsive-type sources. With this kind of source, filtered pulses cannot realistically have any amplitude before zero time of the causal pulse. If the amplitude spectrum of such a filter pulse is specified, there is still an infinite number of pulses that can be produced by specifying the phase spectrum in different ways. *There is, however, one pulse that can be produced in which the phase is linked to the amplitude spectrum. This is called a minimum phase pulse. In this pulse, the energy occurs as close as possible to zero time, while still satisfying the amplitude spectrum.* This is a most important concept for, given any amplitude spectrum, it is possible by rather sophisticated mathematical methods to calculate the form of the minimum-phase pulse. Since it has now been shown that the pulse resulting from a dynamite shot has a minimum-phase form (White and O'Brien, 1974), the procedure takes on real significance. Furthermore, the filtering of such a pulse by a layered system has also been shown to be minimum-phase. This is important, since multiple reflections of the primary pulse in the near-surface layers have a significant effect on the shape of later reflections. These are all part of the filtering effect we are seeking to eliminate.

The objective of deconvolution is to determine, from the record traces themselves, the filter that has been applied (as a result of passage through the earth, the recording system, etc.) and then to determine an inverse filter which will correct the effects of the previous filter(s). This means that the inverse filter must have the effect of bringing the phase back to zero and rendering the amplitude spectrum flat. The amplitude spectrum is rendered flat by multiplying all amplitudes by their inverses to give a constant resultant of unity. In so doing, very large multipliers are sometimes obtained for some frequencies. Since the amplitude spectrum is partially determined by noise, these very high multipliers may not be correct, and a noisy inverse results. Consequently, it is necessary to restrict the bandwidth of the inverse to the range in which the signal spectrum has appreciable amplitudes. The inverse is obtained in any convenient range but is then subjected to a flat-amplitude, zero-phase filter to eliminate noise outside the signal band.

There are, however, some problems. In the first place, the spectrum of a seismic trace containing reflections is not just the (smooth) spectrum of the initial pulse itself but contains ripples due to the interference effects of reflections occurring at different times. Any two positive reflections interfere constructively when their times of arrival are integral multiples of the period of a particular frequency. There is not just *one* resonant frequency, but the spectrum contains a ripple whose wave peaks are separated (in angular frequency).

$$\Delta\omega = \frac{2\pi}{T} \qquad (6.22)$$

where T = distance between the events

Note that events *far apart in time* give rise to a *rapid ripple* on the spectrum and those *close together* give a *slow variation* in the spectrum. Thus, as illustrated in Figure 6.12 (Ware, 1964), it should be possible to smooth the amplitude spectrum of the trace and, to a first approximation, arrive at the amplitude spectrum of the convolving pulse.

This effect can be seen in the time domain if we perform an autocorrelation of the seismic trace, which is done by a convolution operation of the trace with the trace reversed in time. The autocorrelation shows a peak (obviously) at zero delay, followed by a decay and oscillation at nearby time delays—the shape of which is controlled by the average frequency content of the trace—followed by a series of peaks and troughs which mainly correspond to reflections correlating with each other. The near,

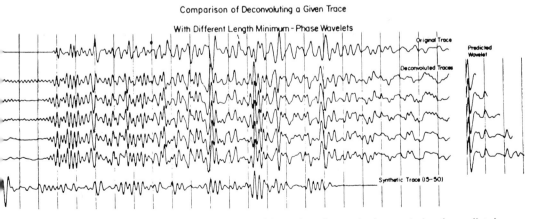

Figure 6.12a Original record trace (top) followed by various deconvolved traces (using the predicted wavelet shown at their right). A synthetic record trace is shown below for comparison.

Corresponding Amplitude Spectra

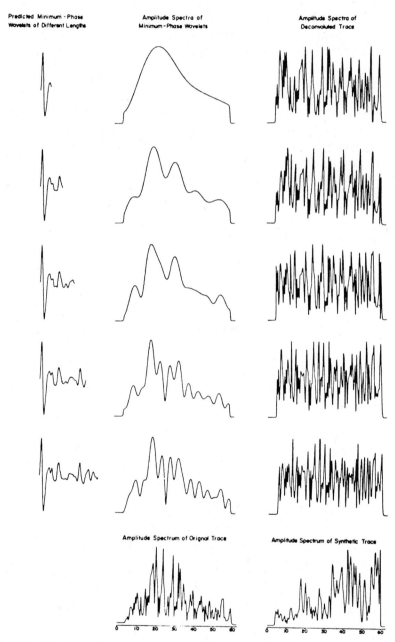

Figure 6.12b The effect of predicting minimum-phase wavelets of different lengths (left), the corresponding spectra (center), and the spectrum of the deconvolved trace (right). [After Ware (1964). Courtesy of the Continental Oil Company.]

truncated, shape of the autocorrelation waveform contains the information needed for estimating the amplitude spectrum of the initial pulse. If the assumption is then made that the initial pulse is a minimum-phase pulse, its phase can be determined from the shape of the amplitude spectrum, and the problem of determining the initial pulse is formally solved.

The actual method of calculation has been described many times in the literature, for example, Burns (1968), Clark (1968), Ford and Hearne (1966), Foster, Sengbush, and Watson (1968), Robinson and Treitel (1967), Robinson (1967a), and Treitel and Robinson (1966, 1969). Matrix algebra and Z-transform representation are involved and are not pursued further here.

Judgment of the degree of truncation of the autocorrelation function is usually based on its characteristics. If a large ghost pulse occurs at some delay τ and smaller ones at 2τ, 3τ, and so on, this is indicative of multiple reflections which, since they are of minimum delay, can be removed by making the inverse pulse long enough (at least greater than 2τ). However, this sometimes violates the original concept of smoothing the trace spectrum, and it is better to use a short operator first to give a spiked deconvolution, followed by another operator which has the characteristics of eliminating a single multiple reflection path.

Such operators are called predictive or gapped, and they consist of a unit spike at zero time, followed by a sequence of zeros, followed by a sequence of calculated values to the end of the operator. The gap and sequence of values thereafter are chosen in such a way that the near-in shape of the autocorrelation wavelet (hence the amplitude spectrum of the data) is preserved, but the ghosts or long-period reverberations occurring in the zone of the calculated values are minimized. As a sequence of sample values, the gapped operator looks like

$$1, 0, 0, 0, 0, \ldots, 0, 0, 0, a_1, a_2, a_3, \ldots, a_n$$

These operators are again derived from the autocorrelation pulse by a method described by Peacock and Treitel (1969) and by Kunetz and Fourman (1968), and reference should be made to these papers for further information.

6.11   DECONVOLUTION OF VIBROSEIS® RECORDS

As mentioned in Section 3.3, the Vibroseis® method of exploration differs from all others in that the signal introduced into the ground is a long train of quasi-sinusoidal waves of the general form

$$S(t) = A(t) \sin\left(at + \frac{b}{2}t^2\right) \qquad 0 < t < T \qquad (6.23)$$

where $A(t)$ = a relatively slowly varying envelope function

   a and b = constants related to the initial frequency and the rate of change in frequency, respectively

Following the methods of Section 4.6, it is to be expected to a first approximation (neglecting attentuation) that, if $I(t)$ is the impulse response of the earth, the received reflection record will be given by

$$R(t) = I(t) * S(t) \qquad (6.24)$$

In other words, each impulse in the impulse response is replaced by the swept frequency signal (or control signal). In order to generate a record $R_c(t)$ that can be interpreted in terms of geology, it is necessary to change the control signal into a compressed form, the most convenient form of which is the autocorrelation function corresponding to the power spectrum of the control signal. This is done by correlating the field trace with the control signal. Correlation is nothing more than a convolution filter using a time-reversed or transposed control signal. So

$$R_c(t) = I(t) * S(t) * S^T(t) \qquad (6.25)$$

and the last two filters can be combined to give

$$R_c(t) = I(t) * \phi_{ssT}(t) \qquad (6.26)$$

where ϕ_{ssT} = the autocorrelation function

At this point, practice does not correspond to theory, and a temporary diversion is necessary. In the above discussion, it was presupposed that, if a vibrator is operated with a control signal $S(t)$, this is the form of the wave being produced in the earth. More specifically, we want to know the wave shape of the particle velocity produced in the earth because, with the geophones commonly used, it is a particle velocity that is being measured. When a vibrator is placed on the earth and the control signal $S(t)$ is fed into it, the output signal is

$$S'(t) = B(t)S(t + \theta) \qquad (6.27)$$

where the phase angle θ is a function of both rock type and frequency, hence varies with both the position of the vibrator along the line and the time within the sweep. $B(t)$ is an amplitude gain function.

In order to introduce into the ground at least a phase-consistent signal under varying surface conditions, a feedback loop (called a phase compensation loop) is normally used. In practice, a signal from the baseplate (from an accelerometer whose output is integrated) and the control signal are both fed into a phase comparator. Normally, this circuit has an output unless the two signals are exactly 90° out of phase (a practical circuit consideration), and this output then acts on a phase shifter to control the input phase to the electrical servo input. Thus the baseplate velocity output to the earth is 90° out of phase with the control signal.

While this 90° phase change is convenient at the vibrator because of instrument simplicity, in the final seismic traces it must be eliminated. Although it could be eliminated directly in the correlation process, it has been found more convenient to proceed as follows.

1. The correlated trace with a 90° phase shift is obtained by correlating the control trace with the field trace:

$$R_c(t) = I(t) * S(t) * S(-t) \qquad (6.28)$$

2. A deconvolution (minimum-phase) pulse is obtained from the autocorrelation of the field trace. This pulse is Fourier-analyzed, using a FFT, and the amplitude spectrum is retained, while a constant $-90°$ phase shift is substituted for the phase angle resulting from the analysis.
3. The deconvolution (constant $90°$ phase) pulse is resynthesized using a FFT.
4. This pulse is convolved against the Vibroseis® correlated traces.

Vibroseis® deconvolution has recently been reexamined by Gurbuz (1972) and by Ristow and Jurczyk (1975), and they have shown how to effect deconvolution entirely in the time domain after correlation by the use of an antifilter:

$$\frac{1}{A_0(Z)} = kS_0(Z)S_0(Z)C_0(Z) \tag{6.29}$$

where k = an unimportant constant
$C_0(Z)$ = a spike deconvolution operator whose amplitude spectrum is that of the field trace
$S_0(Z)$ = a minimum-phase pulse having the same amplitude spectrum as the *control* trace, and it is applied twice

These operators and pulses have been stated as Z transforms, which are treated in an elementary way in the Section 6.12. For the present, it is understood that they are complex filters which can be specified either in the time domain as convolution pulses or in the frequency domain as complex spectra.

The advantage of doing the deconvolution after correlation is that it is possible to make a deconvolution operator that varies with record time, thus compensating, to the extent possible with the noise present, for attenuation.

6.12   RECURSIVE FILTERS AND TIME-VARYING FILTERS

Because of the large number of multiplications and additions involved in computing the output of a digital filter, filtering can be time-consuming and expensive in the overall processing of seismic recordings. Special CFE that allows many of these multiplications to be performed in parallel has really been the prime cause of the economic feasibility of digital processing. We introduce here, very briefly, the use of the Z transform for the representation and manipulation of sampled data. A sampled trace which can be represented by a series of amplitudes, taken at some sampling interval Δt,

$$T \equiv \{a_0, a_1, a_2, a_3, \ldots, a_{n-1}, a_n\}$$

can also be *formally* represented by a power series in the variable Z (which represents a delay factor $e^{i\omega\Delta t}$ corresponding to the constant sampling rate Δt):

$$T(Z) = a_0 + a_1 Z + a_2 Z^2 + a_3 Z^3 + \cdots + a_{n-1} Z^{n-1} + a_n Z^n$$

Two different traces, $T_1(Z)$ and $T_2(Z)$, can be convolved by multiplying their Z transforms to give

$$T_3(Z) = T_1(Z) \times T_2(Z)$$

and one trace, $T_1(Z)$, can be deconvolved with $T_2(Z)$ by dividing the Z transforms:

$$T_4(Z) = \frac{T_1(Z)}{T_2(Z)}$$

Now, by a suitable transformation of the desired convolving pulse—as a rational fraction derived from the Z transform, in most instances we can derive an equivalent computing algorithm which is significantly more efficient than digital convolution. Here we present the main features of the recursive filter, without going into much detail, since this method has an *economic* base rather than a scientific or geological base. It does nothing different from a normal convolution but enables this to be done quickly and economically. For more details, the reader is referred to the excellent tutorial article by Shanks (1967).

In the Z-transform domain, a digital filter can be given as the ratio of two polynomials in Z:

$$F(Z) = \frac{A(Z)}{B(Z)} \tag{6.30}$$

$$= \frac{a_0 + a_1 Z + a_2 Z^2 + \cdots + a_n Z^n}{b_0 + b_1 Z + b_2 Z^2 + \cdots + b_m Z^m} \tag{6.31}$$

This cannot be applied directly as a convolution filter but, by the process of long division (Jury, 1964), it can be transformed into a single, infinite polynomial:

$$= f_0 + f_1 Z + f_2 Z^2 + \cdots$$

If the filter is stable, the values of the coefficients will converge to zero and, at some point, the series can be truncated to give an approximate filter.

The rational fraction, as given in (6.30), can, however, be applied directly. We consider the simple filter

$$F(Z) = \frac{a_0 + a_1 Z}{1 + b_1 Z + b_2 Z^2} \tag{6.32}$$

If $X(Z)$ is the input to the filter, the output $Y(Z)$ is

$$Y(Z) = \frac{a_0 + a_1 Z}{1 + b_1 Z + b_2 Z^2} X(Z) \tag{6.33}$$

or

$$Y(Z) = [(a_0 + a_1 Z)(Z) - Z Y(Z)(b_1 + b_2 Z)] \tag{6.34}$$

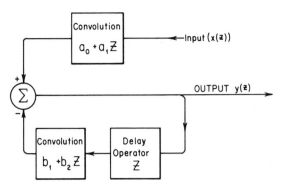

Figure 6.13 The feedback system equivalent to the recursive filter, $F(Z) = (a_0 + a_1 Z)/(1 + b_1 Z + b_2 Z^2)$.

This equation says that the output $Y(Z)$ is equal to the input convolved with the discrete series (a_0, a_1) minus the *output* delayed by one sample interval convolved with the series (b_1, b_2). This is a feedback system, as shown in Figure 6.13; it can also be realized in a digital computer, using a suitable feedback algorithm.

There are numerous ways to compute the coefficients of rational filters designed for specific filtering operations. These are given in detail in Shanks (1967).

From experience and from intuition, it is known that recorded seismic traces have the appearance of generally decreasing in bandwidth (by losing high frequencies) as the record time increases. It can be said that the later reflection events are usually of lower frequency than the early ones. This remark has to be interpreted in the following sense.

If intervals of the seismic record (windows) are chosen to have the same characteristics but centered at gradually increasing record times, and if the seismic traces within these windows are Fourier-analyzed, there will be a gradual, but not necessarily constant, decrease in the high-frequency content of the amplitude spectra as the time of the window increases. Usually, in making these analyses, a time window is chosen that is long enough to provide the necessary frequency resolution while, at the same time, overlap of the windows should be reduced to a minimum.

Based on this experience, it has proved desirable to design a filter system that changes its characteristics with time of application in order to optimize the signal/noise characteristics of the seismic data. While we are willing to optimize the signal power/noise power ratio the foregoing discussion of deconvolution should have convinced us that we should not interfere with the phase characteristics of the record trace after deconvolution. Therefore any time-varying filter we design should be a zero-phase filter.

The design of time-varying filters has been accomplished in several ways but, for filtering seismic traces, use has been made of the simple scheme shown in Figure 6.14 and described below.

This figure shows an unfiltered trace which is going to be filtered in a time-varying manner, filter 1 applying to a time t_1, a mixture of F_1 and F_2 to t_2, a mixture of F_2 and F_3 to t_3, and so on.

Four filtered versions of the (relevant portions) signal are produced and then combined with the weights given on the lower graph. It is obvious that the convolution

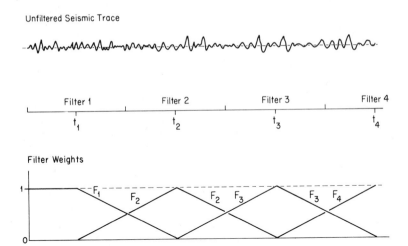

Figure 6.14   A simple method of achieving a time-varying filtered seismic trace.

with F_1 need not extend beyond t_2, since its weight from that time on is zero. Similarly, the trace is filtered only with F_2 between t_1 and t_3, being given zero weight elsewhere.

F_1, F_2, F_3, and F_4 all consist of convolution operators which are zero phase. Some care must be taken in scaling these operators, since they may have different bandwidths and therefore different total energies if the spectrum amplitude is kept the same. One method of approaching this is to compute the sum of the squares of the amplitude of the convolving pulses, for example,

$$\sum_{i=1}^{N} a_i^2 = A^2$$

$$\sum_{i=1}^{N} b_i^2 = B^2$$

(6.35)

and then scale the amplitudes of the pulses by the square root of these energies. The new

$$a_i^1 = a_i \frac{C}{A}$$

$$b_i^1 = b_i \frac{C}{B}$$

(6.36)

where C = a convenient constant

6.13   HOMOMORPHIC DECONVOLUTION

The restriction of only being able to predict the pulse shape from the seismic trace when the pulse is minimum-phase is an irksome one, since there may be some filters we seek to neutralize that are not minimum-phase. The minimum-phase assumption has worked out reasonably well for impulsive-type sources The material in this section is given mainly for the benefit of readers who may wonder if any thought has been given to removing the constraints associated with the minimum-phase assumption. Little routine use has been made of the method in exploration reflection seismology data processing.

The homomorphic method of deconvolution (Ulrych, 1971) has found some favor because it is supposedly free of the minimum-phase restriction. (Homomorphism is mapping from one domain to another.) In this particular case, we start off with the standard frequency domain definition of a filtered trace (neglecting noise):

$$F(\omega) = I(\omega)P(\omega) \qquad (6.37)$$

which states that the filtered trace spectrum (complex) is the spectrum of the impulse response $I(\omega)$ multiplied by the spectrum of the convolving pulse $P(\omega)$. We can transform this to another domain of the logarithm of amplitude by noting that

$$\log F(\omega) = \log I(\omega) + \log P(\omega) \qquad (6.38)$$

It is to be noted that the pulse and impulse response log spectra are now *added* instead of being multiplied as in (6.37).

It is presumed that it is easier to separate two factors that have been added together than two that have been multiplied together. Common sense tells us that this is untrue, unless some further criteria are involved. We give an example and then comment on the fallibility of the process.

Ulrych (1971) shows (Figure 6.15) the case of a mixed-phase pulse (*a*) that is, a pulse whose phase spectrum cannot be deduced from a knowledge of the amplitude spectrum, convolved with a simple delay operator to give a pulse combined with a single echo (*b*). The process calls for digitization, that is, obtaining sample values at various multiples of the delay ΔT to obtain

$$(a_0, a_1, a_2, a_3, \ldots, a_n)$$

and forming the Z transform by associating each amplitude with the appropriate power of Z

$$X(Z) = a_0 + a_1 Z + a_2 Z^2 + a_3 Z^3 + \cdots + a_n Z^n$$

Form $\hat{X}(Z) = \log X(Z)$.

Now take the inverse Z transform of $\hat{X}(Z) \rightarrow \hat{x}(n)$. The latter is called the complex *cepstrum* from the work of Bogert, Healey, and Tukey (1963), who termed the power spectrum of the logarithm of the power spectrum the cepstrum by inverting the order of the first four letters. The word "complex" was added by Oppenheim, Schafer, and Stockham (1968) to emphasize the point that the cepstrum is formed by taking into account both amplitude and phase information contained in $\hat{x}(n)$.

In the specific case of the single pulse and echo shown in Figure 6.15, (*c*) shows the complex cepstrum of (*b*). Note that the original pulse has been replaced by the large spike at zero time *n*, followed by smaller pulses of alternating sign. Ulrych considers these to be well separated from the contribution of the main pulse, and therefore the two effects can be separated.

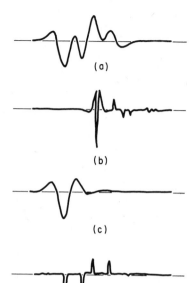

(a)

(b)

(c)

(d)

Figure 6.15 (*a*) Minimum-phase input. (*b*) Complex cepstrum of (*a*). (*c*) Low-pass output—the constituent pulse shape. (*d*) High-frequency output, or impulse response. [After Ulrych (1971). Reprinted with permission from *Geophysics*.]

One is struck by the sudden appearance, in the complex cepstrum, of high frequencies (of *n*) by means of which the times of occurrence are suddenly rendered discrete. The mathematics is too complex to analyze how these high frequencies occur, but we endeavor to show that they are really manufactured in the process by the use of computer-generated examples or by the assumption that there exists a specific origin of time and that all events must occur at a particular sample value.

First, we remark that, if a seismic trace is a continuous function of time and is the convolution of a continuous pulse and a continuous impulse response, the amplitude spectrum of the trace will be the product of the two amplitude spectra and, even though the impulse response has very high-frequency components, these are reduced very much in magnitude since the seismic pulse is band-limited in the low-frequency region. Of course, in theory, the pulse may still have very small high-frequency components (or these may be *manufactured* as a result of a digital analysis of finite accuracy data). It is axiomatic therefore that, in the continuous case, it is never possible to recover the impulse response at frequencies outside the effective bandwidth of the convolving pulse. The spectrum of the log spectrum is essentially a pseudo time trace, and the location of the times of the echoes, or impulse, cannot be determined closer than the bandwidth of the pulse allows. The best that can be done is to position an autocorrelation pulse of a flat-frequency pulse of the requisite bandwidth at the position of each echo, but this is precisely what is done in spike deconvolution working at its best.

In the digital examples, the computer manufactures the output trace from the impulse response and from the pulse (utilizing finite error measurements of the pulse and therefore giving it frequencies up to the Nyquist frequency associated with the sampling rate). It is then possible to turn around and, again using the very small generated values of high frequency, reanalyze the entire system (which has a very high bandwidth), and this leads to the high-frequency output of the impulse response.

Second, we must look at the question of the exact location of the sampling points. It is possible to move the sampling point zero (which is usually at a point of zero-amplitude output) by a small amount in the time domain. This causes all the samples to change in amplitude and in turn forces all the impulse response values to be located at different points in the time domain. This process can be imagined as being carried out at all possible points for the origin, so now the impulse responses will generate a continuous output. What we are saying is that, in the real world, there is uncertainty about the location of the impulse responses and therefore uncertainty in all the phase information for all frequencies of importance. As far as the computer world is concerned, the information read in is exact and an exact answer is generated (according to the instructions given to the computer).

The introduction of noise into the system renders the high-frequency output even more unstable, since the phase of small high-frequency components is changed considerably.

This method of homomorphic deconvolution remains interesting from the point of view of the possibility of determining mixed-phase input pulses. Even this aspect is clouded by the admission that it is only in the case of minimum-phase pulses that the separation of pulse and impulse response components is complete.

6.14 TWO-DIMENSIONAL FILTERING OF SEISMIC CROSS SECTIONS

The foregoing survey dealt with data processing techniques that are applied, trace by trace, to the constituent traces of a seismic cross section. Apart from the stacking of 100% traces taken at different offsets, there was no assumption that any trace of a cross section had any relation to any other seismic trace on the cross section. Hope is always present of course that the combination of seismic traces, one example of which is given in Figure 6.16, will show only aligned seismic events that correspond to reflections from the geological interfaces below. In many cases, this hope is not realized, and disturbing events cross over the cross section, which must be removed before the geological events can be seen clearly. This is particularly true on a 100% record, such as is shown in Figure 6.17, where interference from air waves and near-surface waves often conceals the reflection information. They appear as lineups of pulses which travel across the record with a slow apparent velocity. Two such interferences are shown in Figure 6.17. One of these was to be removed. We anticipate the action of the two-dimensional filter to show the second record with the unwanted event removed. In full seismic sections, unwanted events may originate because of scattering from linear weathering and elevation changes, for example, scarps which almost parallel the line of section, or from linear outcrops, linear gas seeps, and other changes in conditions of the water layer in marine prospecting. There are enough causes; the trick is to find out that there is false reflection information and then to devise a system to remove it.

A restricted form of velocity filtering (pie slice) was originally offered by Geophysical Services and has been described by Embree, Burg, and Backus (1963) and Fail, Grau, and Layotte (1964). The method is as follows.

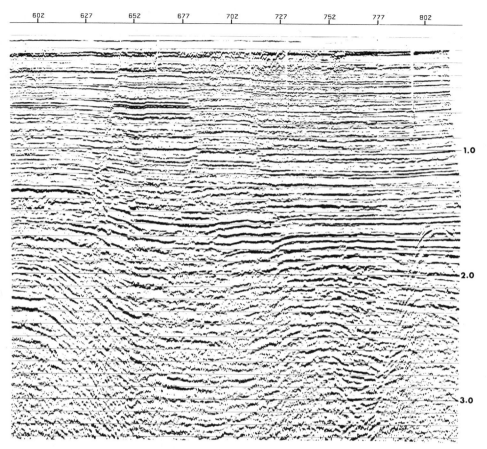

Figure 6.16   An example of a processed seismic section. The hope that this contains only reflected events is not realized in this case, as can be seen from the large diffracted event on the lower right. This event is probably scattered from an object in the water about 1.28 km (4000 ft) to the side of the line. [Courtsey of Seiscom-Delta, Inc. (1975).]

Just as it is possible to describe a one-dimensional seismic trace in the time domain by an equivalent amplitude and phase description in the frequency domain, it is also possible to extend this imagery to two (and more) dimensions. Thus a wave

$$F(x, t) = F(kx - \omega t)$$

which has a wavelength $2\pi/k$ in the x direction as well as a period $2\pi/\omega$ in the time domain can be equally well represented as a function of k the wave number and ω the angular frequency. Transforming between the two is done by a two-dimensional Fourier transform and, as in the case of a single-dimensional function, the inverse Fourier transform also exists.

The transform is done first by making Fourier analyses of the traces in the time direction and then by analyzing these complex amplitudes in the x direction. We are concerned with a form of filtering that will reduce *amplitudes* of particular events as nearly as possible to zero, so we are not concerned with the complex values but only

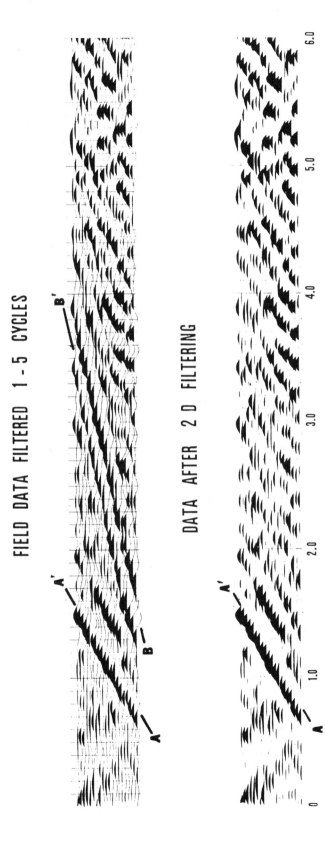

Figure 6.17 Illustration of the efficiency of two-dimensional filtering in removing a particular event (*BB'*) due to interference. Other events are unchanged by this filter action.

177

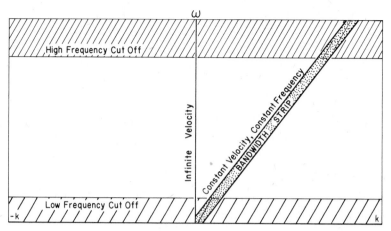

Figure 6.18 The ω-k plane, showing various regions which, if zeroed in amplitude, perform the designated filter operations. The boundaries need not be sharp and, by this means, ringing characteristics can be partially avoided.

the amplitudes $A_i = \sqrt{a_i^2 + b_i^2}$. These are then displayed, much as the time traces are displayed in the form of a map in the ω-k plane, such as is shown in Figure 6.18. On this map, a line drawn from the origin in any direction represents all points for which ω/k, or the velocity, is a constant. The elimination of a sloping band in this ω-k plane therefore eliminates all energy that propagates across the record at a particular velocity.

In the pie-slice process, the attitude adopted was that reflection events usually show up as events with very high apparent velocities and can therefore be left behind if a small wedge of energy on the ω-k plane is conserved (about the ω axis) and the rest of the ω-k plane is given zero amplitude. Figure 6.19 illustrates this. In the case shown, the pie-slice method aids considerably in cutting out low-velocity energy.

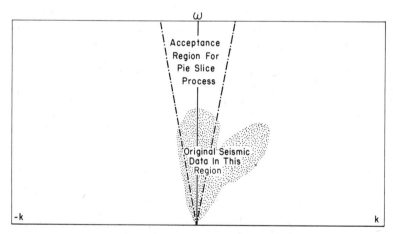

Figure 6.19 The acceptance in the ω-k plane of reflection data having a limited range of acceptable phase velocities ($-V_c < V < V_c$). The velocity boundaries used in the original Pie-Slice® process were sharp and caused a cross-hatched appearance on noisy sections processed this way.

Unless this is caused by very steep reflections on the cross section, the reflection data will be materially improved. If it is due to steeply dipping reflections, the ω-k filter is poorly designed.

The pie-slice process has several characteristics:

1. It is a relatively simple filter and can be carried out in the time domain. Before the FFT, this was a distinct economic advantage.
2. The boundaries of the triangular acceptance region are sharp (e.g., there is full acceptance inside and full rejection outside the acceptance region). This characteristic leads to energy surviving the filter being displayed on the time cross section in the form of crossing events having the velocity cutoff. These are caused by noise (e.g., energy outside the pass band being cut off with an infinitely fast rejection rate). (This is a characteristic of frequency filters, too—they tend to ring at sharp cutoff points.)
3. Because of the symmetrical nature of the filter, it is inflexible as far as passing reflected events of other than zero dip.
4. At points on the cross section where a reflection event suddenly terminates (the end of the section, some dead traces, or for any other reason), the energy from neighboring traces mixes in to give continuity to the interrupted event.

Some of these characteristics are objectionable and have led to a much more general two-dimensional filter system, since the FFT has made large transforms economical. The time domain cross section is of course a large rectangular matrix of points spaced at distance ΔX apart and Δt in time on each trace. This, after Fourier analysis, gives rise to a rectangular grid of points separated by $\Delta \omega$ and Δk. The same problems are experienced in sampling, as aliasing can occur in either direction. It is then necessary to adopt a program that provides not only for zeroing irregular regions in the ω-k plane but also provides a gradual cutoff—either a linear ramp or some more sophisticated scheme, such as the hanning scheme. The reader, by this time has recognized the essential similarity in the treatment and characteristics of two-dimensional transforms and those found for the one-dimensional transform.

Finally, after filtering by multiplying by the transfer function of the two-dimensional filter pulse, an inverse Fourier transform is performed to restore the data to the time domain.

It was by making use of this type of operation, using a constant-velocity, constant-bandwidth filter, that the results in Figure 6.17 were obtained. For very low-velocity rejection, it is most important that the sampling in the x direction is adequate.

Note that this method treats the entire record section in the same manner. The energy for all the sections for a particular ω-k is lumped together. If horizontally varying or time-varying frequency or velocity filters are required, it will be necessary to remain in the X-t domain and to treat the section by means of rectangular convolving filters, such that

$$A^1(m, n) = \sum_{k=-K}^{K} \sum_{l=-L}^{L} P_{k,l} F_{m-k, n-l}$$

where F = cross section to be filtered

The method of two-dimensional filtering has been extensively applied to magnetic and gravity maps (Bhattacharyya, 1972) but has not achieved its full potential in seismic work. There are several reasons why this is so, among them the fairly recent development of a two-dimensional FFT and also the fact that seismic data are sometimes noisy by virtue of random uncorrected time errors due to weathering and other causes. If this potential is to be realized, it must be done through a system, using geological input as well as geophysical input to reach the goal.

6.15 SUMMARY AND CONCLUSIONS

In the last 10 years, there has been a revolution in the treatment of seismic data, due largely to the consistency and flexibility of the digital computer. Data recorded in continuous and reproducible form on magnetic tape now have to be represented by a series of discontinuous samples. These are usually multiplexed, because of the need for sampling across all input signals as they come in—almost simultaneously. Very high-speed analog-converters are now available, which have no difficulty in sampling up to 50 traces per millisecond. Recently, automatically ranging binary gain amplifiers have made recording in a floating-point system possible, and this has decreased the need for care in scaling that was formerly necessary.

For treatment of the data, it is more convenient to have the numbers (amplitudes) arranged in trace sequential form. Then each trace can be treated on its own merits. For proper treatment, auxiliary data from land surveying is incorporated in a line description program which then allows the calculation of trace offsets and primary near-surface corrections, which are then carried with the trace in a header—a few auxiliary words of computer information.

Primary near-surface corrections and a correction for the geometrical effects of offset are then made for each trace, the relevant velocities being taken from the data. These are strictly first-estimate corrections, but they are sufficient to allow the proper data parts to be stacked together in order to have a first look at the stacked CDP reflection cross section. Expansion of the amplitude as a function of time or a slow-working digital automatic gain control is sometimes incorporated to allow both early and late reflections to be seen.

Generalized filtering is much more flexible using digital computer programs. One of the most useful of these is called deconvolution—a program for undoing filtering that has previously taken place. The input pulse, which may have been complicated, or the effect of shallow multiples, can be compensated and effectively made into a zero-phase effect, provided that the distortions in the first place arose through minimum-phase operations. It is the assumption here that, for a given amplitude spectrum, the energy in the pulse is optimally concentrated near the beginning of the pulse. It can be shown that multiple reflections within the shallow layers fall into this classification.

Deconvolution operators are obtained from primary data by first smoothing the amplitude spectrum (to eliminate the effects of reflections) and then calculating the minimum-phase operator from this amplitude spectrum. The correction operator (the deconvolution operator) is then obtained by spectrum inversion (within a given frequency range so that noise values are not enhanced) or by matrix inversion techniques. In the latter, small amounts of white noise are added to improve the stability of the matrix inversion process. The deconvolution process for Vibroseis® records has to be treated in a special manner because the correlation step uses reflection pulses that are not minimum-phase in the restricted sense.

A form of deconvolution called homomorphic has been proposed to deal with situations in which the minimum-phase criterion is not met. Although it has been employed with some success in other fields, there is little present (1976) usage in reflection seismology.

Once seismic reflection results are presented in the form of cross sections, several events can be seen which, from the pattern of arrival on successive traces, will be judged not to be reflections. They can arise from instrumental problems (e.g., cable noise in marine systems) or from diffracted or scattered events. In some cases, they can be removed by means of two-dimensional filters in which the basic assumption is that reflections are coherent, nearly constant time, events. These filters work well provided that the discrimination between other events and reflections map into two distinct regions in the frequency-wavelength plane and if they can be separated out by filters that have no very sharp cutoffs.

Only simple, routine uses of digital computation have been discussed in this chapter, but these uses alone, taken with the reliability of a digital process once it is freed from procedural errors, have really constituted a revolution in the production of reliable seismic reflection data.

APPENDIX 6A: ADAPTIVE FILTERS

In the earlier days of data processing, the final bandwidth of the processing filter (possibly varying as a function of depth) was determined largely by trial and error. The geophysicist in charge of the work would try different combinations of high- and low-frequency cutoffs until satisfied that an optimum signal/noise ratio had been obtained overall. All filters were of course zero-phase in order not to disturb the phase relationships in the seismic traces.

More definite information is obtainable by employing narrow-band filters, using subjective judgement to decide the total number of narrow bands to be kept at each depth. An example is given in Figure 6A.1.

There has been a gradual trend toward more objective analyses of seismic traces to determine the bandwidth to be retained. This is useful, since computer examination of the original data is involved and decisions on the final filters can be made within the processing stream and without the delays caused by trial and error or by the extra observational step involved.

One such process has been described by Branisa (1975). Essentially, a window of a seismic trace is analyzed to obtain the amplitude and phase at each frequency. When two or more traces are examined, the phase can be compared from trace to trace to establish the degree of similarity, or phase coherence. Figure 6A.2 shows how unit length vectors, plotted at the indicated phase angles, tend to add when the phases are nearly the same (mostly signal) and tend to give a small final summation vector when noise is being analyzed. The probability of a given amplitude of the resultant vector, in the case of random noise, was analyzed by Rayleigh (1945) (Figure 6A.3), and Branisa found that experimental noise analysis follows the theory very closely.

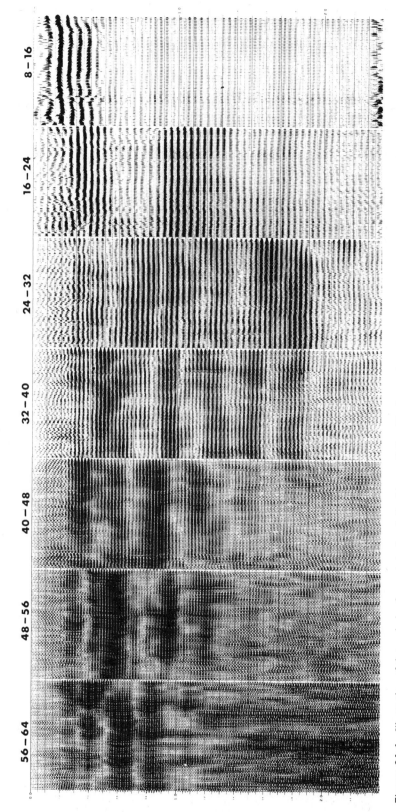

Figure 6A.1 Illustration of the analysis of records by bands, in this case 8 Hz wide. In practice narrower bands, about 3 Hz wide, can be used. The 24 to 32 Hz and 32 to 40 Hz bands appear to have reflection energy down to nearly 2.0 sec, but both higher- and lower-frequency bands cut off at shorter times.

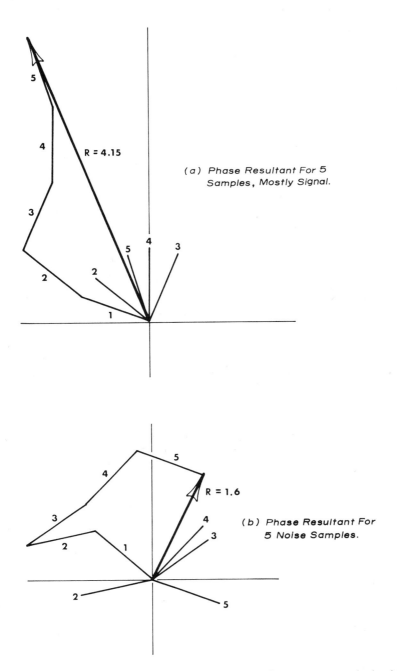

(a) Phase Resultant For 5 Samples, Mostly Signal.

R = 4.15

(b) Phase Resultant For 5 Noise Samples.

R = 1.6

Figure 6A.2 Results of adding five unit vectors derived from coherent traces, mostly signal (a), and from noisy traces, mostly noise (b). The resultant R is the vector addition of the five unit vectors.

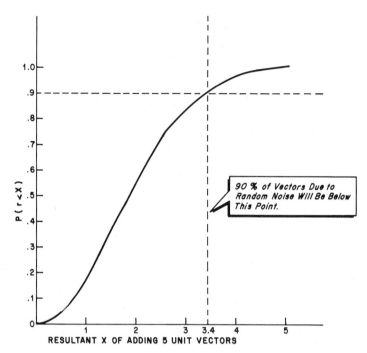

Figure 6A.3 Rayleigh's criterion for the probability of random addition of five unit vectors. Strictly, the theory is applicable only for N large. In practice, it is applicable for five or more unit vectors.

It is now possible to choose a phase coherence level believed to give the optimum compromise between high resolution (a wide-band, flat-amplitude spectrum) and a good signal/noise ratio. Finally, the flattened-amplitude spectrum of the original signals is adjusted (weighted) by the phase coherence to lower the amplitude of the more incoherent frequencies. In practice, five or six consecutive (stacked) traces are used to determine the coherent spectrum at a series of points along a line of section. An average acceptable spectrum can then be computed and used for the entire line.

Time variant limits for adaptive filters can evidently be calculated with this procedure using windows centered on a series of increasing reflection times. A compromise has to be made between having windows of sufficient length for resolution in frequency and rate of change in the filter with record time. In practice 1-sec windows are adequate for most purposes.

REFERENCES

Bergland, G. D. (1969), "A Guided Tour of the Fast Fourier Transform," *IEEE Spectrum*, Vol. 6, No. 7, pp. 41–52.

Bhattacharyya, B. K. (1972), "Design of Spatial Filters and Their Application to Aeromagnetic Data," *Geophysics*, Vol. 37, No. 1, pp. 68–91.

Blackman, R. B., and Tukey, J. W. (1958), *The Measurement of Power Spectra*, Dover, New York.

Bogert, B. P., Healey, M. J., and Tukey, J. W. (1963), "The quefrency analysis of time series for echoes; cepstrum, pseudo-covariance, cross cepstrum and saphe cracking": Proceedings of Symposium on Time Series Analysis, M. Rosenblatt, Ed., John Wiley, New York.

Branisa, F. (1975), "An Automated Signal and Resolution Enhancement Filter," paper presented at the 45th Annual International Meeting, Society of Exploration Geophysicists, Denver, Colorado.

Brigham, E. O. (1974), *The Fast Fourier Transform*, Prentice-Hall, Englewood Cliffs, N. J. This book gives a readable account of the transform properties between the time and frequency domains, as well as a description of the FFT and algorithms for implementing it on a computer.

Brigham, E. O., and Morrow, R. E. (1967), "The Fast Fourier Transform," *IEEE Spectrum*, Vol. 4, pp. 63–70.

Burns, W. R. (1968), "A Statistically Optimized Deconvolution," *Geophysics*, Vol. 33, pp. 255–263.

Clark, G. K. C. (1968), "Time Varying Deconvolution Filters," *Geophysics*, Vol. 33, pp. 936–944.

Collins, R. E. (1968), *Mathematical Methods for Physicists and Engineers*, Reinhold, New York.

Cooley, J. W., and Tukey, J. W. (1965), "An Algorithm for Machine Calculation of Complex Fourier Series," *Mathematical Computation*, Vol. 19, pp. 297–301.

Dix, C. H. (1955), "Seismic Velocities from Surface Measurements," *Geophysics*, Vol. 20, No. 1, p. 73.

Embree, P., Burg, J. P., and Buckus, M. M. (1963)," Wide Band Velocity Filtering—The Pie Slice Process," *Geophysics*, Vol. 28, No. 6, pp. 948–974.

Fail, J. P., Grau, G., and Layotte, P. C. (1964), "Amelioration du Rapport Signal—Bruit a l'Aide du Filtrage en Eventail—Etude d'un Cas Concret," *Geophysical Prospecting*, Vol. 12, No. 3, pp. 258–282.

Ford, W. T., and Hearne, J. H. (1966), "Least Squares Inverse Filtering," *Geophysics*, Vol. 31, pp. 917–930.

Foster, M. R., Sengbush, R. L., and Watson, R. J. (1968), "Use of Monte Carlo Techniques in Optimum Design of Deconvolution Process," *Geophysics*, Vol. 33, pp. 945–949.

Gurbuz, G. M. (1972), "Signal Enhancement of Vibratory Source Data in the Presence of Attenuation," *Geophysical Prospecting*, Vol. 20, pp. 421–438.

Jury, E. I. (1964), *Theory and Application of the Z Transform Method*, John Wiley, New York.

Kunetz, G., and Fourman, J. M. (1968). "Efficient Deconvolution of Marine Seismic Records," *Geophysics*, Vol. 33, No. 3, pp. 412–423.

Levin, F. K. (1971), "Apparent Velocity for Dipping Interface Reflections," *Geophysics*, Vol. 36, No. 3, pp. 510–516.

Meiners, E. P., Lenz, L. L., Dalby, A. E., and Hornsby, J. M. (1972), "Recommended Standard for Digital Tape Formats," *Geophysics*, Vol. 37, No. 1, pp. 45–58 (Digital Format C).

Musgrave, A. (1962), "Applications of the Expanding Reflection Spread," *Geophysics*, Vol. 27, No. 6, pp. 981–993.

Northwood, E. J., Weisinger, R. C., and Bradley, J. J. (1967), "Recommended Standards for Digital Tape Formats," *Geophysics*, Vol. 32, No. 6, pp. 1073–1084 (Formats A and B).

Oppenheim, A. V., Schafer, R. W., and Stockham, T. G. (1968). "Non Linear Filtering of Multiplied and Convolved Signals," *Proceedings of the IEEE*, Vol. 65, pp. 1264–1291.

Peacock, K. L., and Treitel, S. (1969). "Predictive Deconvolution: Theory and Practice," *Geophysics*, Vol. 34, pp. 155–169.

Rayleigh, Lord (1945), *Theory of Sound*, Vols. I and II (originally published in 1877), Vol. I, Dover, New York. p. 41.

Ristow, D., and Jorczyk, D. (1975), "Vibroseis Deconvolution," *Geophysical Prospecting*, Vol. 23, No. 2, pp. 363–379.

Robinson, E. A. (1967a), "Predictive Decomposition of Time Series with Application to Seismic Exploration," *Geophysics*, Vol. 32, pp. 418–484.

Robinson, E. A. (1967b), *Multi-Channel Time Series Analysis with Digital Computer Programs*, Holden-Day, San Francisco.

Robinson, E. A., and Treitel, S. (1967), "Principles of Digital Wiener Filtering," *Geophysical Prospecting*, Vol. 15, pp. 312–333.

Sattlegger, J. (1965), "A Method of Computing True Interval Velocities from Expanding Spread Data in the Case of Arbitrary Long Spreads and Arbitrarily Dipping Plane Interfaces," *Geophysical Prospecting*, Vol. 13, No. 2, pp. 306–328.

Shanks, J. L. (1967), "Recursion Filters for Digital Processing," *Geophysics*, Vol. 32, No. 1, pp. 33–51.

Taner, M. T., and Koehler, F. (1969), "Velocity Spectra—Digital Computer Derivation and Applications of Velocity Functions," *Geophysics*, Vol. 34, No. 6, pp. 859–881.

Treitel, S., and Robinson, E. A. (1966), "The Design of High Resolution Filters," *IEEE Transactions, Geoscience Electronics*, Vol. 4, pp. 25–38.

Treitel, S., and Robinson, E. A. (1967), "The M.I.T. GAG Reports," *Geophysics*, Vol. 32, No. 3. This entire volume is dedicated to selected topics from the GAG reports and gives a convenient summary of most of the work.

Treitel, S., and Robinson, E. A. (1969), "Optimum Digital Filters for Signal to Noise Ratio Enhancement," *Geophysical Prospecting*, Vol. 17, pp. 248–293.

Ulrych, T. J. (1971), "Applications of Homomorphic Deconvolution to Seismology," *Geophysics*, Vol. 36, No. 4, pp. 650–660.

Ware, J. A. (1964), personal communication.

White, J. E., and O'Brien, P. N. S. (1974), "Estimation of the Primary Seismic Pulse," *Geophysical Prospecting*, Vol. 22, No. 4, pp. 627–651.

SEVEN

The Calculation and Measurement
of Auxiliary Information

7.1 INTRODUCTION

In the first six chapters, the principles of exploration seismology were explained, and methods of data gathering and data processing described. These, when fully implemented, allow seismic data to be obtained and processed to yield an optimum cross section which is the basis for a first look at the stratigraphy below the line of section. In other words, it constitutes a picture, in a limited sense, of the true geological cross section.

On this cross section, lines can be drawn which are illustrative of structural geological features, and conclusions can be drawn with respect to the trapping potential as seen from this line of data. There is however, no information on rock types, on the probability of higher-than-normal porosity, or on the probability of oil or gas filling the pores in the rocks at any point in the section.

Other parameters can be derived from these seismic data, and some of these parameters help to fill some of these gaps in our knowledge. In this chapter, these measurements are described as purely objective parameter measurements, and their (subjective) use to derive geological knowledge is deferred until later.

7.2 AVERAGE AND INTERVAL VELOCITY MEASUREMENT

In Section 6.5, the measurement of average velocity from the surface to the depth of a particular reflector was described. This velocity determination (stacking velocity) was made to enable a proper correction to be made for the normal (geometrical) increase in reflection time (NMO) due to offset of the receiver from the source. Many different offset traces are combined, after NMO removal, to give a single CDP or stacked trace.

Although the seismic velocity of a compressional wave through a rock is not a definitive determinant of the rock type by itself, it is certainly an important clue and a partial classifier. For this reason, much effort has been expanded to find ways of obtaining accurate velocity values. Before we discuss some of these in more detail, it is necessary to remember some of the properties of rocks, particularly as they affect the measurements being made on velocities. A rock is not the homogeneous, isotropic medium of theory but an aggregate of mineral particles which have been compressed and cemented and subjected to heat and stress cycles which are not well known. As a consequence, the rock may (and almost always does) show different velocities for P waves, depending on their direction of travel, and waves can be attenuated because of nonlinear stress-strain cycles or because of scattering. In most rocks, the dependence of velocity on frequency is very small and is negligible in the seismic frequency range. As a consequence, a considerable amount of work has been done to measure both P- and S-wave velocities at ultrasonic frequencies, using small samples taken from cores (Wyllie, Gregory, and Gardner, 1958; King, 1962; and Banthia, King, and Fatt, 1965). These experiments have shown that the measured velocities, both P and S, are highly dependent on both the external pressure P_e applied to the matrix of rock particles and also the internal pressure P_i of the fluid content. For sandstones, it has been shown both experimentally and theoretically (Hicks et al., 1956; Brandt, 1955) that the velocity depends on a net effective pressure

$$P_n = P_e - nP_i \tag{7.1}$$

where n is usually close to unity.

The most common way of measuring P-wave velocities has, however, been by ultrasonic logging in a hole. The method is simple in concept. Figure 7.1 is a schematic of the logging tool, which consists of a source S (usually magnetostrictive) which

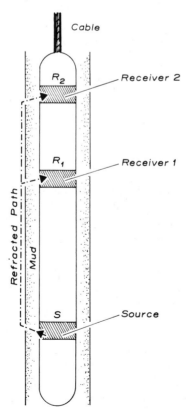

Figure 7.1   Schematic of P-wave-velocity logging tool.

changes volume rapidly on application of an artificially produced magnetic field, the pressure pulse so created being transmitted through the borehole fluid to the wall of the hole. There energy is transmitted by a refraction path to two other points on the wall of the hole and from these points is transmitted to two pressure receivers, R_1 and R_2, separated by a known distance (usually about 1 m). The difference in time of reception of the first impulse at the two receivers is measured electronically, the trigger in each case being the first arrival of energy exceeding a set gate level. With this system, velocities can be measured continually as the system is pulled up the hole. A log is made (usually calibrated in microseconds per foot).

These velocities appear to be reasonably representative of the velocities of seismic waves through the corresponding formations, except when:

1. Invasion of the porous formation by drilling fluid is great, or the formation is altered by drilling fluid—such as in montmorillonite-containing shales—and the source is at an insufficient distance from the receivers. When this happens, the fastest time path has to be through the altered formation, and the velocity is not representative of the true formation velocity.
2. The hole diameter is very large, or very irregular, which either results in unequal times through the fluid at receivers S_1 and S_2 or causes the level of the signal to become so low that there is insufficient amplitude in the first arrival to trigger the electronic counters and cycle skipping results; that is, the trigger is operated by a later, larger arrival, such as the Rayleigh wave or the water wave, and incorrect velocities are obtained. The problem of different mud delays at the two receivers S_1 and S_2 has now been effectively solved by making the P-wave velocity-logging tool symmetrical. An additional source above the receivers, at the same distance as the source below the receivers, allows alternate pulsing of sources. The average time difference is then taken for two successive pulses. In some cases, the problem of cycle skipping has been solved by employing stronger sources but, when the formation velocity is near that of the borehole fluid, nothing can be done.
3. The formation velocity is lower than the P-wave velocity through the drilling fluid (about 1524 m/sec to 5000 ft/sec).

These CVLs are available over some portion of the total drilled footage for thousands of wells and are cataloged by log service companies, along with electrical, neutron, gamma density, and other logs. In some cases, as shown in Figure 7.2, a check on total arrival time is made by placing a suitable receiver in the hole and arranging for shots to be fired in shallow holes near the well. The difference between shot time and arrival time—the transit time of the wave—is corrected for elevation of the shot, offset distance, and other relevant parameters and can then be compared with an integration of the elemental transit times as measured by the CVL.

Dix (1939, 1945, 1946) has dealt in detail with the problems of velocity measurement by shooting in holes. The experimental method is not as foolproof as it sounds, and care has to be taken when it is used to calibrate a CVL.

However, the effects of water invasion, particularly in long shale sections, can be very serious. Special long tools have been used so that the first arrival corresponds to a wave that has traveled for at least part of its path in unaltered shale. This has resulted in agreement with the check shot method and has also caused the synthetic seismograms to correspond better with the field records taken in the same location.

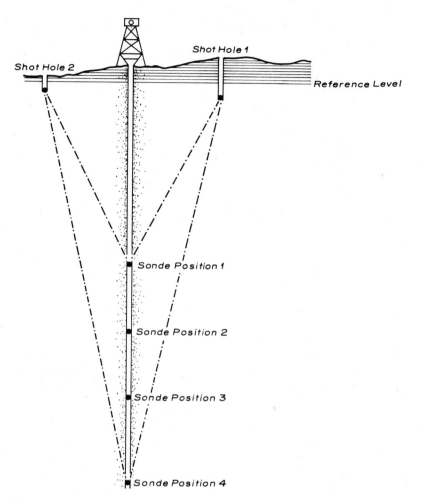

Figure 7.2 Schematic of well shooting method. Usually at least two shot holes are used on opposite sides of the well to check for well deviation from the vertical. Sonde positions are 100 to 160 m apart (300 to 500 ft).

Hicks (1959) showed that the invasion effect occurred between two- and six-hole diameters into the shale and that source-receiver spacings of 2.5 m upward are necessary to obtain the true shale velocity.

No specific tool has been designed to measure shear wave velocities from a borehole. However, an indication of approximate shear wave velocity has been obtained by arranging the P-wave continuous-velocity logger so that the entire received signal can be recorded at each receiver. Then it is found (see Figure 7.3) that the sequence of P, S, and Rayleigh waves can often be seen. The P-wave arrival is a small, discrete event, and the arrival time can be estimated by an automatic trigger circuit set with a threshold just above the noise. The S wave is sometimes buried in noise. The Rayleigh wave can usually be seen, because it is a much larger event and its time can be estimated by an automatic trigger circuit set with a much higher threshold so that it is not triggered by either P or S (or post-P noise). Thus Rayleigh wave velocities are

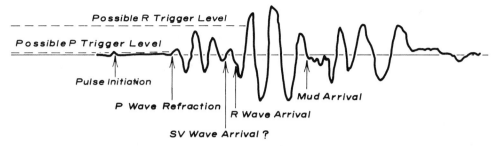

Figure 7.3   Idealized trace of the acoustic wave train in a borehole due to a pressure pulse in the mud.

estimated from the difference in time between the R-wave occurrences at the two receivers.

Since the Rayleigh wave occurs at a velocity slightly less than that of the shear wave in the rock but the fraction varies with rock type, the accuracy of S-wave velocities obtained from these measurements was not good. Remarks previously made about the variation in hole diameter and alteration of rock characteristics with fluid invasion are equally valid as far as shear wave characteristics are concerned. From the practical point of view, the shear wave log has chiefly been used as an indicator of the quality of the cement bond behind casing.

Check shots, using vibrators, have been used to calibrate or obtain both P- and SH-wave velocities, the vertical and horizontal component geophone outputs, respectively, being correlated with the control signal of the appropriate vibrator. There appears to be no necessity for coupling the downhole geophone rigidly to the hole. The hole is usually off-vertical enough that the geophone sonde lies in contact with the hole, and this coupling is sufficient to give an adequate signal. However, a positive coupling is preferred. Examples, for the shear wave case, have been given by Erickson, Miller, and Waters, (1968).

The precision with which these times can be measured is still a matter for conjecture. An absolute travel time is dependent on the characteristics of the source and, since shots are usually detonated below the weathered layer which allows more high frequencies to be generated and transmitted, they are preferred. It is doubtful if accuracies greater than ± 0.002 sec can be expected. Somewhat better precision should be available, however, for differential times either if the same shot is used, with two different receivers separated by several hundred meters (Kokesh, 1956) or if the source characteristics can be held consistent for different receiver locations in the hole. If a surface source does not move, this can be accomplished with modern servocontrolled vibrators. The differential time accuracy should be obtainable to ± 0.001 sec. The limit of accurate interval velocities is therefore controlled by the time interval to be measured. For example, 400 ft of limestone may have a one-way transit time of 0.020 sec, and the precision obtained is, at best, 5%.

It is against this background of direct measurement that we must judge the accuracy and value of indirect, reflection methods of velocity measurement. As described in Section 6.5, the average velocity $\overline{V}$ to a reflector should be determinable from the increase in reflection time as the offset distance (source to receiver) increases. The standard methods are either an X^2-versus-T^2 plot, from which $\overline{V}^2$ is obtained, or the velocity can be obtained by an analysis of CDP data.

These analyses, which are often given proprietary names, consist of several discrete steps, some of which must be:

1. Selection of seismic traces, corrected as well as can be done for near-surface time differences, which apply to a single subsurface reflection point.
2. Application to each trace of a geometrical correction factor based on an assumed velocity.
3. Assessment of the correctness of this velocity and the output of a "goodness" factor for each time window of the record or, in some cases, for each window centered on known reflections.
4. A change in the selected velocity over a predetermined range.
5. An output in clearly visible form of the goodness factor versus the selected velocity for a series of window depths or for depths corresponding to particular reflections.
6. A subjective picking technique which selects the appropriate stacking velocity for each depth, record time, or reflection time.

In addition to these steps, it is often desirable and is a programmed feature of the velocity determination to edit the results in different ways. For example, velocities determined at consecutive depths are compared to see if they represent values that will result in reasonable interval velocities. Other features, of greater complexity, have been deemed advisable in some programs.

Two examples (Figures 7.4 and 7.5a and b), have been given to illustrate the diversity in displays. The first of these is described by Taner and Koehler (1969), and the second is a proprietary program of Continental Oil Company called EVEL. In this section, we examine factors that may limit the accuracy of such measurements.

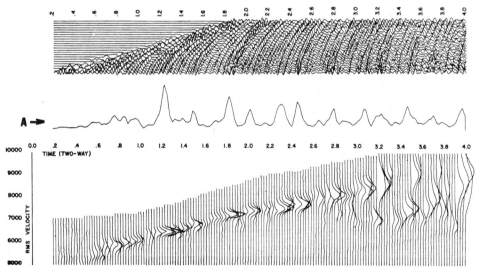

Figure 7.4 Typical velocity spectra display. The peaks on the vertical lines represent the best estimate of stacking velocity. Trace *A* shows the comparative energy of the reflections, hence gives an idea of the reliability of the velocity values. Criteria other than energy are sometimes used. [After Taner and Koehler (1969). Reprinted with permission from *Geophysics*).]

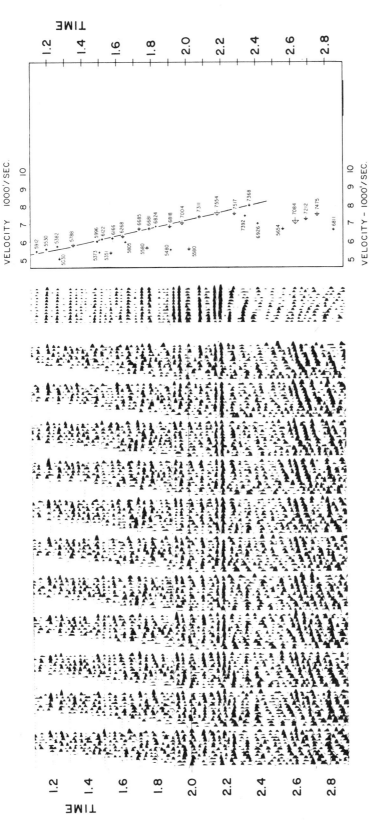

Figure 7.5a The primary records (left), the stacked record, and the velocity associated with a number of events. (After Smith and Diltz. Courtesy of the Continental Oil Company.)

193

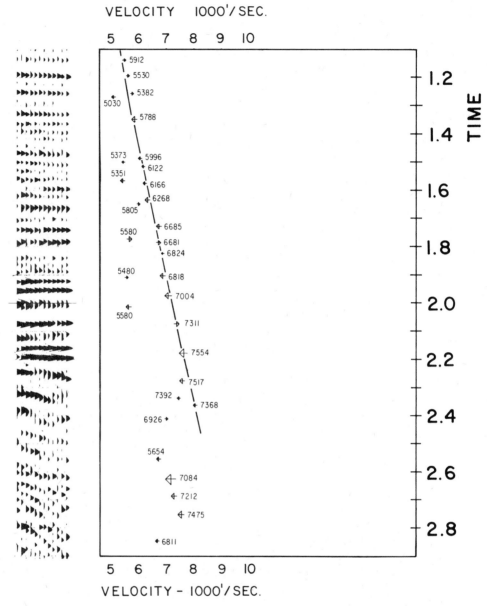

Figure 7.5b The velocity symbols point left or right, depending on whether the event used is a trough or peak. The size of the symbol is related to the energy of the event. The number gives the associated average velocity in feet per second. This is detail from Figure 7.5a.

The first factor is a lack of knowledge of the true path of the energy traveling from the source to the receiver. In Figure 7.6, a series of plane-parallel layers of alternating velocities is shown. This may be a series of sands and shales, or an evaporite sequence such as is commonly found in the Persian Gulf sedimentary section. The ray paths are refracted at the boundaries between the layers and, as the offset is increased, the ratio of the paths in the higher-velocity material to the path in the lower-velocity material

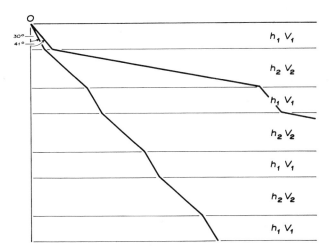

Figure 7.6 Ray paths in a layered medium, showing the pseudoanisotropy caused by differing P-wave velocities in the layers. The example is for $h_1/h_2 = \frac{2}{3}$, $V_1 = 8000$ ft/sec, and $V_2 = 12,000$ ft/sec. Average vertical velocity is 10,000 ft/sec; average velocity at $30°$ incidence is 10,267.9 ft/sec; average velocity at $41°$ incidence is 11,235 ft/sec; critical angle is 41.81°.

changes. Thus the composite material acts as though it were a single material having anisotropy. The average velocity of a composite ray at an angle to the vertical is higher than the velocity vertically through the layered system.

True anisotropy, that is, a velocity in the horizontal direction different from that in the vertical direction, is due to a preferred orientation of the matrix particles and has an effect that changes the pseudoanisotropic effect. In the particular case illustrated, the shale anisotropy is higher than that of the sand, thus causing the shale paths to take a shorter time the higher the angle of incidence. This then reduces the effect of the pseudoanisotropy, since it occurs in the layer with the lower velocity. However, cases can occur as, for example, in the halite-anhydrite example, where the high anisotropic ratio applies to the material with highest velocity and then the pseudo-anisotropic effect is enhanced.

Wave propagation in an anisotropic medium has been the subject of several experimental and theoretical papers (Uhrig and van Melle, 1955; Postma, 1955; Van der Stoep, 1966; etc). The main effects are examined here.

The wave fronts in a homogeneous anisotropic medium are nearly elliptical (Postma, 1955), being characterized by the ratio of the major (horizontal or parallel to the bedding) axis to the minor axis, which is the ratio of the horizontal to the vertical velocity in the medium—the anisotropic ratio. Figure 7.7 shows how Huyghens' principle leads to elliptic wave fronts of the same shape in which the ray path is not in general along the normal to the wave front.

The polar equation of an ellipse, with pole at the center and semiaxes a and b, is

$$r^2 = \frac{a^2 b^2}{a^2 \sin^2 \theta + b^2 \cos^2 \theta} \tag{7.2}$$

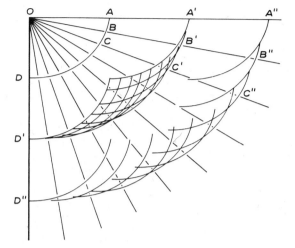

Distances of Progression
Per Unit Time Along Any
Given Directions is the Same
Therefore, the Envelope
A" B" C" D" etc. Has the
Same Shape as the Elemental
Ellipse ABCD etc.

Figure 7.7   Use of Huyghen's principle to show that an elliptical wave front propagates as an elliptical wave front of the same major/minor axes ratio.

so that the velocity in a direction θ from the horizontal is

$$V_\theta = \sqrt{\frac{V_p^2 V_Q^2}{V_p^2 \sin^2 \theta + V_Q^2 \cos^2 \theta}} \qquad (7.3)$$

where V_θ = velocity at angle θ

V_p = velocity parallel to the bedding planes

V_Q = velocity perpendicular to the bedding planes

Note: This applies strictly only to a homogeneous anisotropic material but will apply, in the large, to the pseudoanisotropic material if the layers are thin compared with a wavelength.

Another way to relate the time for a undirectional inclined path to the vertical average velocity, the anisotropic ratio, the depth, and the offset distance is (Uhrig and van Melle, 1955)

$$T^2 = \frac{X^2}{A^2 V^2} + \frac{Z^2}{V^2} \qquad (7.4)$$

For reflections from a constant depth reflector therefore we have

$$\frac{T_R^2}{4} = \frac{X_R^2}{4A^2 V^2} + \frac{Z^2}{V^2} \qquad (7.5)$$

where T_R = reflection time at offset X_R

A plot of T_R^2 against X_R^2 must therefore have a slope of

$$\frac{1}{A^2 V^2}$$

and the anisotropic ratio A is not divorced from the vertical velocity. Only the *horizontal* velocity is determined from this procedure. This conforms to Postma's observation that the anisotropic ratio can never be obtained from surface work alone, a knowledge of the actual depth of the reflector being necessary as well. Since a knowledge of the depth can be obtained only by drilling a well, which also determines the rock type, it appears that anisotropy is of dubious value in characterizing rocks, since it can be measured only after the rock type is known.

The second factor limiting velocity measurements from reflection data is the change in phase of a reflected beam, compared with an incident beam, as the angle of incidence is changed. This manifests itself in a change in wave pulse shape for different offsets and occurs at offsets that correspond to the critical angle when first the P-refracted wave occurs and later when the S-refracted wave occurs. These phase changes apply at a single interface, but there is a more common reason for a *gradual* phase change with offset in the case of layers. In reflections from thin layers, it is the phase of the combined reflections from the upper and lower boundaries that is observed—strictly speaking, it is the combined phase of all multiple reflections within the layer.

In Figure 7.8, these paths are illustrated, and it is shown how the path differences between the upper and lower reflections change. It is particularly noticeable at the

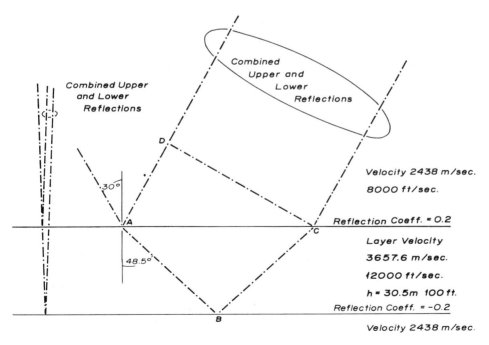

Figure 7.8   Change in phase difference between upper and lower reflections as the angle of incidence is increased.

longer offsets. In the case illustrated [layer thickness 30.5 m (100 ft) and velocities 2438.4 m/sec (8000 ft/sec) and 3657.6 m/sec (12,000 ft/sec)], a difference of phase at 50 Hz of 153.6° (0.0085 sec) is shown, while at 25 Hz the difference is only one-half of this phase angle, or 76.8°.

Of course, this effect is most noticeable for high-velocity layers, where the refracted ray is bent away from the normal and the distance in the high-speed layer increases very rapidly near the critical angle.

Knowledge of effects due to conversions of energy within layers, or simply due to delays in thin layers, is very limited, and no satisfactory general computer method of generating wide-angle synthetic seismograms has yet been given any publicity. These effects are, however, vital and should be an important topic to be researched if more accurate velocities are needed.

A third factor is the dip of the reflecting horizon. A simple geometrical construction shows that the velocity determined by statistical methods based on CDP field work always gives a velocity that is the true velocity divided by the cosine of the angle of dip. The following short table shows that this effect is usually important only at angles of 10° or above, but the velocity calculated is always greater than the true velocity if the medium can be regarded as homogeneous.

Angle (deg)	Ratio, V_{dip}/V_{flat}
0	1.0000
5	1.0038
10	1.0154
15	1.0353
20	1.0642
30	1.1547

Curvature by itself has no effect on the velocity retrieved, since the CDP remains in the same location on the curved surface. It may be noted, however, that in the dipping case the common reflecting point migrates updip with increasing offset and the effects of curvature are then felt. It is difficult to give any general rules on the effect of curvature when dip is present.

7.3 EFFECT OF PORE FLUIDS ON ROCK VELOCITIES

Several different theoretical studies have been made on the effects on elastic wave velocities of different pore-filling fluids (Brandt, 1955; Gassman, 1951; Biot, 1955, 1956; Wyllie, Gregory, and Gardner, 1958; Domenico, 1974; Stolt, 1973; and others), and these have been complemented by numerous ultrahigh-frequency measurements. Recent intense concentration on this subject has been brought about by the knowledge that gas-filled sands are often directly indicated on correctly processed seismic sections by large reflection amplitudes (and other seismically visible effects). These interpretive matters are dealt with later.

In this chapter, only the simpler aspects of this subject are considered. The expressions for the velocity of sound waves in a two-phase medium are usually complicated and generally contain empirical constants which have to be set by comparison

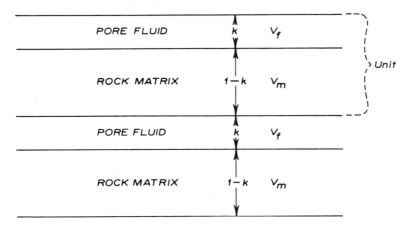

Figure 7.9 The system of layers corresponding to the time-average equation of Wyllie, Gregory, and Gardner (1958). The time through one unit of a repeating sequence is $T = K/V_f + (1 - K)/V_m$.

with measured values. The only equation that is comparatively simple, but is the least effective [see comparison in Kuster and Toksoz (1975)], is the time average equation of Wyllie, Gregory, and Gardner (1958), in which the random medium of matrix material and fluid-filled pore spaces is replaced by a series of plane-parallel layers of liquid and rock and the time through the rock is computed for such a system. Figure 7.9 shows this model and shows how unrealistically it portrays an assemblage of irregular rock particles with fluid in the interstices. The average velocity is then given as

$$\bar{V} = \frac{V_f V_m}{k V_m + (1 - k)V_f} \tag{7.6}$$

where V_m = velocity of sound in the rock matrix
V_f = velocity of sound in the fluid
k = fraction of rock that is pore space

The next equation to be considered (Brandt, 1955) was designed to give the velocity in a sandstone of randomly sized, roughly spherical rock grains filled with fluid. The sound velocity in this type of sandstone is given as

$$\bar{V} = \frac{F(P_r - CP_f)^{1/6}/K^{1/3}}{\sqrt{\phi}\sqrt{\rho_r(1 - \phi) + \rho_f\phi}} \frac{(1 + 17.5B^{3/2}K/\sqrt{P_r - CP_f})^{5/6}}{(1 + 26.3B^{3/2}K/\sqrt{P_r - CP_f})^{1/2}} \tag{7.7}$$

where P_r and P_f = rock and fluid pressures (lb/in.^{2})
ρ_r and ρ_f = rock and fluid densities (g/cc)
ϕ = porosity of the rock
B = bulk modulus (reciprocal of compressibility) of the fluid (lb/in.^{2})
C = fraction of the pore pressure "felt" by the rock, normally $C \approx 0.9$
K = a constant determined by the mechanical properties of the rock, $K \sim 10^{-7}$
F = a constant depending on Poisson's ratio of the rock

For consolidated sandstone, Brandt took F to be 5.75. To allow a description of a shallow, unconsolidated sand, F can be $\lesssim 5.75$.

This equation gives reasonable agreement between laboratory measurements and calculation for porosities higher than 10% (Hicks and Berry, 1956), but it must be remembered that there are three empirical constants which can be set. It does, however, predict an increase in velocity with pressure and a decrease as porosity or fluid compressibility are increased. Sands containing highly compressible fluids, such as oil and gas have lower seismic velocities than water-bearing sands.

If a mixture of both hydrocarbons and water is permitted in the pores, the Brandt velocity equation can be modified as follows. The bulk density ρ_B is expressible as

$$\rho_B = (1 - \phi)\rho_r + \phi(1 - \mu_w)\rho_h + \mu_w \rho_w \tag{7.8}$$

where ϕ = fractional porosity
$\quad \mu_w$ = fractional water content by volume
$\quad \rho_r$ = specific gravity of the matrix
$\quad \rho_h$ = specific gravity of the hydrocarbons
$\quad \rho_w$ = specific gravity of water

The bulk modulus of the two fluids coexisting in the pore space is

$$B_f = \frac{1}{[(1 - \mu_w)/B_h] + (\mu_w/B_w)} \tag{7.9}$$

where B_h = bulk modulus of the hydrocarbons (psi)
$\quad B_w$ = bulk modulus of water (psi)

and the other parameters are as defined previously in (7.7).

Then,

$$\overline{V} = \frac{F(P_r - CP_f)^{1/6}/K^{1/3}}{\sqrt{\rho_B \phi}} \frac{(1 + 17.5B_f^{3/2}K/\sqrt{P_r - CP_f})^{1/6}}{(1 + 26.3B_f^{3/2}K/\sqrt{P_r - CP_f})^{1/2}} \tag{7.10}$$

It can be seen intuitively that, for a two-phase system, the compressibility increases very significantly when even a small amount of *gas* is present. This conjecture has been verified by Domenico (1974), and this fact has very important consequences in the interpretation of reflections of high amplitude from "gas" sands.

Geertsma (1961) has developed from Biot's comprehensive theory (1956) an equation for the velocity of sound of very low frequencies through a fluid-filled porous medium. These frequencies are such that the wavelengths of the sound are very much greater than the pore dimensions. This longitudinal velocity is given as

$$V = \frac{1}{\rho_b^{1/2}}\left[\left(\frac{\beta}{C_s} + \frac{4}{3}G_b\right) + \frac{(1 - \beta)^2}{(1 - \phi - \beta)C_s + \phi C_f}\right]^{1/2} \tag{7.11}$$

where C_s = compressibility of the matrix material
$\quad C_b$ = compressibility of the empty reservoir bulk material
$\quad \beta = C_s/C_b$
$\quad G_b$ = shear modulus of the reservoir bulk material
$\quad C_f$ = compressibility of the fluid
$\quad \phi$ = porosity

This equation was also derived earlier by Gassman (1951).

Curves calculated by Domenico (1974) have been reproduced as Figure 7.10, in which the very sharp change in velocity as small amounts of gas are admitted to the system is seen clearly.

In general, the Biot-Geertsma-Gassman equation gives very good results in comparison with experiment, as indeed it should, since it is based on the solid matrix and the fluid-filled pores comprising two interacting thermodynamic systems.

Other velocity equations have been proposed by Stolt (1973) and by Kuster and Töksöz (1974). These are based on scattering theory and in general give the same velocity properties as the Gassman-Biot-Geertsma equation.

It may be interesting to compare the results for shear waves (based on the Kuster–Töksöz scattering model) with those for P waves.

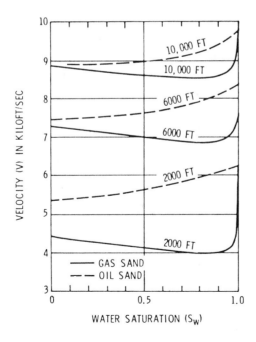

Figure 7.10 Longitudinal velocity V as a function of water saturation S_w for gas and oil sands at depths of 2000, 6000, and 10,000 ft. [After Domenico. Courtesy of the World Petroleum Congresses (1975).]

Their theory shows that, for a shear wave traveling in a rock having a two-phase pore-filling fluid, the velocity is

$$\beta = \sqrt{\frac{\mu^*}{\rho^*}} \tag{7.12}$$

where β = shear wave velocity
μ^* = effective shear modulus
ρ^* = effective density

For spherical fluid-filled pores,

$$\rho^* = \rho(1 - c) + \rho'c \tag{7.13}$$

$$\mu^* = \frac{\mu(1 - c)}{1 + c\left(\dfrac{6K + 12\mu}{9K + 8\mu}\right)} \tag{7.14}$$

where ρ = density of the rock matrix
μ = shear modulus
K = bulk modulus of the rock matrix
ρ' = fluid specific gravity
c = porosity

For any reasonable rock, the shear modulus equation nearly reduces to $\mu^* = \mu[(1 - c)/(1 + c)]$, which relates the effective shear modulus to the shear modulus of the rock matrix and to the porosity. This is in accord with common sense, since the fluid should contribute nothing to the shear stiffness of the rock—it is simply weakened by the holes.

Thus the shear wave velocity can be expressed as a function of porosity:

1.

$$\beta(c) \cong \frac{\beta(0)}{\sqrt{(1 + c)\left(1 + \dfrac{\rho'}{\rho}\dfrac{c}{1 - c}\right)}} \tag{7.15}$$

2. For a water-filled rock, this equation can be written

$$\beta(c) \cong \frac{\beta(0)}{\sqrt{(1 + c)\left(1 + \dfrac{4c}{1 - c}\right)}} \tag{7.16}$$

3. For a partially gas-filled rock,

$$\beta(c) = \frac{\beta(0)}{\sqrt{(1 + c)\left(1 + 4S_w\dfrac{c}{1 - c}\right)}} \tag{7.17}$$

where S_w = fractional water saturation

Thus, while there is some change in shear wave velocity with gas saturation, the effect is chiefly that of the density change and does not change to anything like the extent that the P-wave velocity does. For example, in a rock with 20% porosity, the value of $\beta(c)/\beta(0)$ changes only from 0.908341 to 0.870784 when the water saturation changes from 10% to 99%.

7.4 ATTENUATION OF SEISMIC WAVES IN ROCKS

Observations of seismic data in some oil and gas areas (particularly offshore Gulf of Mexico) has shown that there is some correlation between the presence of gas (and perhaps oil) and anomalous absorption of seismic waves. Mateker et al. (1970) have suggested that attenuation measurements can be made to estimate the sand/shale ratio within a given part of a geological section.

In Section 2.7, a calculation was made which suggested that the classic attenuation from solid friction is not a very large factor in homogeneous rocks, even in the lower-velocity rocks of the Tertiary. It is difficult to imagine the solid friction effects, small as they are over a few hundred feet, to be measurable to an accuracy that would allow differentiation between sands and shales.

From the standpoint of a general relationship, Figure 7.11, which shows measured values of $1/Q$ plotted against velocity (in feet per second), it can be seen that, roughly, a line of slope -2 can be drawn through the data. The consequence of this is that the relation between $1/Q$ and V is

$$\frac{1}{Q} = \frac{C}{V^2} \tag{7.18}$$

where $C \approx 10^6$

It is interesting to speculate that, if this relation is true for shear waves as well as P waves, one would expect, since shear wave velocities in consolidated sediments are about one-half the P-wave velocities, that the attenuation for shear waves would be four times that of P waves (of the same frequency). On a wavelength basis, S waves of the same wavelength should be attenuated about twice as much as P waves.

If it is assumed that there is no dissipation in a material during a purely compressional cycle, it can be shown (Anderson and Archambeau, 1964; Kennett, 1975) that the specific Q's for compressional and shear waves are related by

$$\frac{Q_\beta}{Q_\alpha} = \frac{4}{3}\left(\frac{\beta^2}{\alpha^2}\right)$$

where α = compressional wave velocity
β = shear wave velocity

Although this relation was derived with competent rocks in mind as, for example, in earthquake seismology, it appears that it is reasonably accurate for the measurements in Pierre shale (McDonal et al., 1958) of attenuation for compression and shear waves.

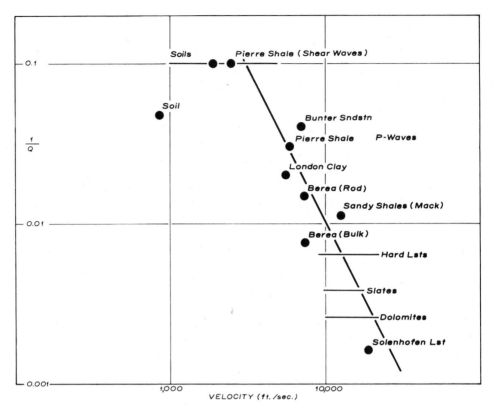

Figure 7.11   Log-log plot of $1/Q$ as a function of velocity for several rocks. The slope of the best fit line is roughly -2.

There is, however, another form of energy loss, due to scattering, which may well be the major cause of attenuation. This results from the loss of amplitude of a seismic wave passing in both a downward and an upward direction through a seismic inter-face. O'Doherty and Anstey (1971) have suggested that this is indeed the major cause of observed attenuation on seismic records. Schoenberger and Levin (1974) made experiments with synthetic seismograms and confirmed the importance of these transmission losses, but their estimate was that the attenuation due to layering accounts for approximately one-third to one-half of the total frequency-dependent attenuation estimated from field seismograms at well locations.

If the primary pulse of unit amplitude is reflected by the nth interface of a series having reflection coefficient r_i, its amplitude will be given by

$$A_n = r_n \prod_1^{n-1} (1 - r_i^2) \qquad (7.19)$$

and it is the continued product part of this that will be of interest. First, we establish some orders of magnitude of the effect. Table 7.1 has been calculated on the basis that all r_i are of the same magnitude. Since this is known not to be true and since this is a nonlinear term in reflection coefficients, we would expect the effect calculated to be less severe than is actually the case. That is, the substitution of an average reflection coefficient gives a conservative answer. The number of layers has been varied in order that the cumulative effect as the number of layers increases can be seen. The drop in power (as well as the amplitude ratio) is given in decibels for those who feel more at home with that logarithmic measure. It is noticeable that the decibel loss of amplitude is proportional to the *number* of interfaces and that even for reflection coefficients of 0.1 (acoustic impedance changes by 5%) the loss is appreciable.

O'Doherty and Anstey have discussed the whole situation in a readable, semi-quantitative manner and have pointed out that this loss of amplitude by the primary event is counterbalanced (to some extent) by the fact that the peg-leg multiples carry the energy for deep reflections. Nevertheless, it is usually the case that the number of peg-leg multiples to be used, which contain most of the energy, is limited to about five, and *all* these paths have to traverse the layers above if they are to have anything to do with the reflections from the layer under consideration.

From the point of view of attenuation therefore, there may be sequences of layers (alternating sedimentation) that attenuate strongly and others (transitional sedimentation that attenuate little. This is very likely the type of attenuation measured through a given geological section. One can see that it is not the sand/shale ratio that is measured, but the number of depositional cycles.

It is not safe to say that the reflection coefficients have to be small in sand-shale sequences, because we have now seen how the presence of a little gas can give rise to

TABLE 7.1

AMPLITUDE OF A UNIT PRIMARY PULSE TRANSMITTED THROUGH *N* INTERFACES IN BOTH DIRECTIONS VERSUS THE REFLECTION COEFFICIENT

	Reflection Coefficient, r					
N	0.01	0.1	0.15	0.20	0.25	0.30
10	0.9990	0.9046	0.7964	0.665	0.524	0.389
	−0.0087 dB	−0.87 dB	−1.977 dB	−3.55 dB	−5.61 dB	−8.19 dB
20	0.998	0.8179	0.634	0.442	0.275	0.1517
	−0.0173 dB	−1.746 dB	−3.953 dB	−7.09 dB	−11.21 dB	−16.38 dB
50	0.995	0.605	0.321	0.1299	0.0397	0.00895
	−0.043 dB	−4.36 dB	−9.883 dB	−17.73 dB	−28.03 dB	−40.96 dB
100	0.990	0.366	0.1027	0.0169	0.00157	0.00008
	−0.087 dB	−8.73 dB	−19.77 dB	−35.46 dB	−56.06 dB	−81.92 dB
200	0.980	0.133	0.0106	0.000285	0.0000025	6×10^{-9}
	−0.174 dB	−17.46 dB	−39.53 dB	−70.91 dB	−112.11 dB	−163.83 dB

large velocity and density changes. Of course, we do not expect that these gassy formations will necessarily be plane-parallel layers, but they will have high *scattering* coefficients and divert much of the energy out of the signal beam.

It is, however, easy to observe the effect of attenuation in seismic record sections that have gas indications (bright spots) on them. The energy carried by reflections beneath the bright spot is sharply reduced. White (1975) showed that, in an unconsolidated sand with partial gas saturation, there is a 20% increase in compressional wave velocity from 1 to 100 Hz and an attenuation of 27 dB/1000 ft at 31 Hz and of 82 dB/1000 ft at 123 Hz. Fluid flow waves are shown to be responsible for the dispersion and attenuation at low frequencies. While this appears to be an important loss mechanism for heterogeneous porous rocks, it can be shown that (say) a 50-ft-thick sand, contributing 100 ft of travel path for reflections from below, would reduce the amplitude of the reflections from below by only 2.7 dB, which is not sufficient to explain the observations. Two reflection boundaries each having a reflection coefficient of 0.4 would provide a further loss of another 3 dB. So the occurrence of small layers within the sand still seems to be plausible. More discussion of this is given later.

It is noted here, however, that the shear wave velocity and attenuation are not affected. Since the reflection coefficients for shear waves are not as large as those for P waves, the attenuation due to layering is also reduced.

A classic case of cyclic sedimentation occurs under Sabkha conditions (periodically flooded and evaporated tidal lagoons) in the Persian Gulf area, and evidence from velocity logs taken in the area shows that there are extensive sequences in the Upper Fars formation where 200 or more interfaces having reflection coefficients of 0.3 (between anhydrite and halite layers) can exist. Here is a built-in mechanism for the known difficulty of obtaining reflections from below this level in this area. The multiple reflections that exist because of this layer and reflection above simply compound their dips to give deep multiple reflections apparently having a structure different from that of the shallow events—but it is not the correct structure for formations at this depth.

Levin and Lynn (1958), using dynamite in shot holes, conducted a series of experiments in deep holes. The character of the entire wave train was recorded and showed graphically how the high frequencies are lost in the downward path. They show, however, that the downhole, clamped geophones also record the upward traveling pulses reflected from layers below the geophone. With this type of a recording system and modern data processing methods, much knowledge of all forms of attenuation could be obtained. Levin notes that the degree of high-frequency attenuation varies very significantly from well to well. Since this experimental work was done before the publication of synthetic records, with multiples and transmission losses, the effects of the boundaries were not taken into account properly. The value of this work lies in the illustration of experimental difficulties, in some cases, the constancy of pulses at different depths, and the description of general methodology.

McDonal et al. (1958) very carefully measured the attenuation constant in Pierre shale (a near-homogeneous shale) for both P and S waves. No allowance was made for the possibility of boundaries (or scattering of any kind). It is now commonly considered that their work forms a basis for considering rocks to have internal friction which gives rise to an attenuation proportional to the first power of the frequency, rather than considering them viscoelastic solids which cause an attenuation proportional to the second power of the frequency. Their attenuation constants for

Pierre shale are given in terms of α (decibels per 1000 ft) times the frequency. This is not the α defined in Section 2.7, and it may well be advisable to clear up this lack of consistency. The α used in this book is defined as being the symbol in the equation

$$A = A_0 e^{-\alpha x} \sin 2\pi f \left(t - \frac{x}{c} \right) \tag{7.20}$$

which is for a plane wave of frequency f traveling at a velocity c.

If we take the ratio A/A_0 and the logarithm of this, we will have

$$\log_{10} \frac{A(f)}{A_0(f)} = \log_{10} e^{-\alpha x} = \frac{-\alpha x}{2.303}$$

However, the power loss in decibels is $20 \log_{10}[A(f)/A_0(f)]$, hence is $-8.684\alpha x$.

The attenuation per foot is therefore -8.684α dB/ft, and the attenuation per 1000 ft (McDonal's α) is $8684\alpha = \alpha_M$.

Now we can take McDonal's value α_M for both shear waves ($\alpha_{MS} = 1.05f$ dB/1000 ft) and P waves ($\alpha_{MP} = 0.12f$ dB/1000 ft) and compute equivalent Q values.

For shear waves,

$$Q_s = \frac{8684\pi}{C_s} \left(\frac{f}{\alpha_{MS}} \right) = 9.99$$

For P waves,

$$Q_p = \frac{8684\pi}{C_p} \left(\frac{f}{\alpha_{MP}} \right) = 32.02$$

The discrepancy between this value and the one given in Table 2.2 apparently arises through averaging the values given in Bradley and Fort's original tables.

Perhaps the chief value of studying McDonal et al. (1958) is to see how meticulously attenuation studies have to be done in order for the values to have some meaning. It emphasizes the difficulty encountered in making *differential* attenuation measurements using normal reflection seismograms.

7.5 CALCULATION OF SIMILARITY, CORRELATION, OR SEMBLANCE COEFFICIENTS

If the source and receiver characteristics remained constant and if the geological section did not change as the measuring points moved over the surface of the earth, it would be expected that the seismic reflection traces would be constant also. That is to say, there would be at the maximum a constant scale factor difference between corresponding values on successive seismic traces. However, we know that this does not happen in practice, and each trace is different from the preceding or the following one. The degree of coherence, or correlation coefficient, is defined as being related to the cross-correlation coefficient between equivalent time windows of the traces:

$$C_{xy} = \frac{\phi_{xy}}{\sqrt{\phi_{xx}\phi_{yy}}} \tag{7.21}$$

where $\phi_{xy} = \sum_{i=1} a_i b_i$

and the traces are

$$x(i) = a_1, a_2, a_3, \ldots, a_i, \ldots, a_N \tag{7.22}$$

$$y(i) = b_1, b_2, b_3, \ldots, b_i, \ldots, b_N \tag{7.23}$$

The terms

$$\phi_{xx}^{1/2} = \sqrt{a_1^2 + a_2^2 + a_3^2 + \cdots + a_i^2 \cdots + a_N^2} \tag{7.24}$$

$$\phi_{yy}^{1/2} = \sqrt{b_1^2 + b_2^2 + b_3^2 + \cdots + b_i^2 \cdots + b_N^2} \tag{7.25}$$

are normalizing factors to facilitate comparing two traces without regard to the amplitude at which they are displayed. Normally, the samples are taken within windows of the original traces which start and stop at the same absolute (corrected) time; that is, no delay is allowed between the traces. (This procedure is contrasted with that used in the Vibroseis® system, in which the correlation value, unnormalized, is played out for two time series as the delay is progressively increased.)

We plan to use a coefficient of this type to quantify, and possibly display prominently, portions of the seismic section in which the geology is changing, as opposed to portions in which the layer thicknesses, layer velocities, and constituent rocks are constant. This, however, necessitates allowance being made for inadvertent shifts between traces. Insensitivity to straight shifts between consecutive traces is made by allowing the delay to change over a small range and then choosing the maximum value of the correlation coefficient. This situation is shown in Figure 7.12.

$$\gamma_{xy} \doteq (C_{xy})_{max} = \max \frac{\phi_{xy}(\tau)}{\phi_{xx}^{1/2}\phi_{yy}^{1/2}} \qquad -\tau_0 < \tau < \tau_0 \tag{7.26}$$

Naturally, it is desirable to minimize the effect of the window used, but it is also required that the changing of end values does not influence the correlation coefficient unduly. For this reason, a window shape is often used that gives small comparative weight to the end values of the window. Foster and Guinzy (1967) have given a table of some windows and their properties.

It should be noted that the coherence is equivalent to the correlation coefficient but is a function of frequency. It is therefore difficult to deal with for different time delays. In spite of the prevalence of this term to denote relationships between seismic traces, it is really applicable only as a phase relationship between one or more constant-frequency wave trains.

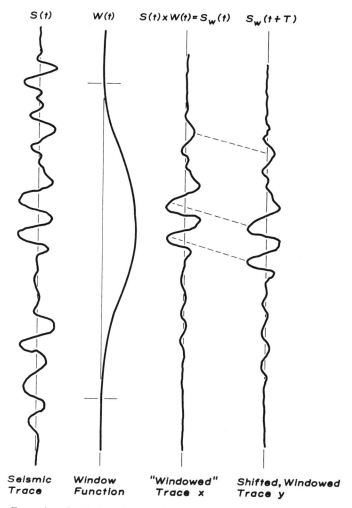

S(t)   W(t)   S(t) x W(t) = S$_w$(t)   S$_w$(t + T)

Seismic   Window   "Windowed"   Shifted, Windowed
Trace   Function   Trace x   Trace y

Figure 7.12   Formation of a windowed trace and the shifted, windowed trace for which γ_{xy} is a maximum when the cross-correlation has a lag of τ.

Semblance is another measure of multichannel correlation in the time domain. It is used, specifically, in measuring the degree to which multiple traces are aligned when they are to be stacked after NMO removal. It is the normalized output/input energy ratio resulting from stacking several channels and is again normally used as a measure of similarity of time of a limited number of phases of seismic events.

If $f(i, j)$ is the jth sample of the ith trace, the semblance coefficient S_c is

$$S_c = \frac{1}{M} \frac{\sum [\sum_{i=1}^{M} f(i,j)]^2}{\sum_{j=K-1/2N}^{K+1/2N} \sum_{i=1}^{M} (f_{(i,j)})^2} \tag{7.27}$$

where M channels are summed. The coefficient is evaluated over a window of width N, centered at point k on the original time-sampling scheme. It is equivalent to the zero-lag value of the autocorrelation of the sum of the traces divided by the mean of the zero-lag values of the autocorrelations of the component traces (Neidell, 1971). As with the correlation coefficient, this formula has to be modified to allow delays between traces.

Our purpose here is to describe the application of correlation coefficients (or other measures of similarity) to finding areas of the seismic cross section where there is lack of similarity and to do this on a quantitative basis. We are seeking geological answers and therefore have to modify our mathematical methods to correspond to geological reality. For example, even if we allow shifts between traces to compensate for the fact that there may be uncorrected shifts originating in the incorrect assessments of the near-surface corrections, the only time we will obtain perfect correlation (maximum correlation coefficient $\gamma_{xy} = 1$) is when all the seismic events suggest that the geological section consists of parallel layers of constant-velocity sediments. Note here that we have said "suggest" rather than "prove," since the seismic spectrum is not wide enough to "see" all of the detail in the earth's lithological variation. This question of resolution of detail by seismic methods is discussed at greater length in Chapter 8.

The systematic computation of γ_{xy} between pairs of traces, for windows of constant length and shape but different centers, for an entire cross section, allows the development of a matrix of values which can be contoured (see Figure 7.13). On this figure, lines have also been drawn indicating the time differences between the traces that give the maximum value γ_{xy} at the centers of the windows. It can be seen that this system can be a basis for the automatic picking of seismic cross sections. Merdler and Paulson (1968) have developed a system which incorporates several additional rules to deal with the problems of ambiguity that obviously arise.

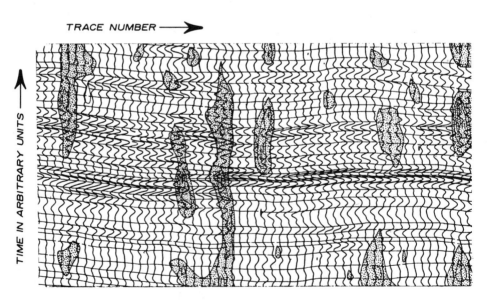

Figure 7.13 Portion of a record section showing similarity coefficient contours superimposed on the section. The nearly horizontal lines show the time relation between traces for maximum similarity. [Courtesy of the World Petroleum Congresses (1975).]

One other problem remains, nevertheless. The trapping of hydrocarbon accumulations occurs almost exclusively on the sides of geological basins, where experience tells us that the sedimentary layers are very likely to be increasing in thickness toward the center of the basin. The correlation coefficient is very sensitive to the increase in time between individual reflection events, and this may therefore vitiate its usefulness in discovering other types of horizontal lithological gradients more favorable for oil trapping and accumulation. This problem can be circumvented by a transformation which can be explained by reference to Figure 7.14a and b.

A sequence of impulses a_i and random delay times τ_i gives rise to an amplitude spectrum whose shape is determined by the amplitudes and time separations of the impulses. The amplitude spectrum contains events e_j which occur at frequencies f_j.

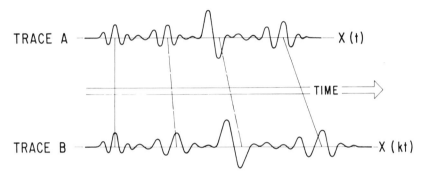

Figure 7.14a   Trace B is similar to trace A except that the time scale of all events has been multiplied by the factor k.

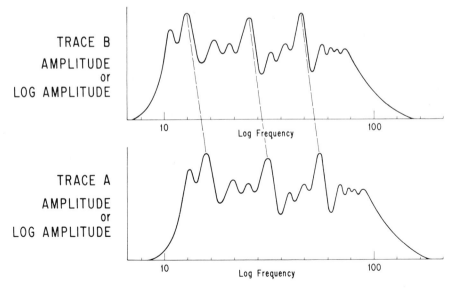

Figure 7.14b   Illustrating the shift in the frequency spectrum due to convergence of the reflecting horizons. [Courtesy of the World Petroleum Congresses (1975).]

If the original sequence is now expanded uniformly in time by the factor k such that all previous delay times τ_i become $k\tau_i$, it will be obvious that all the events e_j on the spectrum occur at frequencies given by f_j/k.

If the original and expanded trace spectra are plotted on graphs having the logarithm of frequency as the horizontal axis,

$$\log \frac{f_j}{k} = \log f_j - \log k$$

so that the events e_j on the expanded spectrum are shifted to the left of corresponding events on the original spectrum. It is now possible to treat the spectra with the maximum correlation coefficient concept, except that $\log f$ replaces time as the independent variable. This revised concept of similarity under expansion gives rise to the similarity coefficient—which is equal to unity when all events are stretched out in a uniform manner. The amount of expansion can be estimated.

It is obvious that, if the spectra were taken just as they come out of the Fourier analysis program, there would be a considerable influence of the low and high cutoffs. It is therefore necessary to remove the average amplitude of the spectra and then taper the ends of the spectra before the similarity coefficient is determined. In order to obtain a fine enough division of the spectrum, the windowed portion of the trace may have to be extended by the addition of supplementary zeros.

7.6   SUMMARY AND CONCLUSIONS

It is evident, from measurements made in boreholes, that seismic velocity is a valuable clue to the identification of rock type. P-wave velocities and approximate S-wave velocities can be measured over small intervals of a few feet in a continuous manner. Such velocities are, however, subject to errors when the rocks are altered by the drilling fluid either chemically or by invasion of the pores. In addition, the velocities are often incorrect when the hole diameter changes rapidly or when the actual rock velocity is near that of the drilling fluid. Sometimes this can be discovered by the use of check shots—but this procedure can give rise to considerable error. The invasion, or alteration-of-rock problem, can be overcome by the use of long-interval logging devices, and these are to be preferred from the strictly seismic point of view.

The use of CDP reflection data allows the computation of velocities—average velocities computed from the surface to the depth of each reflector that stands out above the background noise. It has been shown that this stacking velocity is in reality the average *horizontal* velocity through an anisotropic rock sequence. The anisotropic factor cannot be obtained separately from the velocity by the use of surface reflection data only. The anisotropic factor therefore cannot be used as a rock-type indicator.

The velocities associated with porous rocks depend to a very significant extent on the fluids filling the pores. With fluids as diverse as water and gaseous hydrocarbons, the velocity varies over a wide range. This is the basis for the bright spot change in reflection amplitude which, although there are changes in amplitude when oil replaces water, is chiefly associated with gas in the reservoir. Such changes in reservoir properties are explained quite well by theoretical treatments, such as the Gassman-Biot-Geertsma equation for P waves and the Kuster–Töksöz treatment for shear waves. Much care needs to be taken in assigning quantities of gas to a bright spot reservoir because of the theoretical, and experimentally verified, fact that small quantities of gas are quite effective in changing the elastic properties of a reservoir.

As expected, the change in fluid in a reservoir rock has only a marginal effect on the shear wave velocity—a distinction that may have considerable utility in the future.

The attenuation of waves propagated through consolidated rocks is usually not large enough to prevent exploration, in the widest sense. Theory suggests that it is much greater in rocks containing gas but, with the thickness of gas layers usually encountered, the effect is much smaller than the experimentally observed change in the amplitudes of deeper reflections. An additional cause of change in amplitude is the transmission losses that occur at each boundary between rocks. In some cases, notably in sequential gas rock-shale layering and in anhydrite-halite layering, it may be large enough that penetration through such a system is impossible with known methods.

Attenuation variations in sand-shale sequences are insufficient to be accurately measured and used as a sand/shale ratio indicator. The largest attenuation effect is likely to be the transmission loss mechanism, and this is responsive to the number of interfaces or the thinness of the alternating beds.

Similarly between successive seismic traces, or multiple sets of traces, can be calculated in several ways. With the use of proper windowing techniques, the similarity coefficients become relatively free from end effects, and the similarity coefficient as a function of horizontal position along the profile and the time or depth on the cross section can be contoured and used as an indicator of rapid changes (departures from the normal) of the sedimentary pattern—locations favorable for the occurrence of stratigraphic traps.

APPENDIX 7A: MAXIMUM ENTROPY SPECTRAL ANALYSIS

We have remarked several times that the process of taking the spectrum of a portion of a seismic trace (or of the whole trace, for that matter) involves splitting out (or windowing) sample values of the portion of the trace being investigated. If the values are just separated out, it is as though the trace has been multiplied by a rectangular function

$$W(t) = 1 \quad (\tau_1 \le t \le \tau_2)$$
$$= 0 \quad \text{(elsewhere)} \tag{7A.1}$$

The isolated portion of the trace may be too short to give the needed resolution in the frequency spectrum. For example, a 0.5-sec portion of a seismic trace allows amplitudes to be computed only at frequency differences of 2 Hz. In this case, the custom is to extend the windowed portion of the trace to make it as long as is necessary for the required resolution by adding additional values, all equal to zero, inside the needed interval T.

If we let $F(t)$ be the function to be examined in the interval between τ_1 and τ_2, this portion is obtained by multiplying $F(t)$ by $W(t)$:

$$F_W(t) = F(t)W(t) \tag{7A.2}$$

and, when we obtain the spectrum, it is that of $F_W(t)$. The effect of the window can be determined from the general rule that multiplication in the time domain is equivalent to convolution in the frequency domain, so

$$F_W(\omega) = F(\omega) * W(\omega)$$

The spectrum needed $F(\omega)$ has been convolved with the spectrum of the window $W(\omega)$.

It is possible to choose a different window $W(t)$ so that, when this convolution is done, a minimum effect is produced on the desired spectrum, but in any event, there will always be some smearing of the desired spectrum by the convolution procedure.

Burg (1967) suggested an alternative procedure. The conventional methods mentioned above sometimes generate power spectra, some values of which are negative, or the spectral estimates do not agree with the known autocorrelation function. Since these conditions must not occur, the new approach was adopted.

From the given limited portion of the time series to be analyzed, an autocorrelation function can be obtained:

$$A(t) = F(t) * F(-t)$$

This autocorrelation function of course has a limited number of values, and there is an infinite number of spectra that agree with these known values—but produce additional time-lagged values outside the known range. Maximum entropy spectral analysis (MESA) is based on the choice of all possible spectra that maximize the unpredictability of a time function that agrees with the known values of the autocorrelation function. This assumption corresponds to the idea of maximum entropy as used in the physical sciences.

The characteristics of maximum entropy spectral analysis are:

1. The spectral estimates are nonnegative functions of frequency.
2. The resolution of the MESA is greater than that obtained conventionally.
3. MESA can be calculated from $F(t)$ values in little more time than is required for conventional spectral estimates.

Andersen (1974) has described a fast, simple procedure for MESA estimation for a set of data $\{x_1, x_2, x_3, \ldots, x_N\}$ with equal spacing Δt. It is done by recursive methods which make the computations very much easier. His paper gives both the operative formulas and a flow diagram of the recursive procedure.

As far as is known, there have been few tests of MESA in practical seismic reflection work. Andersen, however, reports that the procedure is superior to the FFT because of its greater resolution for a given series of data. An example is the analysis of 128 points of a sinusoid mixed with 10% white noise. After five iterations, the MESA, in agreement with approximate formulas derived by Lacoss (1971), was better than that obtained by FFT analysis.

REFERENCES

Anderson, D. J., and Archambeau, C. B. (1964), "The Anelasticity of the Earth," *Journal of Geophysical Research*, Vol. 69, pp. 2071–2084.

Andersen, N. (1974), "On the Calculation of Filter Coefficients for Maximum Entropy Spectra Analysis," *Geophysics*, Vol. 39, No. 1, pp. 69–72.

Banthia, B. S., King, M. S., and Fatt, I. (1965), "Ultrasonic Shear-Wave Velocities in Rocks Subjected to Simulated Overburden Pressure and Internal Pore Pressure," *Geophysics*, Vol. 30, No. 1, pp. 117–121.

Biot, M. A. (1955), "Theory of Elasticity and Consolidation for a Porous Anisotropic Solid," *Journal of Applied Physics*, Vol. 26, pp. 182–185.

Biot, M. A. (1956), "Theory of Propagation of Elastic Waves in a Fluid-Saturated Porous Solid. 1. Low Frequency Range. 2. Higher Frequency Range," *Journal of the Acoustical Society of America*, Vol. 28, pp. 168–191.

Biot, M. A., and Willis, D. G. (1957), "The Elastic Coefficients of the Theory of Consolidation," *Journal of Applied Mechanics*, Vol. 24, pp. 594–601.

Brandt, H. (1955), "A Study of the Speed of Sound in Porous Granular Materials," *Journal of Applied Mechanics*, Vol. 22, pp. 479–486.

Brown, R. J. S., and Koringa, J. (1975), "On the Dependence of the Elastic Properties of a Porous Rock on the Compressibility of the Pore Fluid," *Geophysics*, Vol. 40, No. 4, pp. 608–616.

Burg, J. P. (1967), "Maximum Entropy Spectral Analysis," paper presented at the 37th Annual International Meeting of the Society of Exploration Geophysicists, Oklahoma City, October 31, 1967.

Burg, J. P. (1968), "A New Analysis Technique for Time Series Data," paper presented at the NATO Advanced Study Institute on Signal Processing, Enschede, Netherlands.

Burg, J. P. (1972), "The Relationship between Maximum Entropy Spectra and Maximum Likelihood Spectra," *Geophysics*, Vol. 37, pp. 375–376.

Burg, J. P. (1975), "Maximum Entropy Spectra Analysis," Ph.D. Thesis, Stanford University.

Dix, C. H. (1939), "Interpretation of Well Shot Data I," *Geophysics*, Vol. 4, No. 1, pp. 24–32.

Dix, C. H. (1945). "Interpretation of Well Shot Data II," *Geophysics*, Vol. 10, No. 2, pp. 160–170.

Dix, C. H. (1946). "Interpretation of Well Shot Data III," *Geophysics*, Vol. 11, No. 4, pp. 457–461.

Domenico, S. N. (1974), "Effect of Water Saturation on Seismic Reflectivity of Sand Reservoirs Encased in Shale," *Geophysics*, Vol. 39, No. 6, pp. 759–769.

Erickson, E. L., Miller, D. E., and Waters, K. H. (1968), "Shear Wave Recording Using Continuous Signal Methods—Part II," *Geophysics*, Vol. 33, No. 2, pp. 240–254.

Fatt, I. (1958). "Compressibility of Sandstones at Low to Moderate Pressures," *Bulletin of the AAPG*, Vol. 42, pp. 1924–1957.

Fatt, I. (1959), "The Biot–Willis Clastic Coefficients for a Sandstone," *Journal of Applied Mechanics*, Vol. 26, pp. 296–297.

Foster, M. R., and Guinzy, N. J. (1967), "The Coefficient of Coherence—Its Estimation and Use in Geophysical Data Processing," *Geophysics*, Vol. 32, No. 4, pp. 602–616.

Gardner, G. H. F., Wyllie, M. R. J., and Droschak, D. M. (1965), "Hysteresis in the Velocity Pressure Characteristics of Rocks," *Geophysics*, Vol. 30, No. 1, pp. 111–116.

Gassman, F. (1951), "Elastic Waves through a Packing of Spheres," *Geophysics*, Vol. 16, pp. 673–685.

Geertsma, J. (1957), "The Effect of Fluid Pressure Decline on Volumetric Changes in Porous Rocks," *Transactions of the AIME*, Vol. 210, pp. 331–340.

Geertsma, J., and Smit, D. C. (1961), "Some Aspects of Elastic Wave Propagation in Fluid Saturated Porous Solids," *Geophysics*, Vol. 26, No. 2, pp. 169–181.

Hicks, W. G. (1959), "Lateral Velocity Variations Near Boreholes," *Geophysics*, Vol. 24, No. 3, pp. 451–464.

Hicks, W. G., and Berry, J. E. (1956), "Application of Continuous Velocity Logs to Determination of Fluid Saturation of Reservoir Rocks," *Geophysics*, Vol. 21, pp. 739–754.

Kennett, B. L. N. (1975). "The Effects of Attenuation on Seismograms," *Bulletin of the Seismological Society of America*, Vol. 65, No. 6, pp. 1643–1652.

King, M. S. (1962). "Ultrasonic Shear Wave Velocities in Rocks Subjected to Simulated Overburden Pressure," *Geophysics*, Vol. 27, No. 5, pp. 590–598.

Kokesh, F. P. (1956). "The Long Interval Method of Measuring Seismic Velocity," *Geophysics*, Vol. 21, No. 3, pp. 724–738.

Kuster, G. T., and Töksöz, M. N. (1974), "Velocity and Attenuation of Seismic Waves in Two-Phase Media, Part I: Theoretical Formulations," *Geophysics*, Vol. 39, No. 5, pp. 587–606.

Kuster, G. T., and Töksöz, M. N. (1974), "Velocity and Attenuation of Seismic Waves in Two-Phase Media: Part II: Experimental Results," *Geophysics*, Vol. 39, No. 5, pp. 607–618.

Lacoss, R. T. (1971), "Data Adaptive Spectra Analysis Methods," *Geophysics*, Vol. 36, pp. 661–675.

Levin, F. K., and Lynn, R. D. (1958), "Deep Hole Geophone Studies," *Geophysics*, Vol. 23, No. 4, pp. 639–664.

Mann, R. L., and Fatt, I. (1960), "Effect of Pore Fluids on the Elastic Properties of Sandstone," *Geophysics*, Vol. 25, pp. 433–443.

McDonal, F. J., Angona, F. A., Mills, R. L., Sengbush, R. L., van Nostrand, R. G., and White, J. E. (1958), "Attenuation of Shear and Compressional Waves in Pierre Shale," *Geophysics*, Vol. 23, No. 3, pp. 421–439.

Merdler, S. C., and Paulson, K. V. (1968), "Automatic Seismic Reflection Picking," *Geophysics*, Vol. 33, No. 3, pp. 431–440.

Neidell, N. S., and Taner, M. T. (1971), "Semblance and Other Coherence Measures for Multichannel Data," *Geophysics*, Vol. 36, No. 3, pp. 482–497.

O'Doherty, R. F., and Anstey, N. A. (1971), "Reflections on Amplitudes," *Geophysical Prospecting*, Vol. 19, pp. 430–458.

Postma, G. W. (1955), "Wave Propagation in a Stratified Medium," *Geophysics*, Vol. 20, No. 4, pp. 780–806.

Schoenberger, M., and Levin, F. (1974), "Apparent Attenuation Due to Intrabed Multiples," *Geophysics*, Vol. 39, No. 3, pp. 278–291.

Stolt, R. H. (1973), personal communication.

Taner, M. H., and Koehler, F. (1969), "Velocity Spectra—Digital Computer Derivation and Applications of Velocity Functions," *Geophysics*, Vol. 34, No. 6, pp. 859–881.

Uhrig, L. F., and van Melle, F. A. (1955), "Velocity Anisotropy in Stratified Media," *Geophysics*, Vol. 20, No. 4, pp. 774–779.

Ulrych, T. J. (1972), "Maximum Entropy Power Spectrum of Long Period Geomagnetic Reversals," *Nature*, Vol. 235, pp. 218–219.

Ulrych, T. J. (1972), "Maximum Entropy Power Spectrum of Truncated Sinusoids," *Journal of Geophysical Research*, Vol. 77, pp. 1396–1400.

Ulrych, T. J., and Bishop, T. M. (1975), "Maximum Entropy Spectral Analysis and Autoregressive Decomposition," *Review of Geophysics and Space Physics*, Vol. 13, No. 1, pp. 183–200.

Van der Stoep, D. M. (1966), "Velocity Anisotropy Measurements in Wells," *Geophysics*, Vol. 31, No. 5, pp. 909–916.

White, J. E. (1975), "Computed Seismic Speeds and Attenuation in Rocks with Partial Gas Saturation," *Geophysics*, Vol. 40, No. 2, pp. 224–232.

Wyllie, M. R. J., Gregory, A. R., and Gardner, G. H. F. (1958), "An Experimental Investigation of Factors Affecting Elastic Wave Velocities in Porous Media, *Geophysics*, Vol. 23, pp. 459–493.

EIGHT

Resolution and Diffractions

8.1 INTRODUCTION

In optics, an instrument is always qualified by a number which describes, to the initiated, the limit of fineness of detail that can be seen when the instrument is used. In the case of a telescope, this is the angular separation between two points of light that can just be resolved, that is, distinguished as separate. In light microscopes and electron microscopes, the qualification is similar but is sometimes translated into a linear measure. For example, it may be possible to resolve two points (say) 0.001 mm apart. Films used in cameras have a limit of resolution of (say) 200 lines per millimeter. That is to say, if black lines on the film were closer together than this, the area would appear to be a uniform gray color rather than showing a separation between the lines.

It must be remembered that, in optics, we are dealing with visible light having wavelengths from about 4000 to 7000 Å (4 to 7×10^{-7} m), usually very small compared with the detail of the objects being investigated, and the eye is sensitive only to variations in light intensity, not to variations in phase. Now it is known that, even with the most perfect optical system, free of all forms of aberration, the image of a point source is not a point but illumination over a central finite area, followed in a radial direction by successive dark and light rings. These are the famous Fraunhofer diffraction patterns (Born and Wolf, 1959, p. 391), the intensity of which is given, as a function of radial distance from the center, in Figure 8.1a. Although this distribution is not quite a $[(\sin Kx)/Kx]^2$ distribution, it approaches it, and in fact a small rectangular aperture does give rise to such a distribution in both directions parallel to the sides of the rectangle.

In Figure 8.1b, two Fraunhofer diffraction curves have been added together with a separation such that the peak of one curve lies at the first minimum of the other— such a separation occurs at an angle equal to $0.61(\lambda/a)$, where λ is the wavelength of the light being used and a is diameter of the telescope or microscope aperture being used.

There is no need for further detail here, because we have already noted that the eye is sensitive to intensity (amplitude) only, the radiation being used is monochromatic, and there is no possibility of waveform variation. Although optical theory suggests a resolution of two neighboring point sources, which is dependent on the wavelength of the light being used, there are few other analogies we can use.

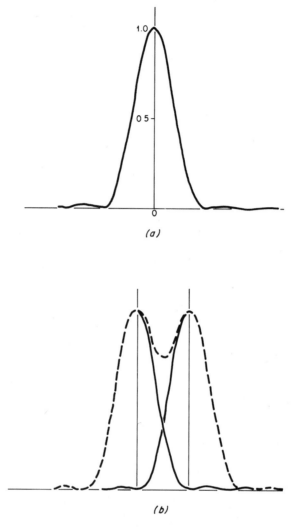

Figure 8.1 (*a*) Fraunhofer diffraction at a circular aperture; the function is $y = [2J_1(x)/x]^2$. (*b*) The sum of two Fraunhofer diffraction curves at separation 0.61 λ/a (Rayleigh's criterion).

8.2 SEISMIC RESOLUTION

The question can be asked, What do we wish to distinguish from what? In a certain (physical) sense, we do not know which part of a seismic cross section or which part of a geological section we wish to examine in detail. Until a specific objective has been stated from other evidence, it can be argued that, in most cases, it is the greater part of the geological section that *may* need to be examined in detail. The aim of all seismic prospecting is to learn as much as possible about a geological section—the manner in which the rocks are laid down, folded, fractured, and faulted, their mineral constitution, and the amount and type of fluid contained in the pores.

The most direct form of measurement uses the various forms of logs obtained in holes drilled through formations, and logs that contain information capable of being determined from seismic measurements are:

1. P-wave velocity logs.
2. S-wave velocity logs.
3. Density logs.

A perfectly logged single hole gives information applicable to a small volume around the hole, so that many holes are necessary to give some of the geological information sought. A seismograph would approach perfection if it could, in a vertical sense, measure the same quantities as the three logs listed, to the same degree of detail.

One answer to the general question about resolution is that we want to maximize the detail with which the vertical variation of the two seismic velocities and the density are obtained. In other words, we would like to be able to determine, as closely as one can using well logs, the depths at which the lithology and connate fluids change.

In another sense, however, the seismic method has been used to acquire information on the lateral variation in some or all of the quantities listed earlier, becoming a cheaper alternative to the drilling of many wells. We must expect therefore to be able to detect horizontal changes in the same elementary parameters—not only to detect them but to place them correctly in space. One can ask, How closely can this be done? And this is another form of resolution about which information must be forthcoming.

Finally, with only three parameters, the bulk and shear moduli of elasticity and the density, that can physically affect the seismic waves, even in the perfect case, how much knowledge of the economic factors associated with hydrocarbon production can be obtained? And what is the precision to be expected?

These are complex questions which do not have easy answers.

8.3   VERTICAL SEISMIC RESOLUTION

Earlier, in Section 4.5, the characteristics of reflections, as related to the velocity and density logs, were given. It was shown in (4.15) that there is a relation between laminar velocity changes in the earth and reflection coefficients:

$$\delta R = \tfrac{1}{2}\delta(\ln V)$$

and this can be extended, if the density also varies, to

$$\delta R = \tfrac{1}{2}\delta(\ln Z) \tag{8.1}$$

where Z is the acoustic impedance of the rock—the product of the proper velocity and the density—which is a function of the depth h or the two-way reflection time t. It is possible to define a piecewise continuous function called the reflectivity:

$$r(t) = \lim_{\delta t \to 0} \frac{\delta R}{\delta t} = \frac{1}{2}\frac{d}{dt}(\ln Z) \tag{8.2}$$

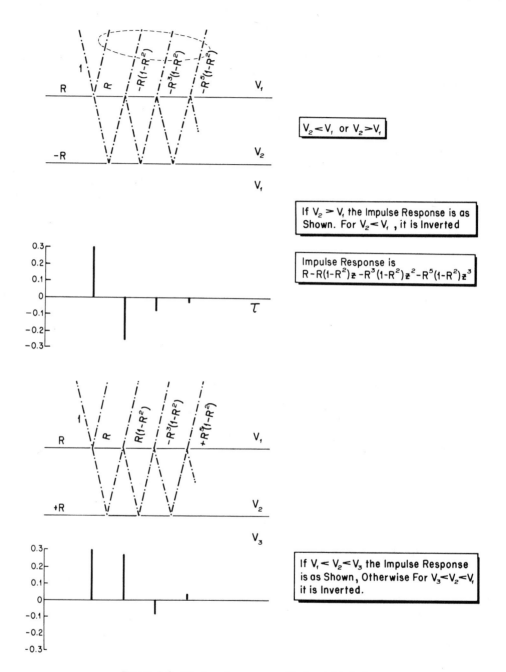

$$V_2 < V_1 \text{ or } V_2 > V_1$$

If $V_2 > V_1$ the Impulse Response is as Shown. For $V_2 < V_1$, it is Inverted

Impulse Response is
$$R - R(1-R^2)z - R^3(1-R^2)z^2 - R^5(1-R^2)z^3$$

If $V_1 < V_2 < V_3$ the Impulse Response is as Shown, Otherwise For $V_3 < V_2 < V_1$ it is Inverted.

Figure 8.2 The impulse response of isolated thin beds.

It may be advisable to point out that the reflection coefficients we have been dealing with have been obtained at constant sampling rates. They really represent the product of the reflectivity and the sampling rate, although this has not been explicitly stated.

Going back now to (8.2), we note that it is not a linear function of the acoustic impedance (a small change in impedance when the average impedance is low has a higher reflectivity than the same change when the average impedance is high). The reflectivity, leaving out some complications for the moment, is the quantity that gives rise to the seismic reflection record, but it is not as easy to interpret, geologically, as would be the actual rock property—the impedance. Thus we transform (8.2) by two steps:

$$2 \int_{t_0}^{t} r(t) \, dt = \int_{Z(0)}^{Z(t)} \frac{d}{dt} (\ln Z) \, dt = \ln \frac{Z(t)}{Z(0)}$$

and   (8.3)

$$\frac{Z(t)}{Z(0)} = \varepsilon^{2 \int_{t_0}^{t} r(t) \, dt}$$

This gives a relationship which allows calculation of $Z(t)$ as a function of t and an assumed, or known, impedance at the beginning t_0 of the log. It does, however, assume a knowledge of reflection coefficients. There are some difficulties.

1. In Section 4.5, it was shown that it is the impulse response (including all multiples and transmission losses) that is actually responsible for the seismic reflection trace.
2. This impulse response is very wide-band in frequency and, to obtain the seismic record, it has to be filtered with the effective bandwidth pulse received by the seismic system.
3. It is assumed that it is possible to produce a seismic trace as though it has been generated by plane waves (i.e., the spherical divergence has been removed exactly).
4. It is assumed that the *exact* values of the reflection coefficients are known. The process is nonlinear and responds in a different manner to large reflections than to small ones. Thus a knowledge of the reflection coefficient scaled by some unknown constant does not allow the exact acoustic impedance log to be determined.

The effect of all these restrictions is to limit the fidelity with which the acoustic impedance log can be displayed. It is a seismic approximate impedance log (SAIL).

These restrictions are now examined in more detail. It is interesting, since we are concerned mostly with questions involving thin layers, to examine the case of an isolated thin layer, that is, thin compared with the wavelengths of the seismic pulse. The impulse response, as shown in Figure 8.2, consists of a series of rapidly diminishing pulses, equally spaced in time by the two-way transit time for the layer $2d/V$, where d is the thickness and V is the relevant velocity. The signs of the separate pulses are either:

1. A first impulse from the upper surface, followed by a second of the same sign and then later ones of alternate sign (for a layer of intermediate acoustic impedance between two extremes) or

2. A first impulse from the upper surface, followed by decreasing pulses all of the opposite sign (for a layer whose impedance is either less than, or greater than, the impedance on either side).

The sequence of impulses decays rapidly and in practice is important only for very large velocity and density contrasts which occur, for example, with thin coal beds in a siltstone matrix or for a gas sand in a shale-water-sand environment. The values plotted in this and some following diagrams are approximately 0.3—which is not a common reflection coefficient in the earth.

Next, we have to deal with the situation that each pulse in the impulse response has been filtered by the combination equipment-earth filter. It is assumed that either the Vibroseis® system has been used or that the equipment phase has been removed by an inverse filter such as deconvolution or a special filter designed for the purpose. In either case, each constituent pulse is a symmetrical (zero-phase) pulse. For the examples given, it is the autocorrelation pulse corresponding to a particular amplitude spectrum.

We first look at the effect of integrating the wide-band reflection coefficient case. A perfect integrator can be regarded as a process or circuit by which a delta function is transformed into a positive step function. This is easily seen since, if the delta function occurs at a time τ, integration up to that time yields zero, by the definition of the delta function. At time τ, the integral yields a unit value, which is retained for all times thereafter. Mathematically,

$$\int_{-\infty}^{T} \int_{-\infty}^{\infty} \delta(t - \tau) \, dt = 0 \qquad (T < \tau)$$

$$= 1 \qquad \text{(otherwise)}$$

If such an integrator is applied to a sinusoidal wave train,

$$\int_{0}^{t} \sin \omega t \, dt = \frac{1}{\omega} \sin \left(\omega t - \frac{\pi}{2} \right)$$

which shows that the integrated output has an amplitude that decreases inversely as the frequency and is retarded in phase by $\pi/2$ radians. Thus, for constant-amplitude input, doubling the frequency (an increase of one octave) reduces the amplitude of the integrated output to one-half.

Figure 8.3a shows the effect of approximating the delta function by a narrow rectangular pulse to obtain an approximate step function—with a steep ramp replacing the step because of the finite width of digital sampling. The amplitude is in arbitrary units because, not knowing the size of the reflection coefficient, we cannot evaluate the exponential function. In Figure 8.3b the integral of a nominal 5- to 100-Hz, flat-spectrum, autocorrelation pulse is shown. Actually, the spectrum had a taper over 4 Hz at each end of the spectrum centered over the nominal end points. Tapering a spectrum of constant bandwidth at the two ends has the following effects:

1. The high-frequency oscillations tend to die out more rapidly with time, the more slowly the tapering is done at the high-frequency end.

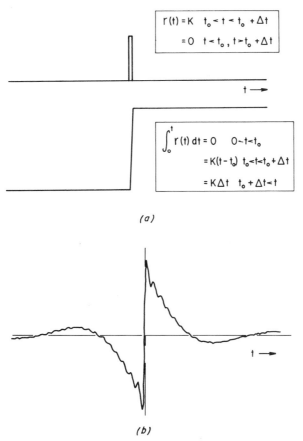

$$r(t) = K \quad t_o < t < t_o + \Delta t$$
$$= 0 \quad t < t_o, \; t > t_o + \Delta t$$

$t \longrightarrow$

$$\int_o^t r(t)\, dt = 0 \quad 0 \sim t < t_o$$
$$= K(t - t_o) \quad t_o < t < t_o + \Delta t$$
$$= K\Delta t \quad t_o + \Delta t < t$$

(a)

$t \longrightarrow$

(b)

Figure 8.3 *(a)* Integration of an approximation to a delta function. *(b)* Integration of a band-limited autocorrelation function. The bandwidth is nominally 5 to 100 Hz, and the ends were linearly tapered from 3 to 7 Hz and from 98 to 102 Hz.

2. The low-frequency oscillations on either side tend to die out more rapidly with time, the more slowly the tapering is done at the low-frequency end.
3. The central breadth of the autocorrelation function tends to increase, the more tapering is done.

Various forms of taper have been used, but in practice the linear taper is almost always employed.

Figure 8.3 shows that, while the integration of a reflection impulse (whose spectrum contains frequencies from zero to some high value connected with the thickness of the pulse) gives an acoustic impedance change over a small interval of time and the change remains after the impulse has passed, it is obvious that the integration of the band-limited zero-phase pulse results in a fast change in apparent impedance (controlled by the high-frequency end of the spectrum) preceded, and followed, by slow decays controlled by the low-frequency end of the spectrum. We are thus forcibly made aware of the need—in order to achieve our goal of acoustic impedance fidelity—of a very wide band of frequencies. Not only are the high frequencies important, but also the lows. This statement is reinforced by several examples later.

We now go one stage further in the process of deriving the practical SAIL trace from the reflection trace, since we know that the absolute values of the reflection coefficients are not likely to be known (exceptions are sometimes possible in certain geological situations in which isolated interfaces between very thick constant-velocity rocks exist).

Instead of the true impedance $Z(t)$, we are now dealing with an approximate impedance function $\tilde{Z}(t)$, and the reflection coefficient trace must be replaced by a special processed record trace $S(t)$. Since the absolute scale of $S(t)$ is not known, the 2 in front of the integral sign has to be replaced by a more general constant A. Thus we have

$$\tilde{Z}(t) = Z(0)\varepsilon^{A\int_{t_0}^{t} S(t)\,dt} \tag{8.4}$$

Provided that $A\int_{t_0}^{t} S(t)\,dt$ is always small which, in practice appears to be nearly true, the exponential term on the right can be expanded, retaining only the constant and first-order terms:

$$\tilde{Z}(t) = Z(0)\left\{1 + A\int_{t_0}^{t} S(t)\,dt + \cdots\right\}$$

or $\hspace{10cm}$ (8.5)

$$\frac{\tilde{Z}(t) - Z(0)}{Z(0)} = A\int_{t_0}^{t} S(t)\,dt + \cdots$$

Thus the fractional change in impedance (as indicated by the SAIL trace) is simply the seismic trace integrated and displayed on an arbitrary scale.

In the remainder of this section, the question of resolution is examined from the point of view of the accuracy with which the variations in the actual acoustic impedance can be forecast (from the surface) by the integration of properly processed seismic traces.

Returning to the isolated thin-layer case, as a suitable starting point, we note that each of the impulses in the impulse response has to be convolved with an integrated symmetrical autocorrelation pulse. In the examples that follow, two pulses are used, having the same low-frequency spectrum (7 Hz, tapered by 4 Hz, centered on 7 Hz) but, to show the effect of the higher frequencies, in one case the upper frequency is 43 Hz and in the other 90 Hz, with a 20 Hz taper, centering on the nominal frequency. These pulses, their amplitude spectra, and the integrated pulses resulting from them are shown in Figure 8.4. Note that, in the case of the 7-43I pulse, the rise time is near 0.020 sec, whereas for the 7-90I pulse, the rise time is approximately 0.012 sec.

As far as the spectra are concerned, we know that integration results in a 6-dB increase in power (double amplitude) for every doubling of the frequency (octave). This is well shown in these figures. One way of integrating the pulse is in fact to make this change in the amplitude spectrum, change the phase by 90°, and Fourier-synthesize the pulse.

In Figures 8.5 and 8.6 are shown the results of a model computer program which computes the reflection response from a generalized layered system. Models corresponding to 10, 20, 40, 60, 80, and 200 ft are computed, the two-way times being

0.002, 0.004, 0.006, 0.008, and 0.020 sec, respectively. The velocities outside and inside the layer were 5000 and 10,000 ft/sec, respectively, giving a reflection coefficient at the layer boundary equal to 0.33.

In these plots, the velocity log is first displayed, showing the different layer thicknesses and the velocities just mentioned. The first column then shows reflection coefficients, positive for downgoing waves to the right. In this display, all vertical logs, except the velocity log, are given their time amplitude relative to one another, although for each layer the maximum excursion is limited to 1 in. Thus we can use the first reflection coefficient trace for scaling purposes. Some features to be noted are:

1. All beds up to the 60-ft bed are thin beds as far as both of the pulses are concerned. The amplitudes continue to grow with bed thickness (although not linearly).
2. For all these thin beds the higher-frequency pulse output is larger than the lower-frequency output, but the shape does not change. Here we have the first difference between the seismic and the light case. For thin beds, amplitude is a measure, in some sense, of the reflecting layer thickness. Resolution, in the light case, does not change with intensity.
3. We do not see any appreciable separation in the Rayleigh criterion sense between the two sides of the SAIL trace until the layer thickness is greater than 80 ft, but it is visible at 200 ft.
4. Between 60 and 80 ft, the higher-frequency pulse output decreases in size, whereas the lower-frequency pulse output increases.
5. The output pulses from the thin layers are very close to the original autocorrelation pulses, before integration. This is because a positive pulse, followed closely by an equal and opposite pulse, effectively differentiates the input pulse. Hence the integration has been countered by the differentiation performed by the impulse response.
6. Both pulses effectively resolve the separate edges of the 200-ft layer, producing a almost square-topped SAIL trace, although it is evident that the layer indication starts from a base very much lower than the reference line. The lack of low frequencies and zero frequency make their effect felt.
7. The impulse response shows that the predicted multiple reflections within the layer cause a slight tail, but their effect is barely noticeable.

In general, a single frequency is reflected from the lower side of a thin layer of this type with a phase change of 180°. Therefore, in order to emerge in phase with the reflection from the upper surface, a phase change of an additional 180° must occur because of the travel path. This happens when the layer has a thickness of one-quarter wavelength for the frequency and velocities concerned. We would expect the maximum amplitude reflection to occur when the layer is one-quarter of the wavelength of the average frequency in a pulse. For the 7-90I pulse, the average frequency must be near 48 Hz, and this, in a medium of 10,000 ft/sec velocity, has a wavelength of 204 ft, and the optimum reflection should be obtained with a layer about 51 ft thick. This corresponds closely to the example given, even though this too simple analysis has been given for a single arithmetic average frequency.

An output almost one-half of the maximum is, however, given by a layer only one-sixth of the optimum thickness, and it is evident that smaller reflection indications can be picked when the signal/noise ratio is good, but are these small events real?

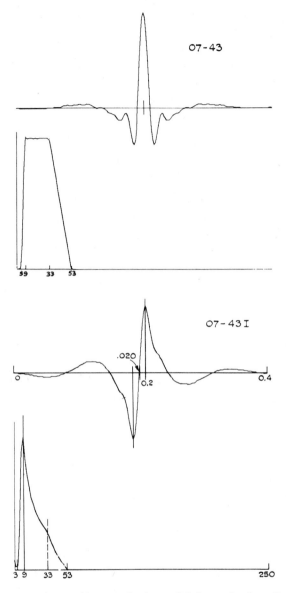

Figure 8.4 Zero-phase and integrated pulses and their associated amplitude spectra.

One other example has been presented (Figure 8.7) in which three different thickness layers are included at different depths in the section. There is no significance to be attached to the actual bed thicknesses chosen, nor to their relative depths. Although the outputs from the three layers appear very similar to those computed earlier (in value, so to speak), the impulse response trace can be used as a clue to see that, as the waves pass through each layer, the downgoing energy pulse becomes more and more complex, resulting in changed reflection outputs from the lowest layer and a series of low-frequency events of greater than expected amplitude trailing the last

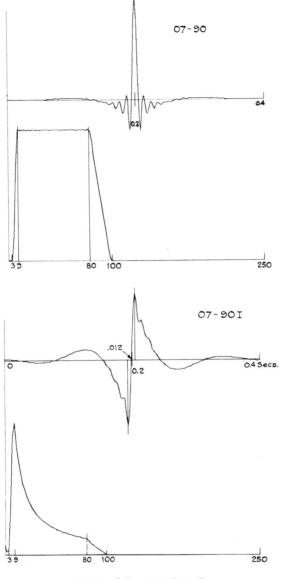

Figure 8.4   (*continued*)

layer output. These can easily be interpreted as thin layers (of alternating velocity contrasts) in their own right. The presence of only a few thin, high-contrast layers can make the SAIL indications ambiguous in meaning.

Discussions of resolution usually are based on the seismic pulse itself, and some additional distinctions are made with respect to *resolution* and *definition*, the former being the ability of an object to give a high-frequency output—which may, however, be ringy if it is derived from a spectrum that is too narrow (the light case can be regarded as an example in which only the envelope of a narrow-band frequency wave is

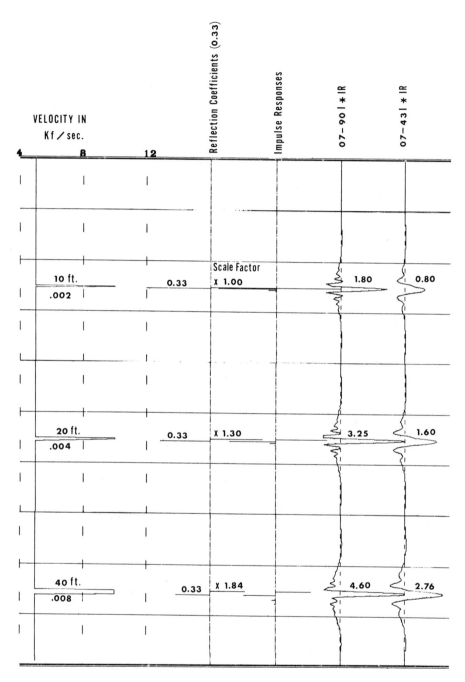

Figure 8.5 Responses of thin beds (10 to 40 ft) to two different integrated pulses.

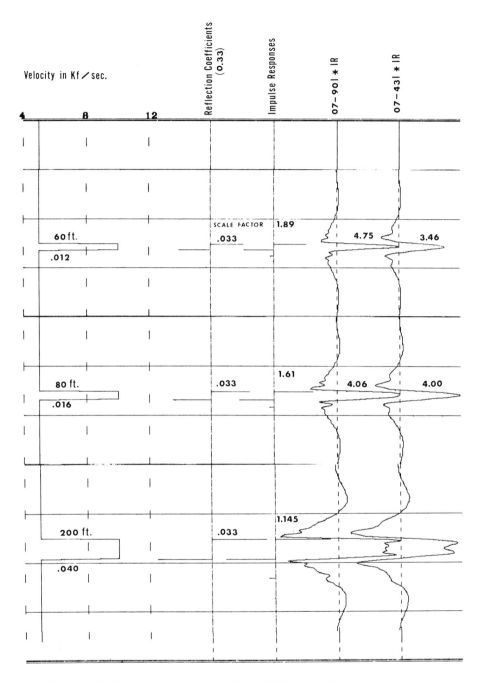

Figure 8.6 Responses of thicker beds (60 to 200 ft) to two different integrated pulses.

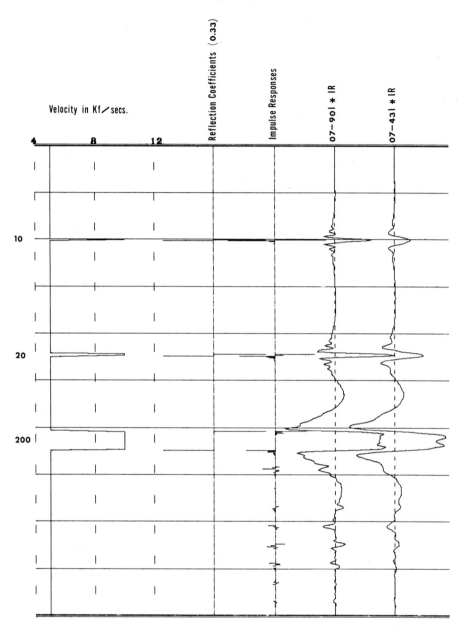

Figure 8.7 The effect of multiple reflections on the response of a succession of thin and thicker beds for two different integrated pulses.

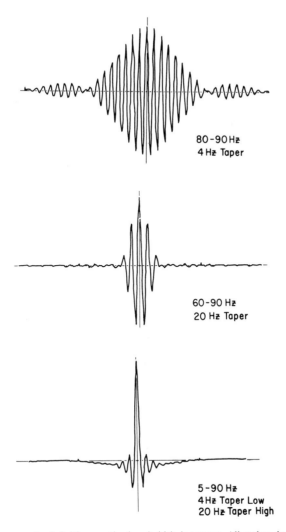

Figure 8.8 The increase in definition as the bandwidth increases. All pulses have the same upper-frequency limit.

effective for resolution; see Figure 8.8—while the latter is a measure of discreteness of the pulse itself in which the waveform is useful in defining two closely spaced objects. The discussions run somewhat parallel, but the direct explanation in terms of the integrated waveform not only involves both these definitions implicitly but also serves the direct interpretation of seismic data problem much more effectively.

So far, these discussions of resolutions have been given in the vacuum of a noise-free environment. It is obvious that changes in amplitude of reflections from thin beds—or indications of layer velocity changes—are interpretable only when random changes of the same nature due to geological or other noise are inappreciable. In Section 8.4 the influence of random noise is discussed.

8.4   VERTICAL RESOLUTION AND RANDOM NOISE

Any seismic trace taken in the real world consists of a mixture of signal and other events which consist of all other energy unrelated to the problem under discussion. On a single seismic trace, there is nothing to indicate which is which, unless some of the noise lies outside the frequency band associated with the reflections. The latter may be falsely generated impulses in the computer processing, but they are usually tracked down vigorously and eliminated. For a single source set of traces, or other genetically related system, identification of false events can often be inferred from their rate of change in arrival time at the trace group—their apparent velocity along the surface and, if these false events have no relation to the reflection process being studied, they can at least be partially eliminated by velocity f-k filtering or, in specific instances, by a method soon to be described.

However, in the case of random noise, the shape, the time of occurrence on any one trace, and the pattern of arrival times are all unpredictable. For most of the remainder of this section, the seismic trace is assumed to consist of the time reflection signal and additive noise:

$$T(\omega) = S(\omega) + N(\omega) \tag{8.6}$$

The coefficient of coherence (Foster and Guinzy, 1967) is given by

$$\gamma_{xy}(\omega) = \frac{|f_{xy}(\omega)|}{\sqrt{f_{xx}(\omega)f_{yy}(\omega)}} \tag{8.7}$$

where $x(t)$ and $y(t)$ are stationary time series with power spectra $f_{xx}(\omega)$ and $f_{yy}(\omega)$, respectively, and cross spectrum $f_{xy}(\omega)$. It must be emphasized that *estimates* of coherence are made, and great care is needed. The coefficient of coherence is related to the signal/noise ratio by the expression

$$\frac{S(\omega)}{N(\omega)} = \frac{\gamma_{xy}(\omega)}{1 - \gamma_{xy}(\omega)} \tag{8.8}$$

Thus, if the coefficient of coherence is estimated, the average noise power can be estimated in terms of the average signal. This provides a tool which we may need in predicting if an event seen on a trace is probably related to geology or whether it is noise. Discussion of the evaluation of this probability and the proper techniques for estimating the coherence are outside the scope of this chapter and should be sought in Foster and Guinzy (1967).

One problem, which occurs very often in all discussions of seismic data, is the proper correction for near-surface corrections. This problem is aggravated if higher-than-normal frequencies are used, since the accuracy of time is related to the pulse duration. Coherency measures in the frequency domain appear to have been unaffected by small time origin inconsistencies, but unfortunately the estimation of the coherence co-efficient requires the establishment of a lag window within which the spectrum is estimated, and the phases of the various frequencies are related to the time origin of the relevant window.

Corrected seismic traces to be composited should have the same geology-related reflections occurring at the same time. They can be added together after being scaled in amplitude by some preselected weighting function. At most, the resultant composited trace has a signal/noise ratio that has been improved by the square root of the number of traces added together. Generally, the weighting constants are equal to unity, but in some cases of high noise level the traces can be weighted inversely as their power. When the noise level is this high, however, it appears inadvisable to pursue the objective of high resolution, that is, the attainment of very high fidelity of the predicted acoustic impedance compared with that attainable by logging.

If a well is available in an area and seismic reflection work is performed in its neighborhood, the coefficient of coherence can be obtained between the synthetic trace resulting from the well log information and the field trace(s) recorded in the vicinity. Since it is evidence of the existence on the field traces of information at a particular depth that is required, the synthetic record made should start from reflection coefficients rather than from the impulse response. The reflection coefficients can be filtered to correspond in power spectrum to the field trace, but Foster and Guinzy (1967) have warned against the dangers with some forms of filter. The minimum requirement for this form of analysis is sufficient correspondence between the synthetic and field traces that a corresponding lag window with consistent starting times can be employed. In some cases, where formation alteration may be a problem, particular attention may have to be paid to obtaining velocity information through the use of a long-offset velocity tool in the hole and correcting the density measurements for any formation alteration.

8.5 VERTICAL RESOLUTION AND DEPTH OF PENETRATION

At some depths, the random noise level eliminates the chance of obtaining adequate fidelity of the seismic acoustic impedance log compared with the real log. The level of random noise is controlled by such factors as wind noise acting either directly on the geophones or indirectly through shaking the ground, scattering of horizontally traveling waves by random inhomogeneities, and so on. The transmission characteristics of the water layer, for example, are excellent, and irregularities of bottom scatterers within the water (fish, gas bubbles, or man-made artifacts) combine to give incoherent scattered energy which persists up to high frequencies.

However, it has been shown that the penetration of high frequencies directly into the earth and the return of high-frequency energy from the target layers is limited by:

1. Attenuation through solid friction.
2. Loss of energy or change in frequency by reverberations and transmission losses due to many layer boundaries.

To a first approximation, the loss of amplitude follows an exponential law:

$$A(z) = A_0 \varepsilon^{-\alpha z}$$

where α = the attenuation constant combining both effects and can be written as
$\pi f / Q c$

f = frequency

$1/Q$ = specific dissipation constant (effective for both effects)

c = velocity

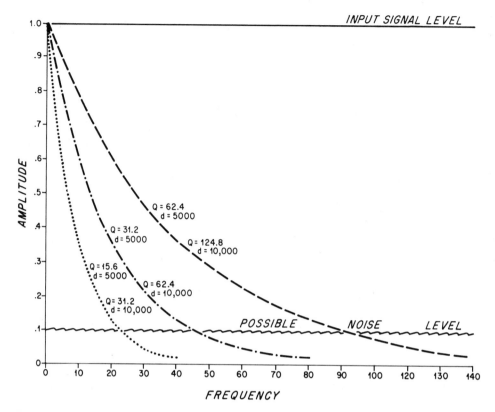

Figure 8.9 The decrease in amplitude due either to Q or to d variation. An increase in Q by a factor of 2 gives the same drop in attenuation as a decrease in distance d traveled by a factor of 2.

Then the loss, in decibels, in traveling a distance d through the earth is

$$\text{Decibel loss} = 8.686 \frac{\pi f d}{Qc}$$

the graph of which is a straight line. Note that the loss is proportional to both d and f. Figure 8.9 illustrates diagrammatically how either of these factors changes the spectrum so that the high-frequency effective output drops below the noise level as the frequency or depth is increased.

As an example, suppose that we wish to propagate 200 Hz to the layers just below Pierre shale, which has a thickness of 5000 ft, a velocity of 7800 ft/sec, and an effective Q of 20. Suppose also that an effective signal/noise ratio of 10 has been established for a frequency of 30 Hz; we would like to know how much higher the signal input must be in order to achieve the same signal/noise ratio at 200 Hz.

The decibel loss at 30 Hz from some reference level is

$$\frac{8.686\pi \times 30 \times 10,000}{20 \times 7800} = 52.48 \text{ dB}$$

The decibel loss at 200 Hz from the same reference level is

$$\frac{8.686\pi \times 200 \times 10,000}{20 \times 7800} = 349.84\,dB$$

A required gain in signal of 297.36 dB, or about 10^{15} in amplitude, would be necessary to accomplish our objective—an impossibility.

Since this section is not meant to be negative toward the accomplishment of high-frequency penetration, it is pointed out that, in a section having an effective Q of 100, the difference in input for 100 Hz and 30 Hz is only about 24.5 dB, an increase in amplitude of only 16.8, an achievable result.

Experiments with shear waves have illustrated how drastic this exponential decay lay can be, and caution is required that an objective is within the realms of possibility before attempting very costly experimentation. At best, the results may be attainable only in some types of lithological sections.

8.6 VERTICAL RESOLUTION AND THE CHARACTERISTICS OF NEAR-SURFACE VELOCITY LAYERING

It was previously stated that near-surface layering, because of the reverberations produced and the effect of these reverberations on the character of the downgoing pulse, can cause loss of high frequencies. This loss is further accentuated by transmission of the reflection pulse upward.

This situation is illustrated by three model studies (Figures 8.10 through 8.12), in which the layering near the surface was chosen to be alternating in characteristics of velocity and the layer thicknesses were chosen randomly between 0 and 100 ft by a suitable computer program. The only difference in input to these three examples is the addition of 10 layers at a time—from 10 to 20 to 30. Reflection coefficients and impulse responses were computed using a zero reflection coefficient at the surface, to avoid extra complications. It can be seen that the additional layers all tend to make the impulse response for the deepest boundary deteriorate as far as high-frequency character is concerned. The impulse response is chiefly a single spike for the 10 layers, although followed by some small consecutive positive values, but the single spike continually diminishes until, at 30 layers, the succession of small consecutive positive spikes becomes dominant.

This impulse response change is reflected in the synthetic reflection records. Whereas for the 10-layer case the 07-90I filter gives a sharper indication of velocity increase at a time of 0.53 sec, by the time the 30-layer case is considered there is no more resolution than with the 07-43I filter. Although this case may seem extreme, it must be remembered that the usual shallow section has many more layers (to represent it properly) than the 30 chosen here, although the reflection coefficients involved may not always be as large.

The synthetic technique of inserting a large, isolated velocity boundary below part of a log is a very useful one for studying attenuation due to just reverberations.

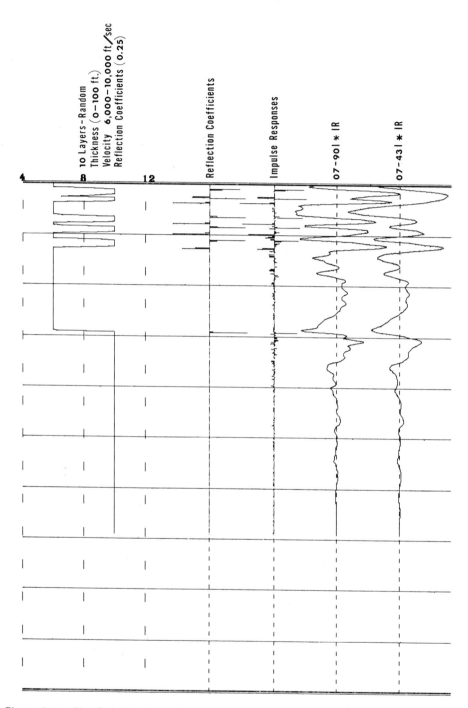

Figure 8.10 The effect of 10 thin, shallow beds of high velocity contrast on the ouptut of a single deeper velocity increase.

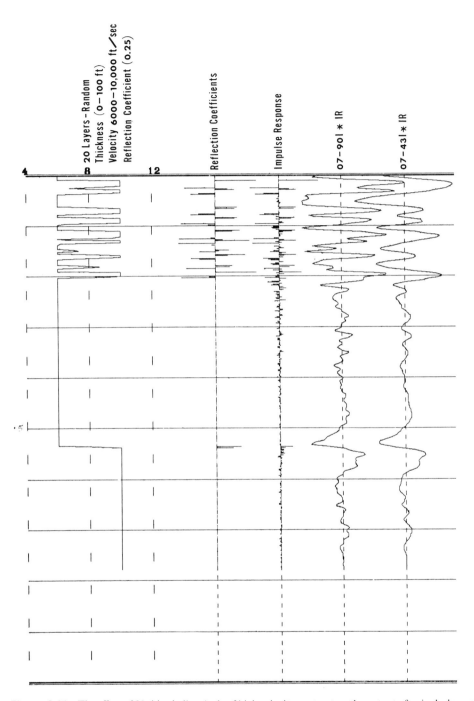

Figure 8.11 The effect of 20 thin shallow beds of high velocity contrast on the output of a single deeper velocity increase.

237

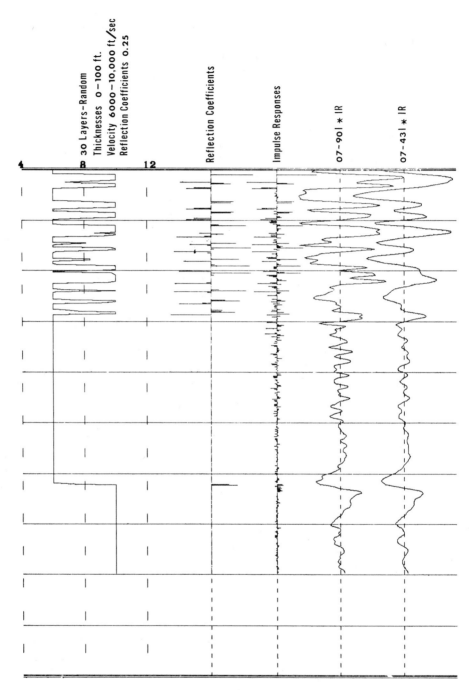

Figure 8.12 The effect of 30 thin shallow beds of high velocity contrast on the output of a single deeper velocity increase.

8.7   SCATTERING AND DIFFRACTIONS FROM REFLECTING DISCONTINUITIES

The task of detecting lateral changes in reflection qualities and placing them in correct relation to one another is another form of resolution. For example, for an unspecified reason, one portion of a reflector may possess a sharp boundary at which the reflection coefficient changes suddenly. If this is a closed boundary or if there is more than one boundary, it may be necessary to tell how large the enclosed area must be before it can be distinguished or whether a boundary of small area can be distinguished from one with a larger boundary. It is a more specific question than the general question of placing all reflectors in their proper places—which is called migration and is dealt with in a later chapter.

We shall see that a discontinuity in the reflection capability of an interface does not mean that there is a discontinuity in the seismic events returning energy to the receivers. In fact, such a discontinuity in energy returns does not occur. The general property of returning energy, or diverting energy from the incident energy beam, is called scattering. An account of scattering of an acoustic wave by a small obstacle in a homogeneous medium has been given by Oliver (1958). The small obstacle gives rise to two different types of scattering. The first, called a simple source, is due to compression and expansion of the obstacle, and this gives rise to radiation uniformly in all directions. The second form of source is a double source and is due to the density difference between the scatterer and the material in which it is embedded. Such a scatterer is then accelerated along a direction perpendicular to the incident wave front, and this gives rise to directional scattering.

The amplitude of the secondary wave is proportional to the incident field, the difference in bulk moduli, the difference in densities, the volume of the scatterer, and the *square* of the frequency of the incident wave. Such frequency dependence is often referred to as Rayleigh scattering. The secondary wave is spherically divergent away from the scatterer. Its amplitude falls off inversely with distance. For the double source, there is an angular dependence of $\cos \theta$, where θ is the angle between the incident beam and the scattered direction.

For more complex problems of scattering, resort often has to be made to numerical calculation. The principles that control such calculations are now given. If a plane wave impinges on a rigid obstacle, each point of the surface, by Huyghens principle, will act as an emitter of seismic waves. As shown in Figure 8.13a, energy is received even though the receiving and source points are not in a relation that is proper for a reflection. This scattered energy, which is organized by the geometry of the scatterer, is called a diffracted wave. It appears that some of the energy comes from the edge itself. A small element of length δl is irradiated, and it is noted that there are pairs of point sources on either side of the center that have equal travel times from the source to the receiver—as long as they are in position for a reflection. If, however, the directions are such $(S'OR')$ that Snell's law is not obeyed, there is a time difference between rays irradiated and emitting at one end and at the other end (see Figure 8.13b). The rate of change in the time delay (along the direction of the obstacle) is then $(\sin \phi - \sin \theta)/c$. If the obstacle is of a more complex form, the problem is to divide it up into a series of small elements whose individual contributions at the receiver are known and then to sum up all the contributions, taking into account the phase due to variation in distance.

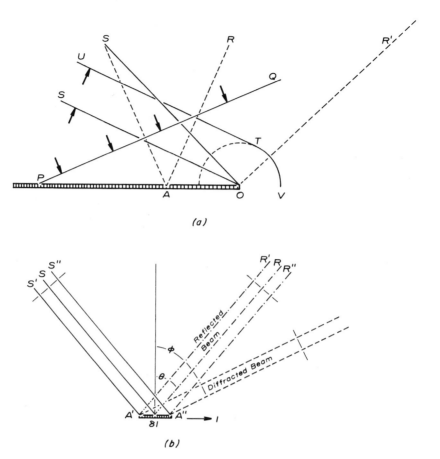

(a)

(b)

Figure 8.13 (*a*) The incident wave front *PQ* on the reflector *AO* is reflected as a plane wave taking positions *SO* and, later, *UT*, but there is a circular portion of the wavefront *TV* with a center at *O*, the termination of the reflector. Energy reaching *R* appears to come from *O*, but in fact comes from elemental contributions from all positions on the reflector. (*b*) Illumination of a small strip (*δl*) and reflected and diffracted beams. The delay per unit length of diffractor per unit distance in the *l* direction is $dt/dl = (\sin \phi - \sin \theta)/c$.

One way of looking at this problem is that it is analogous to a two-dimensional pattern of radiators, each of which is illuminated by a different amount (both in amplitude and time).

The summing procedure often cannot be done formally but has to be done numerically, and several artifices have been introduced by various workers in this field to aid in this task. One of these, used by Hilterman (1970) and Trorey (1970), is to make the source and the receiver occupy the same point. This has the advantage of making the calculations simpler, but at the expense of solving unreal problems, unless the source receiver distance is small, which is seldom the case in the CDP type of reflection profiling. Thus their results, while probably giving correct general results, are not accurate for the general CDP case. Mitzner (1967) has given more accurate methods, but they are expensive in computer time.

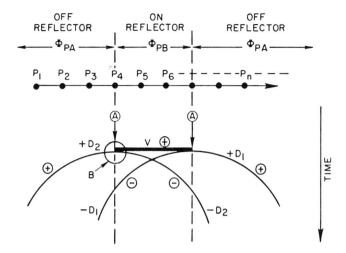

Figure 8.14 A diffraction must undergo a 180° phase change on either side of a diffracting edge. P_1 and P_2 are successive source or receiver locations. D_1 and D_2 are diffractions. The circled signs show the phases of the various parts of the response [After Trorey (1970). Reprinted with permission from *Geophysics.*]

We can, with advantage, however, examine some of Trorey's examples, since they provide general guidelines and some evidence needed later when we finally come to lateral resolution.

The first result is that the sign of a diffracted wave must change in going from one side of an edge to the other. For example, Figure 8.14 shows a strip with diffractions at each boundary. Each branch (right and left) from each edge follows the same geometrical curve in time and distance but, on the right-hand edge, the diffraction is positive to the right and negative to the left. The opposite occurs on the left-hand edge. The reflection itself is positive. It is noted that the diffraction causes the composite reflection and diffraction (they occur at the same time) to have only 50% amplitude right over the edge. An illustration of diffraction from a single edge, in the form of a seismic cross section, is given in Figure 8.15. The drop to 50% amplitude over the edge is clearly shown. Second, the rate of decay of amplitude with the distance of the source and the receiver from the edge is greater than could be accounted for by inverse spreading.

Finally, the shape of the diffraction waveform differs more and more from that of the reflection waveform as the source or receiver point moves horizontally away from the surface projection of the discontinuity. It should be noted that Trorey's solution is for the velocity potential (a scalar value) and that, to evaluate the particle velocity measured at the surface, the derivative in the vertical direction must be taken. It is evident that the vertical component of velocity diminishes at least as fast as the sine of the angle the diffracted ray makes with the horizontal and must therefore diminish even faster than in the solution given by Trorey.

We give a further illustration of how the shape of the pulse is altered when a diffraction from a half-plane is investigated. Some simplifying assumptions are made:

1. The plane is made up of a series of strips as illustrated in Figure 8.16a, each of which can be split up into elements with arrival times differing by one sample time.

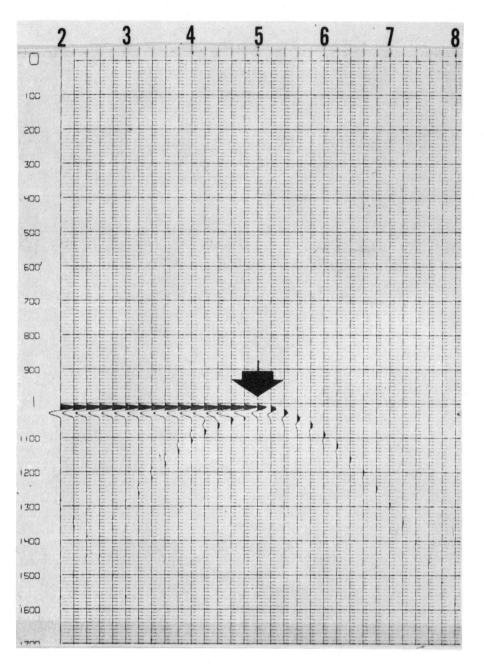

Figure 8.15 The appearance on a seismic record of the diffractions from a single edge. Since correction for spherical divergence has already been applied, the amplitude dies off more rapidly than $1/t$. [After Trorey (1970). Reprinted with permission from *Geophysics*.]

242

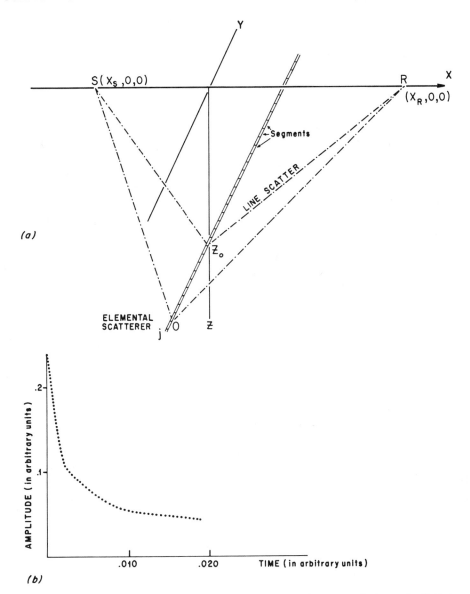

(a)

(b)

Figure 8.16 (a) The geometry associated with scattering from a line perpendicular to the X-Z plane. The line is divided into segments whose time SOR obeys the equation $T_j = T_0 + j\,\Delta t$. (b) Output pulse at R for specific source and receiver locations.

2. The strips are chosen such that their arrivals differ from the nearer strip by one sample time.
3. All elements contribute inversely as the travel distance and inversely as the distance from the source but are otherwise unidirectional in strength. (This assumption causes the pulse shape to fall off in amplitude more quickly than the rate calculated.)
4. All elements emit a delta function of velocity, and this must be multiplied by the sine of the angle of emergence to obtain the vertical amplitude of particle velocity.

The mathematical manipulation is too cumbersome to be included here, but Figure 8.16b shows the shape of the output pulse at R for a specific source-receiver pair. It is assumed here that most of the energy for the diffracted pulse comes from a region on the diffracting plane near the edge. This type of pulse shape for a delta function input can be approximated by a spectrum in which the amplitude falls off with frequency at a 3 dB/octave rate and the phase changes by $\pi/4$. It could be called the result of a "half-integration."

A reflection from the main part of the plane reflector results from convolution of the incident pulse with the delta function assumed in the calculation as the source function —in other words, a reproduction of the incident pulse. The diffraction pulse then corresponds to a convolution of the original input pulse with the wave shape shown in Figure 8.16b. Since it is comparatively richer in low frequencies, the diffracted

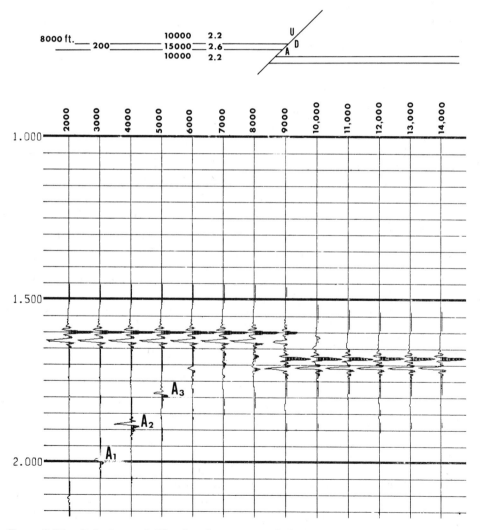

Figure 8.17a Reflections and diffractions from a reverse fault, by computer two-dimensional model. The large diffraction amplitudes A_1, A_2, and A_3 are probably due to the inclined fault face A.

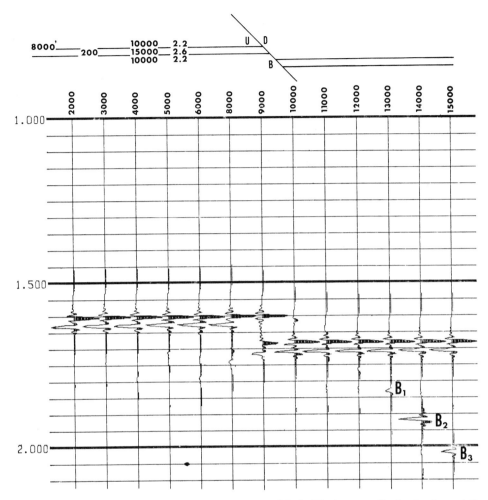

Figure 8.17b Reflections and diffractions from a normal fault. The large amplitudes B_1, B_2, and B_3 are probably due to the inclined face B.

wave gradually becomes more and more low frequency as it is found at greater and greater distances from the reflector edge.

In the earth, real faults displace the ends of the formations, in general both vertically and horizontally. With normal faults, the horizontal motion is such that a vertical line through the formation at the fault location encounters a thinner part of that formation, or none at all; for reverse faults, the opposite is true, and duplication of the section along a vertical line may occur (see Figure 8.17). The effect of the diffractions is to continue the apparent reflected energy laterally past the formation ends. Hilterman's examples show this very well. There is of course a modeling problem here in that the near-vertical wall is often modeled as just being cut in the same rigid material used for the reflector. In the calculations, as in the analog model, the elemental areas of the wall also radiate. In practice, very often the fault is displaced enough that there are different formations laterally lying adjacent to one another. The fault zone is also often fractured and becomes an anomalous zone in porosity (density) and elastic constants.

8.8   LATERAL RESOLUTION

There is an ever-increasing demand for additional fine detail from the seismic re-flection method. Whereas formerly it was used as a reconnaissance tool and as an adjunct to geology in specifying wildcat well locations, it has now become a prime tool for providing evidence for exploitation (i.e., development) wells. The exact lo-cation of faults and the estimation of their hade and throw are now requested. More-over, the seismic reflection method is finding increasing use in other technologies. For example, there are requirements for the location of old, water-filled mines and for the location of fire fronts in the new technology of underground gasification of coal—where seams are too thin for economical mining. While the latter is an esoteric venture, it can provide an example of the new lateral resolution requirement. A coal seam is set on fire along the line of a hole directionally drilled so as to traverse the seam horizontally. Oxygen or air is pumped in through the hole, and the combustion products are obtained by percolation through cracks in the coal seam to neighboring, parallel holes where they are pumped away for use. Figure 8.18a shows the situation diagrammatically. Now, to avoid drilling additional monitoring holes from the surface to the coal seam, a requirement is made that the position of the fire front be monitored (making use of the additional reflectivity of the seam after burning) by seismic stations permanently mounted on the surface. For ease of consideration, surface topography is at first, disregarded here.

It can be seen that, to a first approximation, the burned coal area approximates a narrow strip—more or less parallel to the original oxygen intake hole. The questions to be asked are:

1. At any time where is the boundary of the burned-out zone? Within what accuracy can it be specified?
2. What is the minimum boundary change that can be detected? Can deviations from the thin strip be readily located so as to prevent (or have knowledge of) fire break-through into the collector holes?
3. What are the seismic characteristics needed to make this system feasible in terms of delineation accuracy, seismic frequency bandwidth, seismic surface arrays, and so on?

Evidently, the knowledge required is of the seismic field at the surface for a given source and a given reflector distribution within the coal seam—as well as the rate of change in the seismic field when the burned area changes in size.

The steps involved are, essentially:

1. For a particular source-receiver combination, zones of equal time are drawn on the subsurface. In the case of a separated source and receiver and a constant-velocity overburden, these are ellipses.
2. The points of a grid that fall within each contour zone are evaluated for response at a receiver by use of the general diffraction formulas as given, for acoustic media, by Mitzner (1967).
3. Step 2 is repeated for each contour zone, and the results added—with the appro-priate sample time delay. In this way, the time response at a particular receiver is obtained.

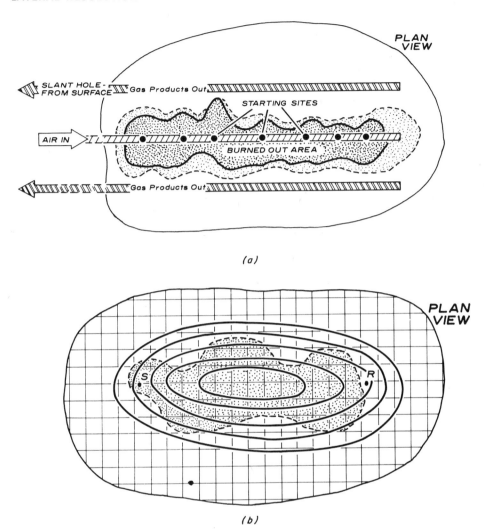

Figure 8.18 (*a*) A present burned-out area (dark boundary) and the new boundary after additional burning. This moving boundary is to be continually determined. (*b*) Contoured subsurface area showing zones of equal *S*-to-*R* time. The hypothetical burned zone is superimposed. Evaluation of the response is done at the grid corners.

4. Steps 1 through 3 are repeated for each source-receiver combination required.
5. The impulse response at each receiver is convolved with the source pulse equivalent. (This may be an autocorrelation pulse in the case of Vibroseis®.)
6. The entire process is redone with a new, or revised, burned-out zone to find the sensitivity or resolution. Note that this process is additive so that, for a given source-receiver combination, previous results can be added to, to avoid redundancy in computer calculations.

As a preliminary to these expensive investigations, however, it is possible to examine the situation from the point of view of time and amplitude of events received from an infinite strip on a line traversing the strip at right angles. Trorey's example in Figure 8.14 sets the stage. For any general trace near the middle of the cross section, there are three events, the reflection and two diffractions of opposite sign and different amplitudes and times of arrival. The lateral resolution of strip width then turns into a problem of vertical resolution of the pulses coming from three different processes. From this point of view, the amplitude spectrum bandwidth and the phase spectrum are important. The depth is also important, since it controls the curvature of the diffracted events, hence the differences in arrival times of the events on any given seismic trace. Details of these events near the strip are shown in Figure 8.19 for a particular strip width and depth. The time scale is controlled by the seismic velocity above the strip. This type of analysis is comparatively rapid and inexpensive, but provides no answer to more refined questions which may be asked, such as those dealing with local fast-burning locations.

Three different seismic impulse responses that could arise from this model are shown in Figure 8.20. Note that, compared with sections 2 and 3, at section 1 the

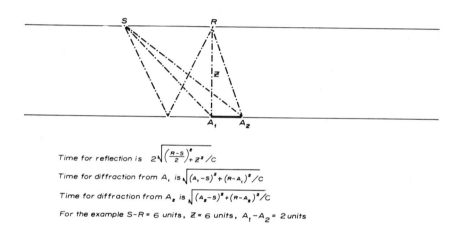

Time for reflection is $2\sqrt{\left(\frac{R-S}{2}\right)^2 + z^2}/c$

Time for diffraction from A_1 is $\sqrt{(A_1-S)^2 + (R-A_1)^2}/c$

Time for diffraction from A_2 is $\sqrt{(A_2-S)^2 + (R-A_2)^2}/c$

For the example $S-R = 6$ units, $z = 6$ units, $A_1-A_2 = 2$ units

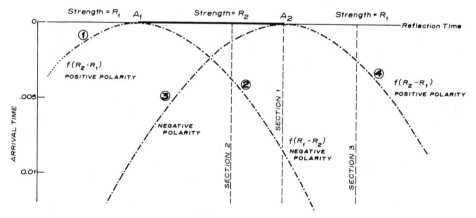

Figure 8.19 Events and times of arrival of reflections and diffractions from an infinite strip. Times are in seconds for 1 unit = 100 ft and velocity $c = 10,000$ ft/sec.

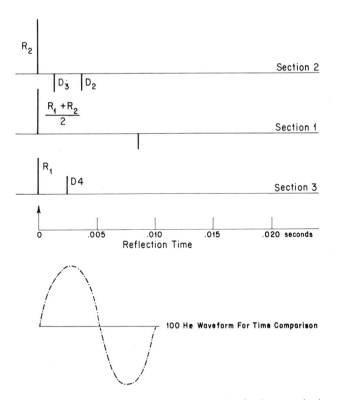

Figure 8.20   The events in approximate amplitude and relative time for the example given in Figure 8.19.

arrival at the reflection time is the mean value. The later diffraction arrivals are plotted approximately to true amplitude, and it is evident that, for low-frequency pulses, their effect is algebraically added to the reflection, while under sufficiently wide-band conditions they can be separated. At section distances up to 200 ft from the edge of the strip, the positive events require time separation of peaks only 0.01 sec apart. Since this corresponds to the first zero of a $(\sin 2\pi ft)/2\pi ft$ pulse, the upper frequency of the pulse has to be 100 Hz.

Now this strip , in comparison with its depth, is quite wide. The amount of energy contributed by the strip is controlled by both the proportion of the energy given out by the source and intercepted by the strip (this is true of all obstacles) and the amount of total energy emitted by the strip, which is intercepted by the receiver. Geometrically, the output is dependent on the solid angle subtended by the obstacle at both the source and the receiver. Thus, as the strip diminishes in width, the ratio of anomalous energy to normal reflected energy received diminishes rapidly (as the width of the infinite strip squared or the area of an obstacle squared in the general case). This is consistent with the Rayleigh scattering principles.

The resolution needed to see all the diffracted events separately from the reflection also increases as the square of the ratio $Z/(A_1 - A_2)$.

It is hoped that the application of principles in this example will allow other cases of particular interest to the reader to be worked out.

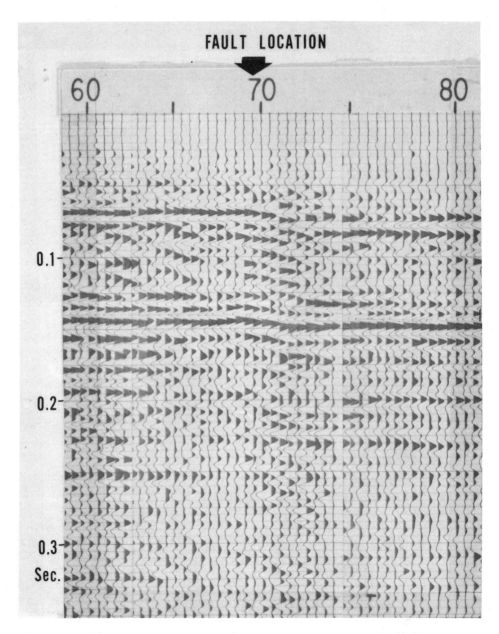

Figure 8.21 Fault evidence on a coal seam reflection cross section (0.15 sec arrival time). The throw is known to be about 8 ft, and the time shift of 0.0035 sec appears to be due to velocity change as well as depth change.

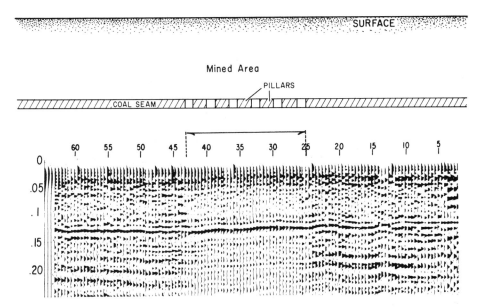

Figure 8.22 High-frequency (40 to 170 Hz) seismic survey over a recently mined area of a coal mine. Mine depth is 550 ft, coal seam thickness is 7 ft, surface shot points are 40 ft apart, and the mined area is air-filled. The section was processed for equal total power in all traces, hence the large reflection over the mined area gives rise to diminished background. (Courtesy of the Continental Oil Company.)

Experimental evidence has been obtained over a shallow coal seam (200 m deep), which suggests that faulting (Figure 8.21) can be located to an accuracy of a few tens of meters and that a mined area can be distinguished from an unmined area within about 20 m from amplitude alone (Figure 8.22). However, the individual mined-out passages cannot be differentiated from the coal pillars left to support the roof, even though these pillars have dimensions of 20 to 30 m. The experimental parameters are:

Trace spacing (subsurface 6.5 m)
Frequency spectrum, 40 to 170 Hz
Fivefold stack
Source pattern, 15 positions over 20 m
Receiver pattern, 20 geophones over 20 m

Diffraction problems, except for those involving the most simple geometries, are difficult enough in the acoustic case without bringing in the added complications of an elastic medium in which two types of waves are possible and in which transfer of energy between the two types occurs at boundaries. Knopoff (1959) has dealt with the case of incident plane P or SH waves impinging on a spherical obstacle of a different material. For problems, such as the one discussed, which essentially involve plane layering with discontinuities, the use of SH waves eliminates any problems connected with energy conversion so they can be treated by the existing scalar wave diffraction theory.

8.9 SUMMARY AND CONCLUSIONS

In seismic reflection work, the distances encountered are always of the same order of magnitude as the wavelength. We observe the whole composite waveform in the case of closely spaced events and, since we are dealing with a linear system, these composite waveforms are simply additions of the individual event waveforms. Thus, while in the optical case, resolution is controlled by the envelope of the intensity curve, in seismic reflection work we have to deal with the composite waveform itself.

Resolution has been dealt with in a form that requires the recovery of an approximate acoustic log from the seismic trace, and it is the (local) faithfulness with which the actual acoustic impedance log can be recovered that constitutes vertical resolution. We need to be able, a priori, to examine a seismic cross section and to determine from it places where oil or gas accumulation is most probable.

The difficulties, some of which may be overcome in time, in achieving this fidelity in reproducing the acoustic impedance log are:

1. The presence of multiple reflections.
2. The difficulty of recovering true amplitude because of attenuation and reverberation in thin-layered, contrasting sections.
3. The transfer function of the near-surface regions on the wave shapes (at both the source and the receiver).
4. The need for extreme bandwidth. This is coupled with the difficulty of introducing low frequencies and the attenuation with depth of high frequencies.
5. The presence of random noise and interference.

Scattering and diffractions from discontinuities constitute the means whereby the discontinuities can be located. The acoustic diffraction theory is reasonably well established but is complex, and only very simple, not very meaningful examples, as far as seismic work is concerned, have been formally solved. However, computer methods can be used to solve specific lateral resolution problems. For problems investigating the size or shape of small scatterers, the time resolution needed (and the high frequency of a very wide bandwidth signal) increases as the square of the ratio of depth to a scatterer dimension. The energy return also diminishes as the square of the scatterer area (fourth power of a scatterer dimension) if shape is maintained. In many cases, this may have to be observed in the presence of a strong reflection. Otherwise, the signal/noise ratio for the whole required spectrum is important.

Field examples of perturbations in the characteristics of a coal seam have been shown. At a depth of 200 m, the edge of a mined-out area can be established to about 20 m, but individual pillars and galleries of these dimensions in the mined-out area cannot be distinguished at an input frequency of 40 to 170 Hz.

While the acoustic diffraction theory is in reasonably good, but complex, condition, there have been few attempts to deal with elastic wave diffraction except under simple, nonrealistic conditions. The use of shear waves, which can be propagated to the distances necessary, should make comparison of acoustic diffraction theory with experiment justifiable for cases involving discontinuities in near-planar reflectors.

REFERENCES

Born, M., and Wolf, E. (1959), *Principles of Optics*, Pergamon, New York.

Foster, M. R., and Guinzy, N. J. (1967), "The Coefficient of Coherence: Its Estimation and Use in Geophysical Data Processing," *Geophysics*, Vol. 32, No. 4, pp. 602–616.

Hilterman, F. J. (1970), "Three Dimensional Seismic Modeling," *Geophysics*, Vol. 35, No. 6, pp. 1020–1037.

Kirchoff, G. (1882), *Berl. Ber.*, 641.

Kirchoff, G. (1883), *Annalen der Physik (Leipzig)*, (2), Vol. 18.

Kirchoff, G. (1883), *Aes. Abt. Nachr.*, 22.

Knopoff, L. (1959a), "Scattering of Compressional Waves by Spherical Obstacles," *Geophysics*, Vol. 24, No. 1, pp. 30–39.

Knopoff, L. (1959b), "Scattering of Shear Waves by Spherical Obstacles," *Geophysics*, Vol. 24, No. 2, pp. 209–219.

Krey, T. (1952), "The Significance of Diffraction in the Investigation of Faults," *Geophysics*, Vol. 17, No. 4, pp. 843–858.

Miles, J. W. (1960), "Scattering of Elastic Waves by Small Inhomogeneities (Shear)," *Geophysics*, Vol. 25, No. 3, pp. 642–648.

Mitzner, K. M. (1967), "Numerical Solution for Transient Scattering from a Hard Surface of Arbitrary Shape," *Journal of the Acoustical Society of America*, Vol. 42, pp. 391–397.

Officer, C. B. (1958), *Introduction to the Theory of Sound Transmission*, McGraw-Hill, New York.

Schwab, F. (1965), "Scattering of Shear Waves by Small Transseismic Obstacles," *Geophysics*, Vol. 30, No. 1, pp. 24–31.

Trorey, A. W. (1970), "A Simple Theory for Seismic Diffractions," *Geophysics*, Vol. 35, No. 5, pp. 762–784.

NINE

Migration—The Correct Placement
of Reflectors and Diffractors

9.1 INTRODUCTION

The fiction of plane, near-horizontal layering has served a useful purpose, but we must now return to reality and take nature as it is found—for this is the condition under which the real seismic reflection method works. Three important elements of structural geology have been ignored:

1. The layers may not be horizontal.
2. Discontinuities exist which give rise to diffracted events.
3. The curvatures of the formations are not negligible.

Nevertheless, all processing described to this point has assumed that these effects are absent and that the seismic reflection record has a one-to-one correspondence with a depth display. It is the purpose of this chapter to trace the development of methods of placing the various events on a seismic time section in their proper place on a seismic depth section. Little time is spent on the older methods, in which the reflections were first picked and then dip bars—lines on a new cross section—were computed and plotted. Only passing reference is made to the aplanatic method, although it is really the progenitor of more modern methods. Some attention is paid to the unsolved problems arising because, with the lithology becoming (at least) two-dimensional, the velocities to each layer are no longer constant. This is a problem, even now.

The advent of more modern methods was triggered by the arrival of the fast digital computer as well as laser plotters and similar devices which can plot two-dimensional digital data of great informational content in a reasonable amount of time.

9.2 EARLY MIGRATION METHODS

Figure 9.1a shows a hypothetical set of seismic traces in which, after correction for near-surface delays and NMO, the reflections constantly increase in time from one trace to its successor. Because of the assumption that NMO has been removed from these seismic traces, it can be taken that these represent traces picked up at the point where the energy was generated. Under the assumption of constant velocity before

254

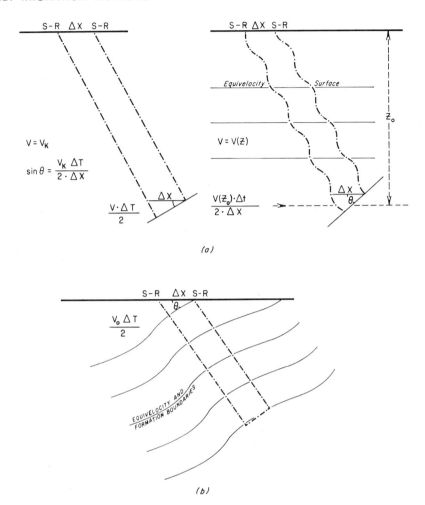

Figure 9.1 (*a*) Examples of angle of dip determination when the equal-velocity surfaces are horizontal or if a constant-velocity medium exists. (*b*) Equal-velocity and reflector surfaces parallel.

the reflector is reached, the wave fronts are circles and the rays are straight lines perpendicular to the reflector. Even under a lesser restriction of a velocity that is a function of depth only, the reflected and incident rays must coincide and be perpendicular to the reflector if the source-receiver coincidence is retained. Then, it is easy to see that the differential time between traces, for these assumptions, is given by

$$\Delta t = \frac{2 \sin \theta \, \Delta x}{V}$$

or   (9.1)

$$\sin \theta = \frac{V \, \Delta t}{2 \, \Delta x}$$

The velocity V used here must be the interval velocity at the depth z of the reflector. By Snell's law, the angle of emergence is $\sin \theta_0 = (V_0/V) \sin \theta$. By making V a function of z only, the assumption is that the formation type does not control the velocity at all and that it is strictly controlled by depth of burial. This is of course nonsense. It would be very unusual geology if all of the formations down to a particular level were horizontal and had the same velocity for a given depth; then this was followed by a gross divergence from these assumptions.

Another possible assumption is that all the formation velocities are constant and independent of depth (Figure 9.1b). In this case, the dip of a particular reflector would be given by

$$\sin \theta = \frac{V_0 \Delta t}{2 \, \Delta x} \tag{9.2}$$

where $V_0 = $ near-surface velocity.

This of course is equally ridiculous. The truth evidently lies somewhere in between (Waters, 1941; Faust, 1951). The equal-velocity surfaces are a function of both x and z and tend to follow the structure to some extent but are not parallel to it.

We are thus in the unenviable position that we know that the dipping reflections shown on the time section do not come directly from below the shot point but have to be moved in a direction updip but, at the same time, we do not know what velocity function should be used to calculate this movement. Since the average velocity is usually intermediate between the surface velocity and the interval velocity, most early forms of migration used the average velocity and many newer forms still do. It is of course possible to grade velocities between CDP or well velocity determinations. Some care is necessary, however, because (unlike the removal of NMO) a change in migration velocity can give rise to structural change, and errors of velocity correspond to errors of structure.

It is wise to keep these uncertainties in mind in the remainder of the discussions in this chapter. Methods have changed, but the problems have not.

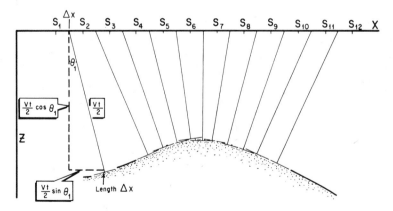

Figure 9.2 A dip bar migrated section.

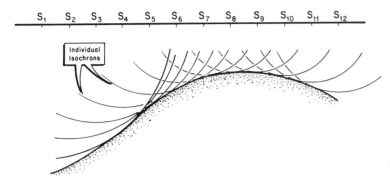

Figure 9.3 Development of an aplanatic surface—and thus the reflecting surface—from individual isochron surfaces.

After choosing a velocity depth function (or selected one from a graded set related to velocity determinations), the first method was to pick the records, compute the angle of dip, swing the center of the reflection sequence updip by the calculated amount, and draw a dip bar of the appropriate length on the cross section (see Figure 9.2). This was repeated for all reflections that could be picked, often grading the reflections for continuity and indicating this by the quality of the line drawn on the cross section. If the cross section was not to be made with equal-distance scales on the X and Z axes, the proper modifications can be made in the angle of dip.

The second method, sometimes adopted, was called the aplanatic surface method. This again was a manual construction. If from any surface pair of source-receiver reflection times a set of constant time lines is drawn (Figure 9.3), these will represent all possible positions of a reflecting surface that will receive energy from the source and return it to the receiver at a particular time. The form of these aplanatic surfaces must be consistent with the velocity function assumed to apply. For example, if the source and the receiver are coincident, either a linear increase in velocity with depth or a constant velocity will give rise to a series of concentric spheres (which are cut by the plane of observation to give circles). The height of the center above the source depends on the rate of increase in velocity with depth. If the source and the receiver are separated, the isochron (equal-time) lines will be ellipses. When a reflection can be identified and correlated from trace to trace, it is possible to draw these isochrons, which correspond to the reflection, at each pair of source-receiver points. In Figure 9.3, they are shown as circles with centers on the reference line at the source-receiver position. The aplanatic surface is the common tangent to all the isochron lines, as shown, and this is the reflecting surface. This construction of course makes use of the stationary time principle of Fermat—in this case, minimum time paths. As a manual construction, this process is time-consuming but, except where there are strongly concave reflecting surfaces, produces a continuous reflecting curved surface. It is now chiefly of interest as the progenitor of the wave front migration method.

9.3   THE WAVE FRONT MIGRATION METHOD

A seismic trace is digitized so that every sample time is represented by an amplitude of the measured parameter (usually particle velocity or excess pressure). Now, if we had the time, we could draw isochron lines on the cross section for any one of these

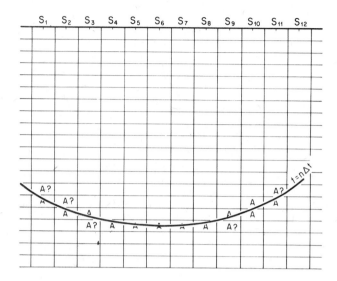

Figure 9.4 One isochron line passing through a gridded system. A weighted portion, dependent on the vertical angle, of the amplitude of the digital sample is assigned to each cell through which the isochron passes. For more than one isochron passing through a cell (or another time sample or another source position) the amplitudes assigned are added.

sample points and for all the seismic traces and then draw the common tangent lines. The following method, which employs a digital computer, results in a migrated seismic cross section.

In Figure 9.4, a single isochron line is shown, which crosses over a rectangular grid formed by vertical lines midway between the individual traces and by the points midway between the sample points. Other definitions could have been used. As indicated, the cells through which each isochron passes are assigned (on some weighting basis) the amplitude of the seismic trace sample. For more isochrons the amplitudes within the cells are added algebraically.

Additional criteria may be added, if desired. For example, not all parts of the isochron line may carry equal weight. It may be decided that dips greater than N_0 milliseconds per trace should not be admitted, in which case, the isochron line is discontinued when its slope reaches that dip limit. Another possibility is assigning a probability of a dip occurring as proportional to $\cos \theta$ (or some more complex function of vertical angle) and weighting the amplitude to be assigned to the cells accordingly.

Finally, after all the isochron sample lines have been used in this manner, the accumulated amplitudes in the vertical columns of cells are played out as new seismic traces in which the events are migrated. Essentially, this has been done by the aplanatic surface method. The justification for the name "wave front method" lies in the fact that the aplanatic surfaces and the wave fronts (for half the reflection times) are coincident if the sources and receivers are coincident, and this is virtually true if all traces have been properly corrected for NMO.

Figures 9.5 through 9.7 show some of the theoretical possibilities, which have the following characteristics:

1. Constant velocity.
2. Events, such as dipping reflectors and diffractions, have been included in the un-migrated data in Figure 9.5.
3. All events have been migrated by using the full half-circle isochrons in Figure 9.6. Note that the hyperbolic diffraction events have been condensed to a minimum area consistent with the band-pass characteristics of the seismic pulse. Amplitudes (except for the diffractions) have been roughly conserved.
4. All events, except the one having the steepest dip, have been migrated in Figure 9.7, in which the aperture for dip migration is limited to $\pm 30°$.

Note the characteristic, when migrating a finite length of reflector, of adding an inverse diffraction at the ends of the reflector—such a characteristic is especially noticeable in practical examples when seismic noise is present (Figures 9.8 and 9.9). These examples are provided by courtesy of the Continental Oil Company.

Some modifications, in the form of editing the original sections before or during the migration process, are obviously necessary, particularly when the signal/noise ratio is not large. This editing consists of applying an algorithm which allows only sequences of dips to be picked that can be followed over several seismic traces. Thus large noise amplitudes are not migrated in the form of curved events of large amplitude. The criteria used as a basis for editing are usually controversial—as are all attempts

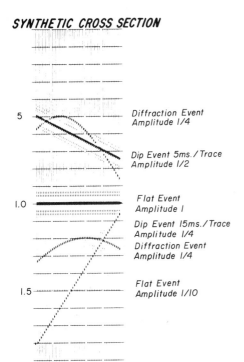

SYNTHETIC CROSS SECTION

Diffraction Event
Amplitude 1/4

Dip Event 5ms./Trace
Amplitude 1/2

Flat Event
Amplitude 1

Dip Event 15ms./Trace
Amplitude 1/4
Diffraction Event
Amplitude 1/4

Flat Event
Amplitude 1/10

Figure 9.5 Synthetic cross section showing various types of seismic arrivals.

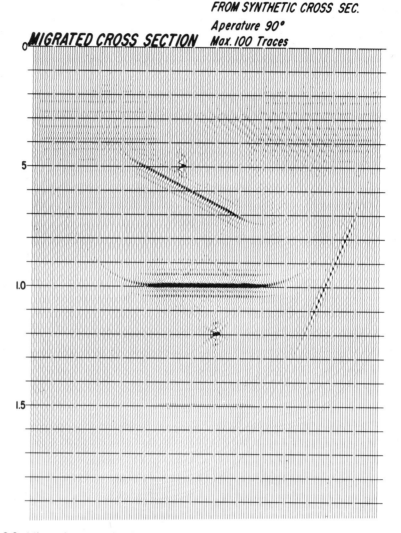

Figure 9.6 Migrated cross section by wave front method from data given in Figure 9.5. (After Heath. Courtesy of the Continental Oil Company.)

to substitute machine picking of events for picking by human interpreters (Paulson and Merdler, 1968)—but even a few simple rules, which can be generally agreed on, help considerably in reducing the overall confusion.

Synthetic results show that, apart from the inverse diffractions, the wave front method performs well, as long as the correct velocity information is used. Since energy from the postulated reflectors and diffractors is spread out all over the cross section plane, where it will tend to cancel if the number of reflectors is large and they are randomly spaced, the noise background increases. For this reason, the migration method should be attempted only with seismic cross sections with a high signal/noise

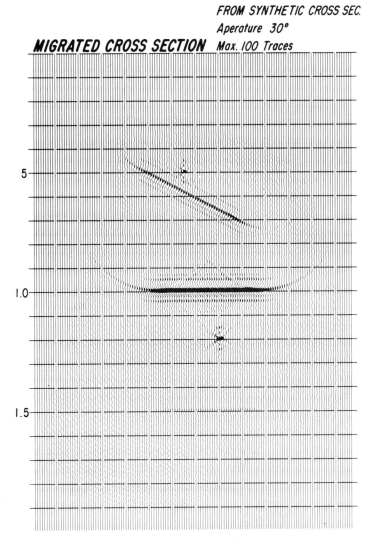

Figure 9.7 Migrated cross section by wave front method (but with limited aperture). (After Heath. Courtesy of the Continental Oil Company.)

ratio. There have been suggestions that migration, by any automatic method that produces a migrated seismic section retaining the seismic wave forms on the traces, can be used on the primary records that make up the CDP records. As a result of the multifold stacking now adopted, these primary *migrated* cross sections can then be added—to reinforce the true data and to reduce the noise level. It is not obvious that this will be the result achieved. The 100% records are much noisier than the CDP record (on the basis of random noise and N-fold stacking they should have a signal/noise ratio of $S_F \sqrt{N}$ if S_F is the final signal/noise ratio achieved after CDP stacking). The stacking of the cross sections may achieve a $\sqrt{N}$ improvement over the 100%

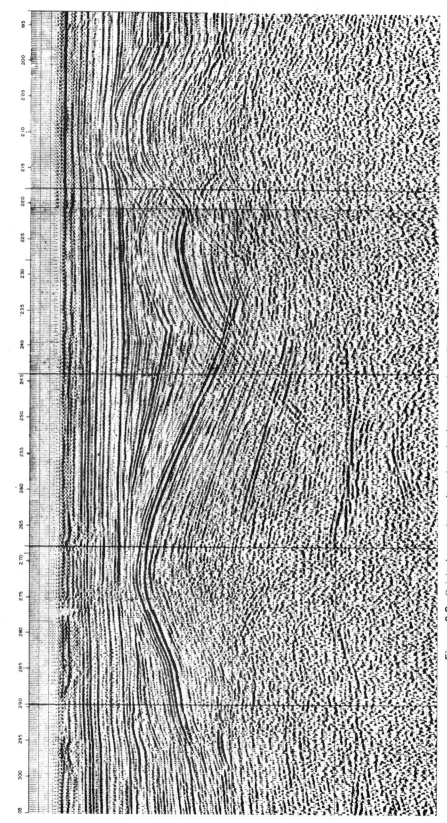

Figure 9.8 Original, nonmigrated marine seismic cross section. (Courtesy of the Continental Oil Company).

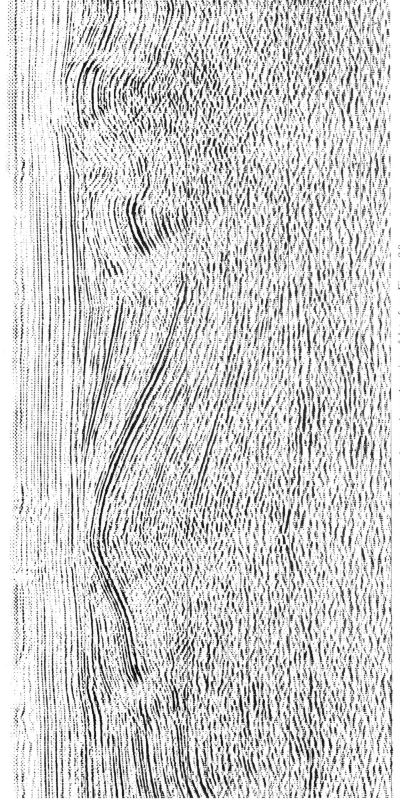

Figure 9.9 Wavefront-migrated version of data from Figure 9.8.

263

cross sections and thus get back to the situation achieved by migrating the CDP traces. Examples tried have not resulted in any appreciable improvement, and the cost is very much higher.

Wave front–migrated sections can be made if the velocity function varies smoothly laterally. That is, each trace is treated using an interpolated velocity function. The shape of the isochron lines under these conditions is not, however, symmetrical with respect to some vertical line, and caution is needed both as to the aperture angle used as well as the rate of lateral change in velocity.

9.4 MIGRATION BY DIFFRACTION STACKING (KIRCHOFF SUM METHOD)

In what appears to be an almost inverse method, the diffraction stacking method relies on the Kirchoff integral method for solution of the wave equation. In essence, each portion of the subsurface is regarded as a scatterer of acoustic waves (this simplification of the elastic wave problem is necessary). In Chapter 8, we saw that this scattering gives rise to an hyperbola-like curve when plotted on a T-versus-X graph, the exact shape depending on the variation in acoustic velocity with depth. It is therefore reasonable to try to perform the migration by summing up the energy that appears on the seismic cross section along these diffraction curves. Such summing then gives a high-amplitude event at the (T, X) position of the scatterer. Based on Huyghens' point of view that a reflector can be treated as a linearly arranged set of scatterers, each of which gives its output independently of the others, it is seen that the procedure of summing along the appropriate diffraction curve should also work for reflectors.

The two different processes can be compared using diagrams (Figure 9.10) due to Schneider (1971). The proper diffraction curves have to be calculated for each sample time in the vertical ray path, and the proper samples to select at offset traces must be determined. For a generalized velocity function, this can be done by the method outlined in Appendix 9A. Examples of diffraction curves for a simple velocity depth function are given in Figure 9A.1.

When velocity functions are measured at different points along the line of section, such velocities can be interpolated, and the diffraction curves can be calculated for migration use at each source point. In the process of stacking by this method, it is evident that for events received on traces laterally removed from the scatterer, the energy is reduced because of the spherical divergence of the wave front (in the case of true reflection). For real scatterers that are lines, the appropriate divergence rate is more nearly cylindrical, although the initial divergence from the source point to the line scatterer is spherical. It is therefore preferable to prepare true-amplitude cross sections before migration, and in fact it is realistic to weight the amplitude of each trace according to a scheme that gives preference to traces near the vertex of the curve. Weighting inversely as the arrival time may be suitable but, as in the case of wave front migration, the aperture must be limited to avoid an increase in noise in low-signal areas.

Figure 9.11 shows an excellent example of a time section and a migrated time section for some onshore data (courtesy of Prakla-Seismos, Gmbh).

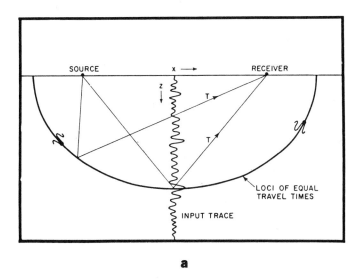

a

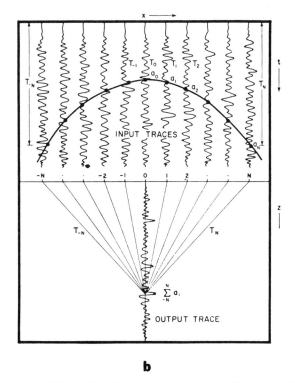

b

Figure 9.10 Illustrating the difference in philosophy between the wave front (*a*) and diffraction curve (*b*) methods of migration. In (*a*) a single trace amplitude is distributed over the proper wave front. In (*b*) many traces are sampled at the times appropriate for a single scatterer and added to give a single output trace amplitude. [After Schneider (1971). Reprinted with permission from *Geophysics*.]

ON SHORE SEISMIC DATA

Unmigrated Time Section

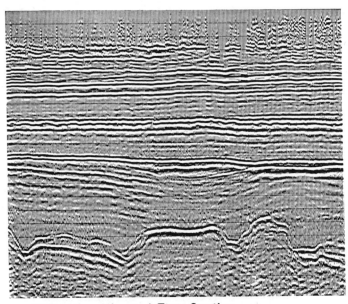

Diffraction Migrated Time Section

Figure 9.11 Original unmigrated onshore seismic data and a diffraction-migrated time section. (Courtesy of Prakla-Seismos, GmbH.)

9.5   FINITE DIFFERENCE METHODS BASED ON THE ACOUSTIC WAVE EQUATION

In Chapter 8, when considering scattering and diffraction, we saw that complex effects are often introduced into the time cross section because of the various paths in the plane of the depth section that can arrive on a particular trace. For example, a syncline which has too great a curvature gives rise to three events, all from the same reflector and, in general, at different times. This situation always occurs when the radius of curvature of the (concave) structure is less than the radius of curvature of the wave impinging on it—that is, the depth.

The situation would be much better if the surface of observation were down in the earth far enough that the radius of curvature of the subsurface was less than the depth (for a particular reflector). The best possible situation would be (from an academic standpoint) when the plane of observation is coincident with the reflector or scatterer in question. Then the distance of travel would be small enough that there would be no migration involved. It is the purpose of this section to show how this can be done by a method related to the wave equation. Most of this work was done by Jon Claerbout and his associates at Stanford University, whose aim was to produce practical methods of data processing of use to exploration seismology. Since the formal solution of the acoustic wave equation in inhomogeneous media is generally an impossibility, resort has been made to numerical methods. In particular, these are related to methods of finite differences—that is, the solution is a finite array of values with the rows representing the x position of the seismic traces and the columns being the record time t just as if the traces represented depth and had been converted to time using the known velocity function. The starting point can be a CDP seismic cross section, an array of the same kind with columns at every observation point and event amplitudes as a function of observation time. The observation time of course consists of the distance traveled by the wave divided by an appropriate velocity over the path taken, which need not be, and in general is not, vertical.

The phrase "downward continuation" has been used for many years in gravity and magnetics interpretation to denote a method whereby the observed fields at the surface can be manipulated in such a way as to generate the field as it would appear if the observation were at a lower level, thus giving more detail and resolution. It is, however, a method dealing with time-invariant potential fields and is much simpler to visualize than the present method. It amounts to a boost of high-frequency components of the signal.

We can imagine the process, however, in a different way and one that is fundamental to the downward-continuation process for acoustic waves. Imagine that each point on a subsurface inhomogeneity is a source, in the manner of Huyghens, and let these sources radiate upward but with only half the velocity normally associated with the geological section. Then, apart from considerations of the directionality of these secondary sources, the ensuing signals received at the surface are very close to those that would have been generated by a simultaneous set of sources at the observation points $(S/R)_1$, $(S/R)_2$, and so on. Thus only the upgoing wave of the solution to the wave equation needs to be dealt with—using one-half of the normal velocities.

Now if it were possible for the observer to travel upward with the wave, an even greater simplification would result, since then the wave field would change only very slowly with time. This can easily be done in the mathematical formulation of the basis

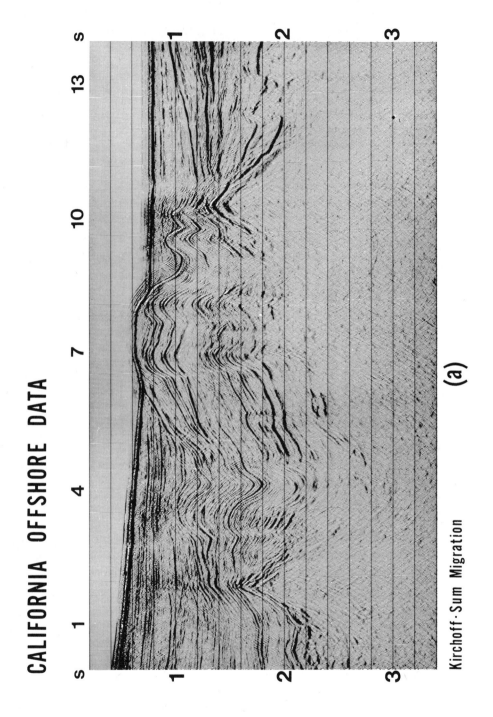

CALIFORNIA OFFSHORE DATA

Kirchoff·Sum Migration

(a)

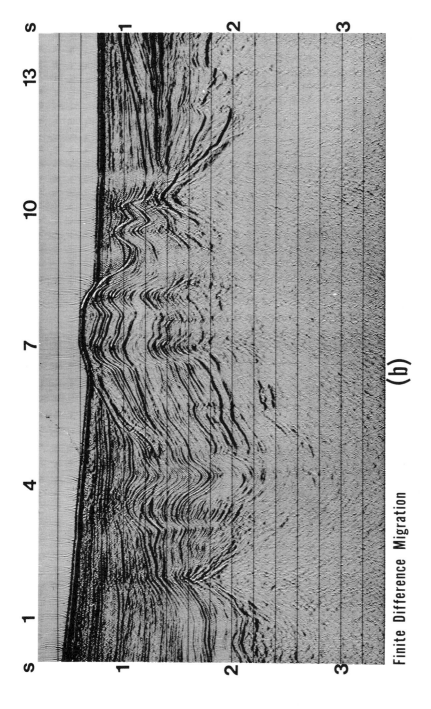

(b)

Finite Difference Migration

Figure 9.12 The similarity of results of migration by the Kirchoff sum and the finite difference methods. All dips are less than 25 . (© 1975. Western Geophysical Company of America.)

for migration by changing the coordinate axes to reference axes that move upward with the wave field.

If one takes two columns of amplitudes, it is possible to predict, using the known physical laws embodied in the wave equation, from any three elements (two in one column and another in the second column) the size of the fourth adjacent element in the second column. In order to do this, the wave equation has to be put in a suitable form—a finite difference form which relates the amplitudes of nearby elements to those already known. Although we have discussed this as though we were going to do it one cell at a time, it has been found possible for grosser approximations (e.g., all dips limited to 15°) to proceed by larger steps. The step consists of convolving the existing data with a calculated pulse or transfer function.

In the first step, transformation from record time t to migrated time τ is done down to a depth Z_1, followed by a similar step down to Z_2, and so on. These steps may be equal or unequal, but the former is usual for reasons of economy of computer effort and simplicity. The transfer function appropriate to the transformation can then be standardized for a given depth. Since this is a finite difference approximation to the wave equation, several errors will be made, which depend on the size of the steps made, the range of dips of the reflectors, the variability of the velocity function, and so on. These errors have been investigated by Claerbout (1971a), Alford, Kelly, and Boore (1974), and Loewenthal, Roberson, and Sherwood (1974).

After the application of each transfer function, the section is migrated fully down to ΣZ_i (the sum of the depths of the range of the transfer functions) but only partially migrated (transformed from t to τ) below this depth.

Claerbout and Johnson (1971) have given the (phase) transfer function across a layer of thickness ΔZ as

$$\phi(Z) = \frac{1 - [aT(1 + Z)/(1 - Z)]}{1 + [aT(1 + Z)/(1 - Z)]} \tag{9.3}$$

where $a = (C \, \Delta t \, \Delta Z / 8 (\Delta X)^2)$
Z = transform operator $\exp(-2\pi i f \, \Delta t)$

Since only shifting of seismic energy is involved in the migration process, the operator involves only the *phase* of the signal and the amplitude is left alone. $T/(\Delta X)^2$ represents some finite difference approximation to the differential operator $-(\partial^2/\partial x^2)$. This may be the simple double difference operator D corresponding to the three-point convolver $(-1, 2, -1)$, or it may be a preferred better approximation such as (Loewenthal, 1974)

$$T = \frac{D^2}{1 - D/6} = D + \frac{D^2}{6} + \frac{D^3}{2136} + \cdots \tag{9.4}$$

A full justification of the transfer function (9.3) is not within the scope of this chapter. Reference may be made to the original works of Claerbout and Doherty for more details.

Although the finite difference method is in use by several companies, the efficacy of its use compared with the diffraction stacking method is open to question. It is probably a question (as in fact are many decisions in exploration seismology) involving the combination of data from a particular area, the original signal/noise ratio, the personal preference of the interpreter, and the economics of the process. Examples given (courtesy of Western Geophysical Company) of some offshore California seismic data (Figure 9.12) show little difference except the noise produced by the Kirchoff sum method above the water bottom.

9.6 THREE-DIMENSIONAL MIGRATION AND ACOUSTIC HOLOGRAPHY

All the methods discussed so far have been two-dimensional. They can of course all be adapted for three-dimensional migration except for two problems:

1. Data are not usually available on a continuous basis for giving the extra experimental information necessary.
2. The extension of these two-dimensional programs to three-dimensional ones would increase the computer time by (at least) an order of magnitude and may increase the storage necessary to uneconomical limits.

Thus, while we may wish for three-dimensional migration, it is unlikely to be feasible except in special cases. It is, however, desirable to know how the lack of three-dimensional migration can affect the appearance of migrated data in a two-dimensional presentation. We deal with this subject first before finally touching on the subject of acoustic holography which, for special purposes, may have some use in defining three-dimensional objects using two spatial dimensions and one time dimension.

French (1974) has given some (model) examples which show the influence of out-of-plane structures on the seismic results obtained by a single line profile. Two-dimensional migration is seen not to be the answer to all these problems. Essentially, the difficulty arises from the fact that the three-dimensional reflecting surface is in a sense concave laterally—that is to say, legitimate reflections are possible from points on the subsurface that are laterally offset from the vertical plane through the line of source and receiver points. If the data are available (see, for example, the three-dimensional swath method described in Section 5.3), the common tangent to ellipsoids or spheres takes the place of that to ellipses and circles or the common surface generated by a set of near hyperboloids of revolution (about the vertical axis) takes the place of the common plane generated by a set of near hyperbolas. The principles are not different—only the difficulty of organization of the data and the time required to place the contributions of each (x, y, t) cell into the appropriate (x, y, z) cell. Three-dimensional programs are now in use by Prakla-Seismos, GmbH., and other geophysical contracting companies.

The term "holography" originated with Gabor in 1948. It was originally proposed as a two-step method of optical imagery (see Born and Wolf, 1959, p. 452; Gabor, 1948, 1949). In this method, the first step is to illuminate the object and a recording plate with coherent light (or electron waves) of a single frequency. The object scatters the electromagnetic energy falling on it so that the photographic plate records the scattered energy as it interferes with the primary energy. This diffraction pattern is called a hologram, and the process, holography. If the plate, suitably processed, is now illuminated by the background wave alone, the wave transmitted through the plate contains information about the original object, and with the proper optical system an image is seen which appears to be three-dimensional.

First, we note the requirement of this original holographic method that the illumination be single-frequency and coherent and that the *phase* be preserved in the form of an interference map in a particular plane. The object itself is extremely large compared with the wavelength of the illumination. As was evident in our discussion of resolution, the problems of resolution for light and for seismic waves are quite different. In the seismic case, the wavelength and the objects to be measured may easily be the same size and, in addition, it is possible to measure the phase of the seismic signal directly. This is done by recording the source output and the geophone output signals on the same recording medium, so that, by comparison, the phase difference can be recovered from the recorded data. This is different from the optical case, because the photographic plate records only amplitudes (or intensities).

"Acoustic holography" refers to a method of using high-frequency acoustic waves (usually for medical purposes) for investigating abnormalities in an otherwise homogeneous medium. If the medium has an acoustic velocity like that of water (1500 m/sec), the wavelengths can be subcentimeter and therefore usually smaller than the object sought.

A further difference, when seismic methods are considered, lies in the type of object sought. In the optical and acoustic cases these are three-dimensional objects, whereas in the seismic case the holographic method has to deal with smooth sheets of material which are almost two-dimensional. These, in terms of the wavelengths used (of the order of 150 m) are mirrorlike on their surfaces. Under these conditions, an image of the source is seen in the mirror—distorted of course by the curvature of the mirror. To see such an image is not desired. The requisite output of any system is to "see" the structure of the rock layers.

Finally, the seismic case is characterized by a velocity function in the earth which, in comparison with other holographic media, is highly variable. In near-surface layers, for example, the velocity often varies by factors of 4, and irregular thicknesses give rise to phase noise which may be many times that being sought to elucidate the structure of subsurface objects. The layered structure and the number of thin layers make the reverberation problem much more difficult.

Having cited some of the difficulties, nevertheless we shall see how a holographic method can be used and how it compares with existing methods. In Figure 9.13a, the general layout of the system is shown. A source S of monochromatic elastic (P) waves is assumed to be present, without generating surface waves or shear waves. Over a large area compared with the wavelength of the seismic waves, a set of receivers (R) is set out. Their output is recorded, in groups or all simultaneously, with the provision that the initial signal generated by S is also recorded. With a vibrator, these recording characteristics are easy to obtain, but the providing of a P-wave-only generator may be extremely difficult (see notes on the output from a surface source in Section 3.5). A source array and receiver arrays may be needed before the signals received are essentially P wave only.

The surface array of geophones receives part of the output from a scatterer O on the object and, for a constant-velocity medium with no weathered layer (Figure 9.13b), the output constitutes a circular pattern of constant-phase lines which are

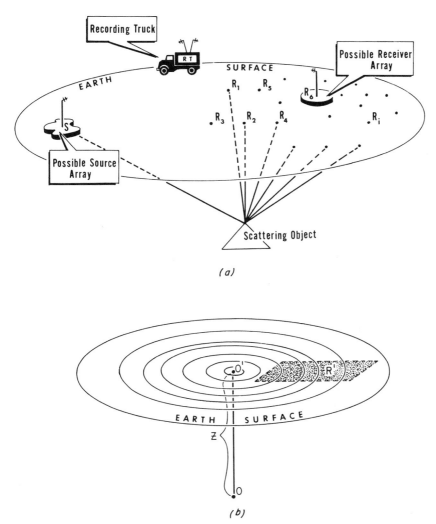

Figure 9.13 (a) A possible field setup for a holographic method using seismic waves. Note that this is a monochromatic two-dimensional surface version of standard Vibroseis ℝ practice. (b) A series of constant-phase lines on the surface due to a scatterer O below the surface. Note that, theoretically, information within a limited portion of the surface R' should be sufficient to establish the location of O.

unique to the scatterer O (since the center is at O', vertically over O, and the rate of change in phase with distance from O' is unique for the depth Z of the scatterer). Thus, in theory at least, a correlation of the output pattern received by the set of receivers with a set of theoretical patterns (similar to the Fresnel zone plate in optics) establishes the position of O in the earth. A set of such correlations has been made for three different levels by Chapman (1967), and these are reproduced as Figure 9.14 (courtesy of the Continental Oil Company). It is seen that the overall response at a point in the earth reaches a peak and diminishes, the peak of the envelope being at the proper depth. At three different levels, the proper one and $\pm 10\%$, the response over a horizontal plane has been computed, to show the lateral resolution. This example is

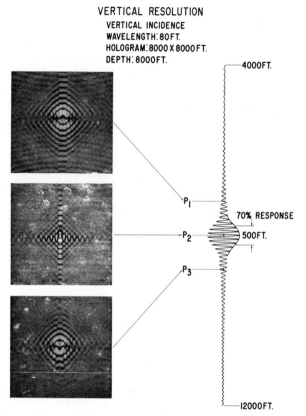

Figure 9.14 Vertical and horizontal resolution achieved with a single-frequency (100-Hz) signal at three different levels (left). On the right is the continuous output down the vertical axis. [After Chapman. Courtesy of the Continental Oil Company.]

somewhat extreme for a field case, since it corresponds to a 100-Hz wave investigated without aliasing over an area equal to the depth squared. For proper detection without out aliasing over an area equal to the depth squared. For proper detection without aliasing 400 × 400 receivers would be necessary. In principle, the more of the amplitude and phase pattern intercepted by the set R of receivers, the more exactly Z can be determined. It is possible, however, to derive an approximate value of Z from an array R' which only partially intercepts the output of O. Let us examine the case R' which is a line of receivers along a radius from O'. In this case, the phase and amplitude can be measured and, provided that the proper phase line can be established from one receiver to the next, either the diffraction method or the wave front method of migration can be used, assuming different values of Z (t_0, time vertically to the surface) until the proper value of t_0 is found when all the indicators cross at a point. In the Vibroseis® method, a continual change in frequency with respect to time allows this method to become more positive. After correlation with the sweep signal which is recorded on the same recording medium, the events accepted by the geophones R become pulselike, their times are uniquely available, and the process becomes the standard migration process. Deviations of time of arrival due to the presence of the weathered layer give, in both the holographic and the direct, migrated Vibroseis®️ method, uncertainty as to the position of the scatterer O.

If now, going back to Figure 9.13a, there is a multitude of scatterers like O, all of them will be found, since they have different centers O' and different diffraction patterns. In the monochromatic case, the resolution is limited both in a vertical and in a lateral sense by the size of the array (aperture) and by the wavelength of the illuminating signal.

Chapman (1967) also investigated the depth resolution that can be obtained using a series of frequencies (50 to 100 Hz), and this is illustrated in Figure 9.15. The ghosts on the trace labeled "Sum, $\lambda = 80'$ to $180'$" are caused by the limited(?) aperture being used, but it is shown that a much greater definition of the proper depth is obtained. This definition would be still further improved by widening the swept frequency band on the low side—as is the normal practice in Vibroseis®️ work. The ghosts may be eliminated, or at least reduced, by tapering of the array.

It is obvious that, with an array of the magnitude described, the noise level is considerably reduced and the effects of phase noise caused by variable near-surface residual errors tend to cancel. However, they reduce the amount of usable high-frequency component.

Other methods have been advocated for making a digital reconstruction of the object. These follow either the lines of the optical theory or are called the method of conjugate phase. The results achieved differ little from those of the correlation procedure, except that they appear to be consistent with the use of absolute amplitudes

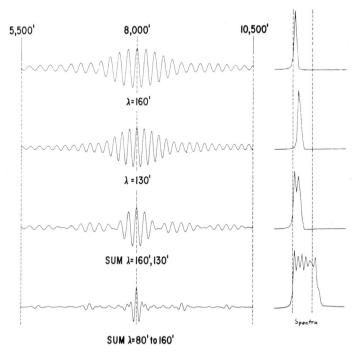

Figure 9.15 The increase in definition obtained at the 8000-ft level in Figure 9.14 by summing six different wavelengths (80 to 160 ft) corresponding to frequencies from 50 to 100 Hz. The corresponding spectra are shown on the right. (After Chapman. Courtesy of the Continental Oil Company.)

only. The envelope of definition is, however, very little changed. All methods make use of the Huyghens-Kirchoff diffraction theory to work backward from the diffracted field on a given surface to the causative scatterer within the medium.

We have seen that seismic holographic methods, when extended to finite bandwidth signals, achieve the same results through the same diffraction methods as those of normal migration. Any apparent advantages accrue from increasing the size of the reception array and making it two-dimensional. These factors can of course be enhanced in normal seismic methods, and holography is seen, in the seismic case, to be another name for reflection seismology with adequate migration.

9.7 SUMMARY AND CONCLUSIONS

It is a mistake to regard a seismic reflection cross section as a one-to-one display of the depths to the reflecting horizons. This is obvious in some cases that display steeply dipping events, since in many cases continuous reflection (diffraction) events cross over each other—an impossibility for solid rock layers.

Subjective methods consisting of reflection picking and individual dip migrations according to a simple velocity function have now been replaced by automatic, digital computer–performed procedures which do not need the picking of correlated events from trace to trace. In many cases, nevertheless, such reflection picking aids in cross-section simplification, but its efficacy is questionable unless all the criteria of computer picking are agreed on and their effects fully understood.

The computer methods available are of three general types. First, the wave front migration method takes the energy at a particular time on a given input trace and distributes it in space (two dimensions) according to the constant-time curves that apply between the source and the receiver. These are near circles for coincident source and receiver (as can be simulated by stacked traces) or are near ellipses for offset sources and receivers. The entire process consists of summing all energy contributions moved into elemental sections of space (usually rectangular elements having the dimensions of sample time rate and trace distance separation).

Second, it is possible to look at the process as one in which the proper elemental energies of a series of traces are summed to give the energy associated with a scatterer in the migrated section space. This addition has to be done with the selection times corresponding to the proper diffraction curves. This Kirchoff sum or diffraction scattering method is applied to every time element on every trace, the final result being the migrated section. It makes use of the Kirchoff-Huyghens diffraction hypothesis that every element of a reflector acts as a separate scatterer.

The final method, to date, is a finite difference method due to Claerbout and his associates, which involves a mapping from the reflection time domain to a migrated time domain. Examples of the same data, treated by different methods, are usually in agreement on the main effects but differ in detail—the treatment (and the generation) of noise and, in some cases, the resolution of faults.

It must be emphasized, however, that seismic records always contain energy which is not generated or scattered from the vertical plane of the seismic cross section. In marine shooting, this is probably more evident, since the water layer is an excellent transmission medium, and the seismic cables used do not have any means of discriminating between events that come from the side and those that come from below.

The presence of the water surface, introducing an image of the scattering point within the water layer (see Figure 9.13), does, however, cause the overall response of the cable to be different for near-horizontal events compared with near-vertical events. The specific difference is hard to define, since it is related to the size of the scatterer, its depth, the roughness of the sea surface, the turbidity or aeration of the water, and many other factors. It is the purpose of this note to warn the interpreter of the need for alertness when unusual events are present on the seismic cross section. Not all these should be migrated and, if a specific cause can be ascertained, the adulterating noise should be removed before migration.

Three-dimensional migration is possible by an extension of the two-dimensional methods—if the data are available. The chief problem to be solved by three-dimensional migration is that posed by "side swipe"—events that come from points on the subsurface reflecting surface, which are laterally disposed with respect to the vertical plane through the source and receiver points.

Seismic holography has been brought up as a possible distinct method for producing three-dimensional images of the subsurface. On careful examination, it is found to rely on the formation of a scattering function distribution in the earth by operating on the diffraction pattern measured at the surface. Since the phase differences, in seismic work, can be measured directly, there is no need for the artifice of mixing in the original signal and the production of a photographic plate, followed by illumination by a coherent light source—as is done in optics. Rather the amplitudes and phases of the samples, spread over a wide area, can be used to reconstruct the image digitally— using the same principles of the Huyghens-Kirchoff diffraction theory. The definition of the subsurface scatterer is inadequate for even the highest single frequency used, and resort has to be made to a series (or a continuum) of frequencies. This brings the method back to standard seismology using Vibroseis® methods if the results are treated by migration subsequent to recording.

APPENDIX 9A: CALCULATION OF DIFFRACTION CURVES FOR A GIVEN VELOCITY DEPTH FUNCTION

We assume that the interval velocity is known as a function of depth. A special case considered later is $V(z) = V_0(1 + kz)$.

The time of travel from S to D to R (R and D coincident) is (see Figure 9A.1)

$$T = \int_0^z \frac{ds}{V(z)} \quad \text{where } ds = \frac{dz}{\cos \theta} \text{ and } dx = dz \tan \theta$$

$$= \int_0^z \frac{dz}{V(z) \cos \theta} \tag{9A.1}$$

Similarly,

$$X = \int_0^z dz \tan \theta \tag{9A.2}$$

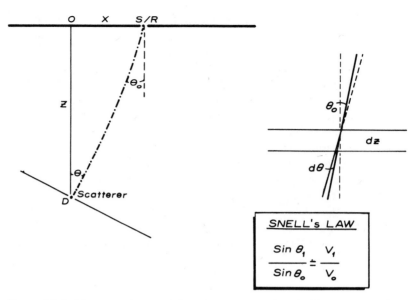

Figure 9A.1 The assumed ray path from the scatterer to the coincident source-receiver.

These two parametric equations, together with Snell's law,

$$\frac{\sin \theta(z)}{V(z)} = \frac{\sin \theta_0}{V_0} = p \qquad (9A.3)$$

determine the T, X curve.

Since $\cos \theta = (1 - \sin^2 \theta) = (1 - p^2 V^2)^{1/2}$

$$V = V(z)$$

$$T = \int_0^z \frac{dz}{V(1 - p^2 V^2)^{1/2}} \qquad (9A.4)$$

$$X = \int_0^z \frac{pV \, dz}{(1 - p^2 V^2)^{1/2}}$$

Usually, these equations are solved by calculating (in the digital computer) the values of T and X for various values of $\theta_0 = n \, \Delta\theta_0$ and then interpolating to give values of T for equal increments of X.

For the relatively simple case of

$$V_z = V_0(1 + kz) \qquad (9A.5)$$

it can be shown that

$$T = \frac{1}{V_0 K} \ln \left(\frac{V_z(1 + \cos \theta_0)}{V_0(1 + \cos \theta_z)} \right) \qquad (9A.6)$$

$$X = \frac{1}{K \sin \theta_0} (\cos \theta_0 - \cos \theta_z)$$

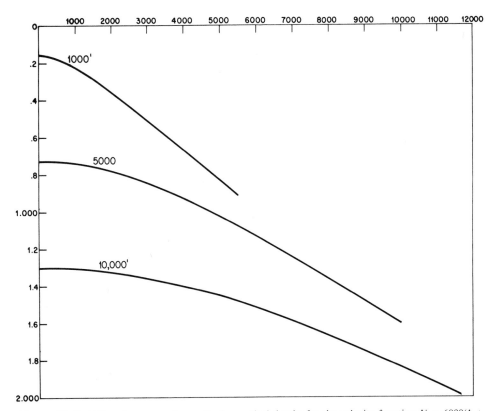

Figure 9A.2 Diffraction curves for particular vertical depths for the velocity function $V_z = 6000(1 + 0.00006z)$.

where V_z is the velocity corresponding to the particular depth chosen and

$$\cos \theta_z = \sqrt{1 - \sin^2 \theta_z} = \sqrt{1 - \left(\frac{V_z}{V_0}\right)^2 \sin^2 \theta_0} \qquad (9A.7)$$

By taking $\theta_0 = n\,\Delta\theta$, the values of T_n and X_n can be determined and plotted against one another. The curves shown in Figure 9A.2 are calculated for the velocity function

$$V_z = 6000(1 + 0.00006z)$$

and are for vertical depths of 1000, 5000, and 10,000, respectively. These values can be in either feet or meters.

REFERENCES

Alford, R. M., Kelly, K. R., and Boore, D. M. (1974), "Accuracy of Finite Difference Modeling of the Acoustic Wave Equation," *Geophysics*, Vol. 39, No. 6, pp. 834–842.

Born, M., and Wolf, E. (1959), *Principles of Optics*, Pergamon, New York.

Claerbout, J. F. (1970), "Coarse Grid Calculations of Waves in Inhomogeneous Media with Application to Delineation of Complicated Seismic Structure," *Geophysics*, Vol. 35, No. 3, pp. 407–418.

Claerbout, J. F. (1971), "Toward a Unified Theory of Reflector Mapping," *Geophysics*, Vol. 36, No. 3, pp. 467–481.

Claerbout, J. F., and Doherty, S. M. (1972), "Downward Continuation of Moveout Corrected Seismograms," *Geophysics*, Vol. 37, No. 5, pp. 741–768.

Faust, L. Y. (1951), "Seismic Velocity as a Function of Depth and Geologic Time," *Geophysics*, Vol. 16, No. 2, pp. 192–206.

Faust, L. Y. (1953), "A Velocity Function Including Lithologic Variation," *Geophysics*, Vol. 18, No. 2, pp. 271–288.

French, W. S. (1974), "Two Dimensional and Three Dimensional Migration of Model-Experiment Reflection Profiles," *Geophysics*, Vol. 39, No. 3, pp. 265–277.

Gabor, D. (1948), "A New Microscopic Principle," *Nature*, Vol. 161, p. 777.

Gabor, D. (1949), "Microscopy by Reconstructed Wavefronts," *Proceedings of the Royal Society, A*, Vol. 197, p. 454.

Gabor, D. (1951), "Microscopy by Reconstructed Wavefronts: II," *Proceedings of the Physical Society, B*, Vol. 64, p. 449.

Laski, J. D. (1970). "Simultaneous Estimation of Parameters of Reflection Events (Depth, Dip, Velocity) and Relative Static Corrections, *Geophysical Prospecting*, Vol. 18, No. 2, pp. 269–276.

Loewenthal, D., Lu, L., Roberson, R., and Sherwood, J. W. C. (1974), "The Wave Equation Applied to Migration and Water Bottom Multiples," paper presented at the 44th Annual Society of Exploration Geophysicists Meeting, Dallas, Texas.

Musgrave, A. W. (1961), "Wave Front Charts and Three Dimensional Migrations," *Geophysics*, Vol. 26, No. 6, pp. 738–753.

Paturet, D. (1971), "Different Methods of Time-Depth Conversion with and without Migration," *Geophysical Prospecting*, Vol. 19, No. 1, pp. 27–41.

Paulson, K. V., and Merdler, S. C. (1968), "Automatic Seismic Reflection Picking," *Geophysics*, Vol. 33, No. 3, pp. 431–440.

Sattlegger, J. W. (1964), "Series for 3-Dimensional Migration in Reflection Seismic Interpretation," *Geophysical Prospecting*, Vol. 12, No. 1, p. 115.

Sattlegger, J. W. (1969), "Three Dimensional Seismic Depth Computation Using Space Sampled Velocity Logs," *Geophysics*, Vol. 34, No. 1, pp. 007–021.

Sattlegger, J. W., and Stiller, P. K. (1974), "Section Migration, Before Stack, After Stack, or In Between," *Geophysical Prospecting*, Vol. 22, pp. 297–314.

Schneider, W. A. (1971), "Developments in Seismic Data Processing and Analysis," *Geophysics*, Vol. 36, No. 6, 1043–1073.

Waters, K. H. (1941), "A Numerical Method of Computing Dip Data Using Well Velocity Information," *Geophysics*, Vol. 6, No. 1, pp. 64–73.

TEN

Near-Surface Corrections

10.1 INTRODUCTION

At this relatively late stage in the description of the seismic reflection method, we have said little about the effects of surface elevation variations and near-subsurface velocity and thickness variations on the reflection results. In Chapters 5 and 6, just enough was said to indicate that proper corrections are highly desirable, but the actual procedures, except in one simple case, were bypassed. The need, strangely enough, was not felt very strongly during all of the early time when use was made of dynamite in holes. For one thing, it was considered necessary to make such corrections accurately only at places where holes were drilled for explosives. It was considered probable that, when holes were drilled through the weathered layer, the uphole time and other relevant data were sufficient to make the corrections, and in a large fraction of the areas they were. Exceptions are of course always available and, in particular, glacial drift areas— in which the holes would have to be so deep as to be impractical and uneconomical— were cases for which special methods (Handley, 1954; Dobrin, 1942; Duska, 1963; Patterson, 1964) were devised.

10.2 SIMPLE METHODS USING SHOT HOLES

In the case of explosives, when only structure was considered, the methods and results were usually straightforward. A hole was drilled through the weathered layer, and the depth of shot, elevation of the top of the hole, and the uphole time for the seismic wave were recorded. In many areas, it was standard practice to drill at least one deep hole and, by a series of shots at different depths, measure the vertical velocity through the subweathering material. These are important pieces of information and are not usually available when surface sources are used. They allowed a time correction to be made to a reference surface. The concepts are simple, and the corrections are (see Figure 10.1): T_r is the uncorrected reflection time for a shot at A_1 received at E_2. $T_r^{(c)}$ is the corrected time for the same shot and receiver but given with reference to the reference plane E_s. It is given by

$$T_r^{(c)} = T_r - \frac{(E_1 - d_{s_1} - E_s + E_2 - d_{s_2} - E_s)}{V_0} - t_{uh_2}$$

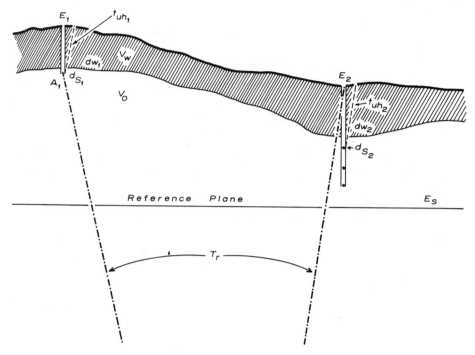

Figure 10.1 Corrections when shot holes are used.

The same reference plane is used throughout the area and is usually taken as near as possible to the average shot elevation to avoid the slant path problems for long-offset traces. It should be obvious that the exact elevation chosen for the reference plane is of little consequence.

There are nevertheless some velocity problems that cannot be treated by these shallow holes. For example, near the Fort Peck Reservoir in North Dakota, the U.S. Corp of Engineers drilled holes in the flood plain of the Missouri River and found recent, unconsolidated sands and gravels as deep as 600 ft. Other discrepancies arose in areas of West Texas and New Mexico, where mesas caused by uneroded areas of resistant rock, several hundreds of feet above the surrounding plains, were found to influence the velocity in the subsurface. This effect, due to the static loading of the surface by the excess weight of the mesa, gave rise to a distribution of velocity abnormalities which existed not only below the mesas but for thousands of feet to the side and in decreasing percentage with depth.

This type of problem usually shows up when the time differences between a deep and a shallow reflection are mapped and contoured. The resulting contours partially correlate with the surface elevation.

A correlation coefficient can be calculated between the variations in the surface and the time variations in the subsurface reflections:

$$k = \frac{\sum_i^N (E_i - \bar{E})(t_i - \bar{t})}{[\sum (E_i - \bar{E})^2 \sum (t_i - \bar{t})^2]^{1/2}} \tag{10.1}$$

If one adopts a geological postulate that the surface topography bears no relation to the subsurface structural variations, corrections can be applied, based on the correlation coefficient and knowing the surface variations, to the reflection times. This method is of course subject to the danger that the postulate may be wrong (in some cases, surface lows correspond to subsurface anticlines), and it must be used with care.

The assumption of a reference plane is obviously incorrect in country with large-scale elevation changes, since the assumed constant-elevation correction velocity V_0 is most likely affected by both loading effects as well as changes in lithology. For time cross sections, a horizontal reference plane is an obvious necessity; however, when reflection times were to be corrected to depth using some velocity function and the horizon levels referred to a sea-level datum, a floating time reference was sometimes used. Many times, the level of this floating surface was a constant plus an estimated fraction of the change in shot elevation:

$$(E_{R_1} - E_{R_2}) = k(E_{S_1} - E_{S_2}) \qquad (10.2)$$

where $\quad k =$ constant $0 < k < 1$
E_{R_1} and $E_{R_2} =$ elevations of the reference surface at positions 1 and 2.
E_{S_1} and $E_{S_2} =$ elevations of the shots at positions 1 and 2

This device, however, begged the question, because it was then necessary to apply a (constant?) velocity down to the reference surface and then convert the corrected reflection times to depth assuming that the entire velocity log (possibly determined from a well survey) moved up and down with the reference surface. It has been remarked previously that the intermediate traces at the various receivers between the widely spaced shot points served only to correlate, to follow the desired reflection from shot point to shot point. There was no redundancy in these data.

As soon as the CDP method came on the scene, there was redundancy, and there was further reason for the correct estimation of static corrections. The data to be stacked together to give a better signal/noise ratio and some multiple-reflection cancelation came from different shot points and receivers and therefore had to be properly corrected before either signal/noise ratio improvement could result or proper velocities could be calculated. In the following sections, various methods of obtaining near-surface corrections from redundant CDP reflection data are examined. They have their dangers, too, and it is not surprising, even to the initiated, that full success has yet to be achieved. These CDP methods can of course be used with shots in holes as sources. It is assumed, for simplicity, however, that all sources and receivers are on the surface. If further information, such as uphole times, is available, so much the better. These data can be used as a further control.

10.3 SOURCES AND RECEIVERS ON SURFACE (CDP)

Although, on the basis of surface elevations and an estimate of the *near-surface* subweathering velocity, it is possible to make corrections for elevation change—and this is supposedly done in the remainder of this section—there is still a necessity for making an estimate of the effect of the weathered layer on the reflection record. This effect is most often regarded as a time-delay constant for all frequencies in the reflection passband. In a later section, we take up the question of whether or not it is possible to treat a frequency dependence in the weathering effect.

The first, and still common, practice was to estimate the weathering time by the *ABC* refraction method described in Section 5.4. This procedure of course adds to the cost of the survey and does not permit any determination of the possibility of change in velocity of the weathered layer. In practice, the velocity in the weathered layer is not constant, as it is very much a function of water saturation of loose, unconsolidated soils and rocks. There is economic pressure, always, to determine the residual static *corrections* (i.e., the surface-consistent extra time delays after the elevation corrections have been made as well as possible) from the redundant data collected during practice of the CDP method. In this section, the principles involved in determining these residual static corrections (by three separate methods) are reviewed.

A typical set of seismic traces is shown in Figure 10.2 for a sixfold stack system. Note that the traces are numbered with the shot point and the receiver numbers, separated by a slash. When arrayed in this manner, we can classify the traces by the manner in which they are aligned. For example, all traces with a common reflection (depth or basement) point are aligned vertically. For example, the columns under SP 16, consisting of the traces 14/18, 13/19, 12/20, 11/21, 10/22, and 9/23, all have a basement point under SP 16.

Similarly, the constant receiver traces are lined up downward to the left, constant shot point traces downward to the right, and finally constant offset traces are along horizontal rows such as 1/10, 2/11, 3/12, 4/13, and so on. Each method to be described makes use of one or more of these arrangements, and all have their weaknesses.

In order to find comparative times, it is usual to cross-correlate selected windows of the traces concerned to find out at what (algebraic) time difference the best match is obtained. This consists of plotting the cross-correlation function

$$\phi_{fg}(\tau) = \int_{t_1 + \tau}^{t_2 + \tau} f(t)g(t + \tau)\,dt \tag{10.3}$$

for a range of delay times (say, -20 to $+20$ msec) and over the range of the window t_1 to t_2, moved by the corresponding shift τ in one trace. It is important both to keep the number of samples within the window constant and to minimize the effects of the end samples, so this is usually done over as long a window as possible, and the window itself is tapered. It is presumed that the following precautions have been taken.

1. The window is taken over prominent reflections.
2. The window is not allowed to include long intervals of noise. It may be better under some circumstances to use two or more separate windows and combine or average the time delays found.
3. The normal movement has been removed as closely as possible (this applies to all but one of the methods to be discussed). Correlation is very sensitive to stretching or shrinking of a trace, as we noted when considering the Vibroseis® method.

It is usually necessary to resample the individual traces at a finer sampling rate in order that the time delay for maximum cross-correlation can be obtained accurately, that is, with better precision than the original sample rate of the trace. Linear, quadratic, or cubic interpolation may be done, but the best method (see Appendix 10A) is to use the $(\sin x)/x$ method.

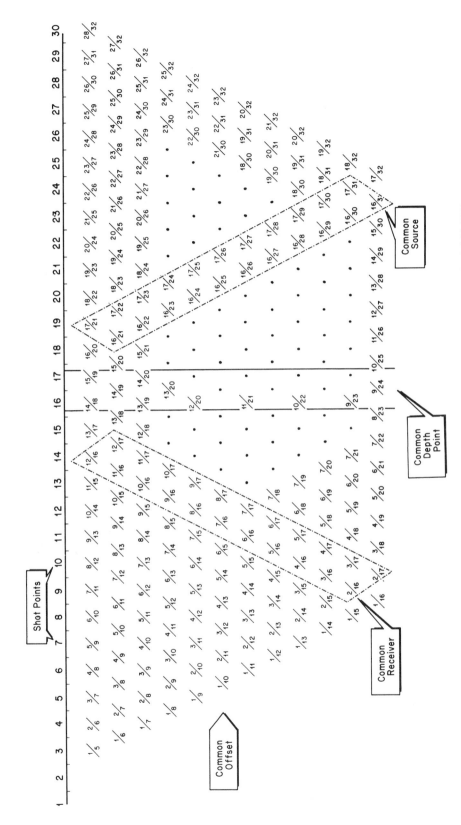

Figure 10.2 The characteristics of depth points, offsets, sources, and receivers in a specific sixfold CDP system.

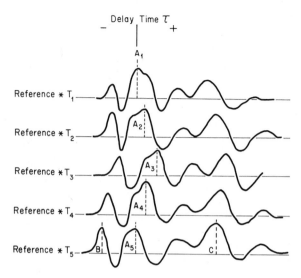

Figure 10.3 Typical cross-correlation functions. The picked peaks at A_1, A_2, A_3, A_4, and A_5 have varying delay times. B and C represent some higher peaks which can be machine picked if the data are noisy or if the criteria are not strict enough. The asterisks indicate cross-correlation.

The outcome of the cross-correlation is a curve, such as is shown in Figure 10.3, with its peak (hopefully this is unambiguous) shifted from zero delay, and this delay can be determined. These are then the time differences that are used, and they can be entered into an array similar to the array shown already, in which every trace has its place.

Note that these determinations of delay time can be subjected to a statistical analysis, to give the mean μ and the variance σ^2 [see, for example, Wine (1964)]. The standard deviation σ is the distance from the mean (for a normal distribution of observations) within which approximately two-thirds of the observations fall. Ninety-five percent of the observations in a normal distribution fall inside the range $\mu - 2\sigma$ to $\mu + 2\sigma$, and it is a commonly accepted practice to throw away, as suspicious, any observations that fall outside this range.

Method 1—Common Reflection Point

Hileman, Embree, and Pflueger (1968) appear to have been the first group to publish this method. The τ discovered from the cross-correlation of two traces is made up of several individual contributions:

$$\tau = \tau_d + \tau_{WS} + \tau_{WR} + \tau_{NMO} + \tau_N \qquad (10.4)$$

where τ_d = dip time difference
τ_{WS} = weathering at shot point
τ_{WR} = weathering at receiver
τ_{NMO} = time difference due to incorrect NMO calculation
τ_N = noise

If a method involving the CDP is chosen, the term due to dip should disappear.

First, however, all traces, for a common depth, are compared with a reference trace. This reference trace can be one of the traces themselves, in which case, $N - 1$ values of τ are obtained. We use only a set of τ's with a zero mean, so this adjustment gives 12 values of τ (including 1 for the reference trace). By this procedure, the entire table can be filled out. Now, the set of values along the *common receiver* line, if averaged, gives a value due to this receiver alone (it is assumed that the shot values average to zero), and the set along the common shot point, if averaged, gives rise to a value which can be taken to be the value for this shot point, by similar reasoning.

If the values are examined that lie along the line of the CDP, the statics may vary in a random manner about a near-zero mean, or they may vary systematically with offset (either an increase or a decrease), and this is a sign that the NMO has been incorrectly removed.

Gross errors in the estimation of either shot point static or receiver static can be perceived easily and removed. These usually arise through the possibility of two possible maxima (peaks of almost the same height in the cross-correlation function), and one or both must be rejected.

Systematic changes due to incorrect NMO are removed by low-order polynomial fitting and retaining the residuals.

Finally, the statics applied to the individual traces making up the stacking set are the sums of the source point and receiver point statics. *Surface consistency* is thus achieved. This means that a particular point has a source or receiver static which is applicable and is independent of the offset, or particular combination of source and receiver, used.

The reference trace should really be the near trace, in which error due to NMO is smallest, and this is probably the best single one to use if the signal/noise ratio is high. Other references have been used, such as the nonshifted average of all the traces. However, if the static shifts between the traces are large compared with a period of any frequency, there is danger that the character of the trace may be so altered that the cross-correlation trace (which has the common spectrum between the two traces and the difference of their phases) will have an altered shape and an incorrect position for its maximum.

It is intuitively obvious that, since the sampling is good within a spread length but the combined shot point-receiver redundancy is poor outside the spread length, this method measures high-frequency statics (i.e., statics that vary rapidly along the surface) well but measures long-period statics poorly. This is in fact what is found and, since there is a limit to the spread length that can be used, statics that affect overall structure are poorly estimated, and false structure can be inserted. Remember that the averages are taken over the spread length for both source point and receiver values. Thus, although CDP traces are used to avoid errors due to dip, errors can be made that affect the dip.

Method 2—Equivalence of Source and Receiver Statics

Another variant of the statistical automated statics system is one in which advantage is taken of the common shot point, or common receivers. The table in Figure 10.2 can be filled in by correlating each trace of a common shot point record with a reference trace. The same remarks as previously given apply to the choice of the reference trace, and in general the same cross-correlation procedure is adopted. The advantage here may be in the common spectrum given off by the source-earth combination—consistency is difficult to achieve in practice, even with the Vibroseis® system. The table is thus filled in, and the same precautions can be taken as before, namely,

1. The receiver τ's must form a zero-mean set.
2. Values that deviate more than 2 standard deviations from the mean may be (at least) questioned at this stage.

The same procedure is of course available when the traces are sorted into sets having different shot points but a common receiver. Again, the justification may be that the near-surface conditions at the receiver are common (i.e., the effect of near-surface reverberation is the same). These values then allow the table to be filled out again. Note that these values fill out the table along the lines down to the right (common source) or down to the left (common receiver). Effects of NMO removal error are still present. Effects of dip are still present in each method, except that it is a different dip for each determination, and there is some hope that τ_{dip} will cancel. Note that there are $2N - 1$ determinations of the common source and $2N - 1$ values of the weathering difference between a given pair of surface points. If an average is then taken (on the assumption that *time differences* are independent of whether the surface point is a source or receiver), surface-consistent values of the weathering can be obtained. This method provides more statistics (since both sources and receivers are treated as

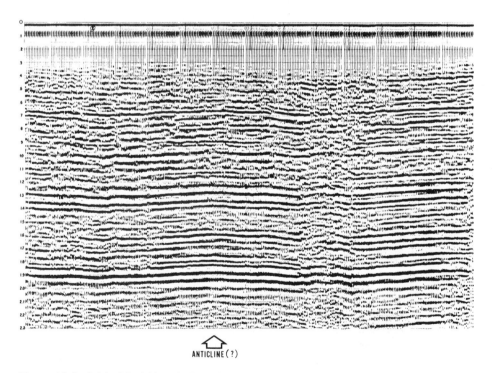

ANTICLINE(?)

Figure 10.4 Original fivefold-stacked test data area A. (Courtesy of the Continental Oil Company.)

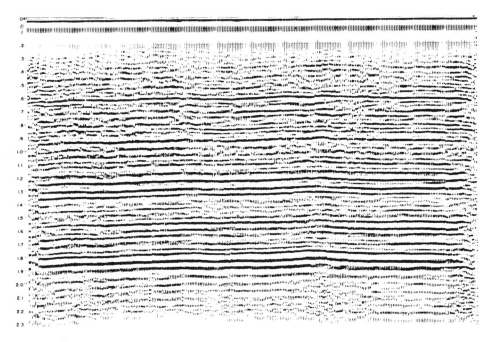

Figure 10.5 Test data in Figure 10.4 treated with statistical correction program 2 described in the text. Note the change in the reflection structural pattern and the deterioration near the ends where the full stack has not been obtained. (Courtesy of the Continental Oil Company.)

equivalent and thus dubious values can be checked), and the statistical error can be calculated more accurately.

However, the difficulties of remnant τ due to incorrect NMO and dip have not been solved, except that one can expect high-frequency statics to be well determined, and those of long surface period, which are the dangerous ones, will be undetermined. Figures 10.4 and 10.5 show fivefold field data both in the original stacked version and after application of method 2 statics before stacking. Also, well illustrated in these two figures is the ambiguity of structure and the dilemma of the interpreter—is the anticline shown by the arrow on Figure 10.4 present or not. The answer is not in either of these sections.

Method 3—Constant Offset Static Corrections

Returning now to (10.4),

$$\tau = \tau_d + \tau_{WS} + \tau_{WR} + \tau_{NMO} + \tau_N \tag{10.4}$$

we can reduce a different term to zero if we take only traces that have a constant offset distance to correlate with one another. Then, as long as the average velocity to the reflection stays constant, $\tau_{NMO} = 0$. In Figure 10.2, these traces occur along the horizontal lines. Reflection seismic traces from adjacent sources having the same offset are cross-correlated:

$$\tau_{i,i+1} = \tau_d + \tau_{WS,i} - \tau_{WS,i+1} + \tau_{WR,i+P} - \tau_{WR,i+P+1} + \tau_{N,i} - \tau_{N,i+1} \tag{10.5}$$

where P is the offset being used. All offsets can of course be utilized. The dip still stays in the time difference determined, as has been characteristic of all methods—even though CDP method 1 *apparently* avoids this complication. It has to be dealt with in an interpretive way, for example, by assuming that a mean of a sufficient number of determinations averages out the dip for the short-period variations. In practice, it is usual to fit a low-order polynomial

$$\Delta t = a_0 + a_1 x + a_2 x^2 + a_3 x^3 + \cdots \tag{10.6}$$

to the τ's computed and then to use only the residuals (values determined less the polynomial calculated values) as the proper weathering corrections. As before, we assume that τ values due to noise cancel. Finally, we state the hypothesis that the source point statics are equivalent to the receiver statics (there is surface consistency whether the point is a source or receiver), and we obtain

$$\tau_{i,i+1} = \tau_d + \tau_{W,i} - \tau_{W,i+1} + \tau_{W(i+P)} - \tau_{Wi+P+1} + \tau_{N,i} - \tau_{N,i+1} \tag{10.7}$$

This is more of a mixture of terms than we have encountered previously, because different receivers and different sources are involved in each determination of $\tau_{i,i+1}$. Since there are $2N$ equations and $2N + P_0$ weathering values to be determined, the values appear to be underdetermined in any set. However, this is by no means the case, since the redundancy caused by shift of only a partial spread at a time makes the chief values overdetermined and the usual procedures apply. Of course, at the beginning and end of lines there are some points with no, or very little, redundancy, and as in other weathering measurements, the determinations are made so that one value may arbitrarily be given some constant value. Figure 10.6 shows that the set of $2N$ equations, if added, gives very nearly $2N$ times the difference between $\tau_{W,i}$ and $\tau_{W,i+1}$, two other weathering values entering in (the nearest offset and the farthest offset) with a weight of only 1. It is usually sufficient to adopt this simple approach, but a correction

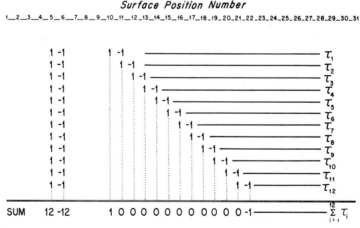

Figure 10.6 Data from cross-correlations of adjacent pairs of common-offset traces (common shot points 5 and 6) arranged to show how the sum gives large weight to the shot point weathering values. Multiples of the weathering time concerned (1 or -1) are shown in the table.

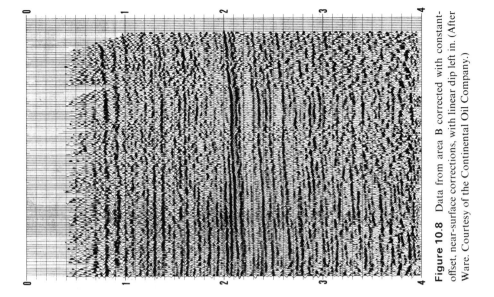

Figure 10.8 Data from area B corrected with constant-offset, near-surface corrections, with linear dip left in. (After Ware. Courtesy of the Continental Oil Company.)

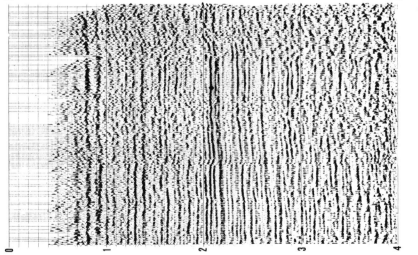

Figure 10.7 Original 10-fold-stacked data, area B. (Courtesy of the Continental Oil Company.)

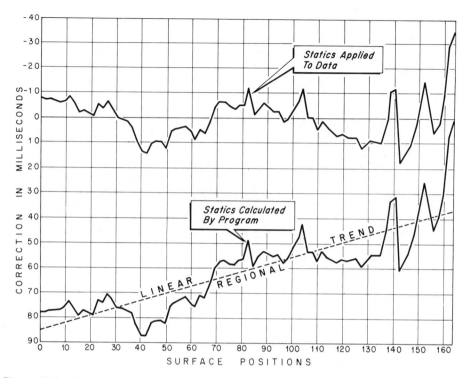

Figure 10.9 Showing the relation between the static corrections calculated by the program and the values applied to the data to leave in linear dip.

may later be made after the first approximation has been established to all of the weathering corrections. The higher the value of N, the less this second approximation is needed.

The calculation of the residuals is then made by a least squares best fit to a low order polynomial (or any other suitable long period function of offset distance). The standard method (pseudo inverse matrix method) is suitable.

Uncorrected and corrected data, together with a graph of the calculated and residual values applied to the data, are given, respectively, in Figures 10.7 through 10.9. Even the residuals are very large, ranging from 17 to −36 msec, but such areas are not uncommon, and these data illustrate the need for exceptionally good near-surface corrections in order to improve data quality. The structure is of course a function of the polynomial fitted. It should be pointed out, however, that the statics calculated here do include elevation corrections, and this procedure is contrary to the recommendation made previously. It was necessary because of changing subweathering velocity.

10.4   SOURCES AND RECEIVERS ON SURFACE— ITERATIVE AND ADVANCED STATISTICAL METHODS

It is evidently necessary—particularly when high-frequency records are being processed—to obtain the near-surface correction times as accurately as possible. An error of (say) 0.005 sec causes the 100-Hz component of one trace to be out of phase with another with which it is being compared or stacked. The chief purpose of great accuracy in near-surface corrections is simply to preserve the high-frequency components of the reflections when the CDP components are added together. Such records better reflect the geological boundary space relationships but naturally give the changes in the individual boundary positions more clearly. Thus, although the methods outlined so far are possibly adequate for most areas and for the routine requirements of a present-day reflection survey, there remain some locations and some final requirements that demand better corrections.

Inaccuracies may be present, in one or other of the methods cited, as a result of

1. Residual NMO.
2. Dip between basement points (particularly of the slowly varying type).
3. Incorrect assessment of time difference between traces (cycle skipping or a change in character of the central peak.
4. Errors due to noise.
5. Different estimates for different frequency bands.

Once an estimate of the static shifts has been made, this estimate can be incorporated in the original data and a new set of velocity determinations made along the line. On the basis of these new velocities (which will be more accurate if the original traces are aligned better after preliminary static correction), the NMOs can be made better and the process of static determination again performed. This is an iterative procedure, but there is no guarantee that it will be convergent. Care is needed in the comparison of successive iterations. Saghy and Zelei (1975) have given a method which includes repeated averaging according to depth point, source point, and receiver point and have found the effectiveness of the original Hileman, Embree, and Pflueger (1968) method considerably improved. It appears possible to examine the variance of the time delays when the velocity determination is made and thus terminate the iterations at the point of minimum variance, as well as give some indication of the expected standard deviation of the weathering determinations.

Wiggins, Larner, and Wisecup (1975) set up the problem as a set of linear equations whose solution parameters consist of spatially varying statics, the NMO and structural terms that best match, in a least-squares sense, the redundant travel time data described previously. It is found that the data are not sufficient to determine all the variables exactly, consequently a least-squares best fit to all observations is sought.

These problems are best handled by matrix methods, which lie outside the scope of this book. However, a few remarks can be made which may give the flavor of the method. First, the data and the variables are spectrally decomposed to give a set of relationships between statics and other parameters that are variable along the surface. Each characteristic function so found is independent of any other characteristic function, so that each can be handled independently. In fact, these characteristic functions are very much like sinusoidal variations along the surface, and they can be thought of as having low to high wavelengths. A property of the decomposition is that, if a gross estimate of the uncertainty of the observations can be given, the accuracy of the statics solutions can be assessed *as a function of the wavelength.*

The solutions are found by an iterative method (of matrix inversion), but now the number of iterations is known in advance. As expected, errors in the shorter-wavelength, more rapidly varying, statics converge rapidly to zero, while those in the longer

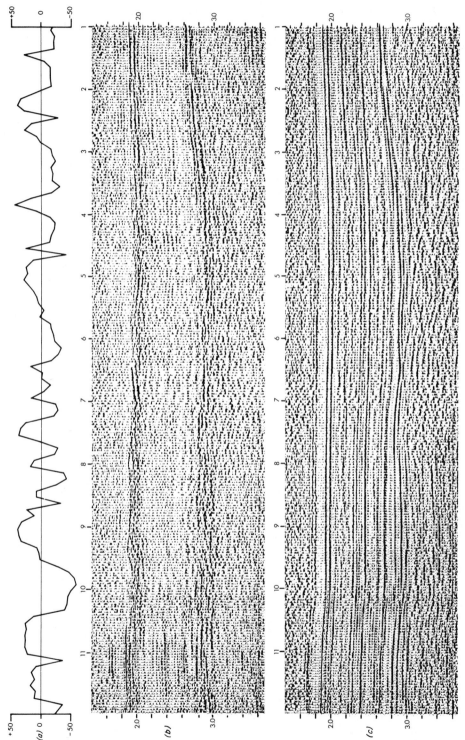

Figure 10.10 (*a*) Original residual static correction. (*b*) Record section corrected with original residual statics. (*c*) Same record section corrected with MISER static correction. (© 1975, Western Geophysical Company of America.)

wavelengths converge slowly or may not converge at all. This is consistent with previous observations that the use of redundant data gives very little control over gradual statics changes, of the order of the length of the spread.

This method (known as MISER, a Western Geophysical proprietary program) produces excellent results such as are seen in Figure 10.10 in which the total receiver static corrections, the field static (elevation corrections only) record section, and the MISER-conditioned section are shown.

10.5 WEATHERING DETERMINATIONS BY SURFACE WAVE MEASUREMENTS

One universal problem that plagues the solutions obtained from CDP data is the inability, except by the introduction of extraneous criteria, to separate out the structure of the near-surface corrections from the structure of the geological formations. Since the latter is of vital importance—a prime part of the oil-finding information being sought—there is still a need for a method independent of the reflection data, which can be used as a reference to check the CDP-derived data at intervals. If only the weathered layer is thought to be the problem, then one method, the ABC refraction method, has already been described.

A second method, which may be useful in some areas, is the use of the pseudo-Rayleigh waves generated in the weathered layer by a surface source. A single source is all that is needed, together with a short spread of single vertical geophones, as shown in Figure 10.11. The waves are usually very strong and dominate the record received. In Figures 10.12 and 10.13, the original correlated Vibroseis® record is shown at the top.

The phase velocity of the pseudo-Rayleigh waves varies with frequency (Dobrin, 1951) and, for a given set of velocity and density parameters, a set of phase velocity-versus-frequency curves can be computed for a series of depths. Such a set plotted with a log frequency scale is given in Figure 10.14. It now remains to determine the phase velocity-versus-frequency characteristics of the field data. This is done by using a series of band-pass zero-phase filters on the field data. As is well known, the narrower the band pass of the filter, the more spread out in time the energy. A compromise therefore has to be made so that the time width of the filter will allow identification of the filtered data with the more discrete event on the primary record, while at the same time allowing a reasonably discrete frequency to be assigned to the phase velocity measured. In the cases shown the filters were nominally 3 Hz wide.

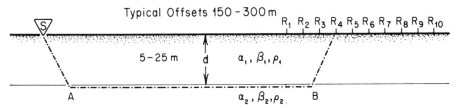

Figure 10.11 Typical spread and source configuration for recording pseudo-Rayleigh waves in the weathered layer.

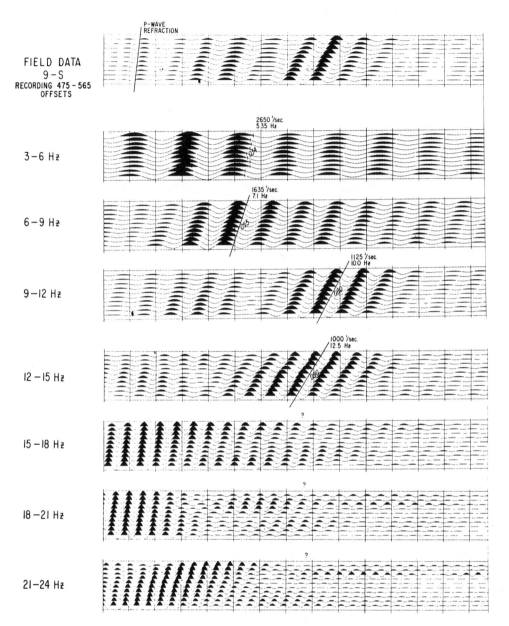

Figure 10.12 Three-hertz-wide band-pass data from a particular set of pseudo-Rayleigh-wave field data.

Provided that this can be done, there is little difficulty in measuring the time differential across the record, hence the phase velocity. Examples of easily interpreted arrivals are shown in the first four filters in Figure 10.12, but higher frequencies tend to be associated either with the P-wave refraction event marked or with other, unidentified, arrivals. A more consistent set of filtered results is shown in Figure 10.13.

These experiments have shown that it is necessary, even for normal depths of weathering and normal velocities, to be able to generate energies in the very low-frequency range (for exploration seismology) of 3 to 20 Hz. In good areas, there is

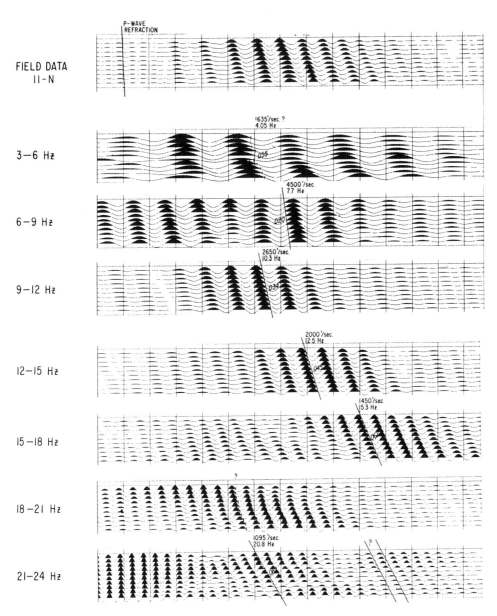

Figure 10.13 Three-hertz-wide band-pass data from a more organized second set of pseudo-Rayleigh-wave field data.

not much difficulty in identifying the corresponding curve of the phase velocity–frequency plot, but it must be remembered that:

1. Phase velocity curves are chiefly sensitive to the shear wave velocity in the weathering and subweathering.
2. The assumption is made that the weathered layer is a constant-thickness layer with plane boundaries.

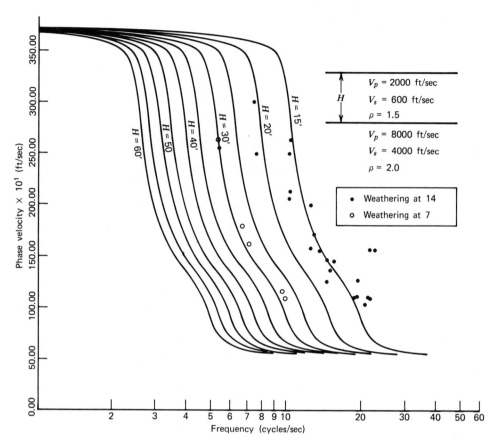

Figure 10.14 Phase velocity–frequency theoretical curves for a model of a single layer over an infinite half-space. Some field data have been superimposed so that typical scatter can be seen.

Programs have been written that iterate to obtain the set of parameters allowing a best fit to be made to the experimental data. Their success depends on the quality of the field data.

10.6   TREATMENT OF THE NEAR-SURFACE LAYERS AS A GENERALIZED FILTER

There appears to be some evidence that the action of near-surface layers, particularly that of the weathered layer, is not simply a delay mechanism. In the worst areas, the best determinations of delay times do not work well, although they are acceptable in most areas. This has led to speculation that the effect on the phase of the constituent frequencies is not just a linear function of frequency with a zero intercept of 0 or π, but a generalized filter. A simple example, showing a condition under which such a generalized filter is generated, is obtained when the reflection coefficient at the base of a single weathered layer and equal delay times for all frequencies are considered, but multiples are included. Figure 10.15a and b illustrates this concept. If the filter at

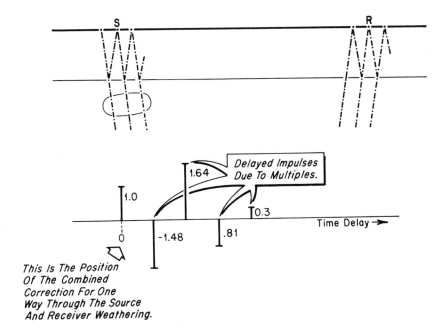

Figure 10.15 (*a*) The original source pulse is reinforced by multiples in the weathered layer. At the receiver all arrivals are augmented by multiple reflections. The combination of these two effects acts on the reflected energy as a filter whose main weight may not occur at the sum of the one-way delay times of the weathering at the two ends. (*b*) The transfer function of a layer (source end and receiver end are included) of density 1.5, velocity 610 m/sec, over a half-space with density 2.5 and velocity 2438 m/sec.

each end is represented by a transfer function pulse, the total filter will be that obtained by convolving the two transfer functions together. It is not difficult to see that, if R_W is the coefficient of reflection at the base of the weathering, the total transfer function may have its highest peak later than the time corresponding to the one-way delay time of the weathering at each end.

Using the Z-transform concept, the transfer function at the source end (for deep reflections) is

$$F_1(Z) = 1 - R_W Z^N + R_W^2 Z^{2N} - \cdots \tag{10.8}$$

and at the receiver end is

$$F_2(Z) = 1 - R_W Z^M + R_W^2 Z^{2M} - \cdots \tag{10.9}$$

both referenced to the base of the weathered layer.

Multiplying these two together, the total transfer function is given by

$$F(Z) = 1 - R_W(Z^{N+M}) + R_W^2(Z^{2N} + Z^{2M} + Z^{N+M}) - R_W^3(Z^{N+2M} + Z^{2N+M})$$
$$+ R_W^4(Z^{2N+2M}) - \cdots \text{higher order terms} \tag{10.10}$$

Let us take the case in which the following parameters hold:

1. The weathering at each end is equal to $N \, \Delta t$, where Δt is the sample time.
2. R_W is derived from the velocities and densities $V_2 = 8000$ ft/sec $\rho_2 = 2.5$ and $V_1 = 2000$ ft/sec $\rho_1 = 1.5$. $R_W = 0.74$ approximately. Then,

$$F(Z) = 1 - 1.48Z^N + 1.64Z^{2N} - 0.8085Z^{3N} + 0.3Z^{4N} \tag{10.11}$$

This transfer function is graphically illustrated in Figure 10.15b, where it is obvious that the center of gravity is not at the combined delay times of the weathering at the source and the receiver, which is the position of the first impulse. The methods related to CDP combined with a straight-delay measurement therefore measure incorrect values. The problem is of course that the characteristics, both velocity and density, of the weathered and subweathered layers are not constant. In the valleys, the amount of sediments above bedrock is usually greater and can be water-saturated, whereas on the hills the aeration of the weathered layer is greater, the layer is generally thinner, and the reflection coefficient R_W is higher than in the valley. The transfer functions therefore constantly change shape.

We can deal with this situation by any of the formerly described CDP methods, but for illustration we choose the one using constant-offset traces. Instead of using cross-correlation to establish a time delay between consecutive traces of constant offset,

$$\tau = \tau_{S_1} - \tau_{S_2} + \tau_{R_1} - \tau_{R_2} + \tau_{\text{dip}} + \tau_N \tag{10.12}$$

we go to the frequency domain in which each trace $f_i(t)$ is represented by a complex function of frequency $g_i(i\omega)$. (See Gol'din and Mitrofanov, 1975.)

Trace $i + 1$ is then divided (in the frequency domain) by trace i to obtain

$$T_{i,i+1}^{(i\omega)} = \frac{g_{i+1}(i\omega)}{g_i(i\omega)} \frac{IR_{i+1}(i\omega)}{IR_i(i\omega)} \tag{10.13}$$

Now we can assume that, over the window chosen, the reflections are constant. That is to say, the impulse response of the earth's layering is constant for two near sources and two near receivers. It must be admitted, however, that time delays due to dip are still included in $g_K(i\omega)$ and the differences in depth of the reflections appear in the transfer function calculated. There seems to be some point, now, in choosing a reasonably shallow window with a good reflection energy/noise ratio and with smaller dip (on an expectation basis) than the deeper reflections.

$$T_{i,i+1}(i\omega) = \frac{g_{i+1}(i\omega)}{g_i(i\omega)} = \frac{f_{S_2}(i\omega) f_{R_2}(i\omega)}{f_{S_1}(i\omega) f_{R_1}(i\omega)} \tag{10.14}$$

where S_2 and R_2 correspond to the trace $i + 1$ and S_1 and R_1 correspond to the trace i.

Now if logarithms are taken (very carefully), we will find an equation

$$\ln T_{i,i+1}(i\omega) = \ln f_{S_2}(i\omega) - \ln f_{S_1}(i\omega) + \ln f_{R_2}(i\omega) - \ln f_{R_1}(i\omega) \tag{10.15}$$

an equation that corresponds very closely to the form of (10.12), except that we have chosen to ignore noise in this development. The noise contributes to each of the terms in (10.15).

The interesting point is that we have one of these equations for every value of ω (every Fourier component), hence have to deal with N times as many equations as would have to be dealt with in the original constant-offset method. We can proceed one stage further.

If complex numbers like $f_{S_1}(i\omega)$ are expressed in the form $|R_{S_1}|e^{i\theta(\omega)}$ and natural logarithms are taken, we can formally write

$$\ln f_{S_1}(i\omega) = \ln |R_{S_1}| + i\theta(\omega) \tag{10.16}$$

and by analogy other terms transform the same way. In this case,

$$\ln T_{i,i+1}(i\omega) = \ln |R_{S_2}| - \ln |R_{S_1}| + \ln |R_{R_2}| - \ln |R_{R_1}|$$
$$+ i[\theta_{S_2} - \theta_{S_1} + \theta_{R_2} - \theta_{R_1}] \tag{10.17}$$

it being understood that all terms on the right-hand side are functions of ω.

Now, the same geometrical scheme adopted for the constant-offset time-delay system can be immediately used to generate successive differences of $\ln |R_{i+}/R_i|$ and

$$\theta_{i+1} - \theta_i$$

Thus, if a source and a receiver at a particular surface point have the same transfer function, the individual surface point transfer functions can be generated, and they can be used in combination to correct all the field traces.

It is noted that, if a shallow reflection zone is chosen, this method should end up by correcting phases and amplitudes of all reflections relative to a *constant-time* shallow zone, so it will combine the effects of the weathering with an isochron method mentioned earlier.

10.7 SUMMARY AND CONCLUSIONS

The correction of seismic reflection records for the influence of near-surface, or low-velocity, layers is one of the most important to be made but is one of the least successful of all data processing methods.

Early corrections, based on uphole times, were successful as long as the sub-weathering velocity remained constant. Geologically, subcrops can change in lithology and elastic parameters. Elevation changes (and consequent stress changes in the subweathering rocks) induce velocity changes. These either have to be monitored closely by drilling and surveying deep holes, or methods have to be devised that take such subweathering velocities into account.

The CDP methods of reflection prospecting produce much redundant data, used in the stacking process for improving the signal/noise ratio of the reflections. However, much near-surface correction time information can be derived from the reflection data themselves. The methods depend on the assumptions made. Three conceptually simple methods have been given which, however, need a great deal of bookkeeping in the computer programs. They are:

1. CDP NMO-corrected traces are cross-correlated to determine time differences between individual traces.
2. Common source or common receiver traces, NMO-corrected, are cross-correlated to determine time differences. Surface consistency of the time delay is assumed no matter whether the surface point is occupied by a source or receiver.
3. Common-offset traces are used to avoid errors due to incorrect NMO removal. Cross-correlations give time differences.

In all these, averages or other devices are used to separate out the surface consistent residual statics. One of the most prominent difficulties is the fact that it is impossible to distinguish between long-wavelength (along the surface) changes in statics from changes in reflection time due to geological structure.

To obtain greater accuracy in both near-surface corrections and in velocity, iterative methods have been devised. The convergence of these methods under all conditions of signal/noise ratio has not been established. A recent method by Wiggins, Larner, and Wisecup expands the input data in the form of orthonormal functions of distance along the surface, and both the accuracy and the number of iterations necessary to converge to a given precision can be established a priori. High precision in near-surface corrections is most important for high-frequency reflection work.

The dispersion of surface waves and the use of the ABC refraction method are two possibilities for establishing an absolute value for the weathering thickness. These values may be of some utility in checking CDP statistical methods. However, in the case of surface wave dispersion measurements, these must usually be made at the very lowest bands of the usual exploration seismology frequency spectrum, and some estimates of the P-wave velocity, S-wave velocity, and density must be available for both the weathered and subweathering layers. The results are most sensitive to S-wave velocities and only secondarily affected by density and P-wave velocity.

A method of treating near-surface effects as a generalized filter has been given, and the procedure outlined.

It appears that good progress has been made in solving the problem of calculation and correction of near-surface effects. More advancement is needed, particularly in high-frequency reflection work, and there appears to be a good change of additional progress being made. It is emphasized that modeling of the weathered layer as a simple constant-velocity, variable-thickness layer is untenable, and the problem of corrections for the near surface is one of the most important and complex of all of the problems of exploration seismology.

APPENDIX 10A: INTERPOLATION OF SEISMIC TRACES USING THE SIN x/x METHOD

We first prove that a time function, known to be limited in frequency to f_N (the Nyquist frequency) on the high end, can be represented by a series of sin x/x-type functions whose amplitudes are the sample values of the trace at the known sampling rate. $f(t)$ is known to be given by $\int_{-\omega_N}^{\omega_N} g(\omega)e^{i\omega t}\, dt$. We sample it at a series of points separated by τ, where $\tau = 1/2\omega_N$. Then the sampled trace is

$$f_s(t) = \sum_{j=\infty}^{\infty} \int_{-\infty}^{\infty} a_j \delta(t - j\tau)\, dt \tag{10A.1}$$

an infinite series of delta functions whose amplitudes are those of the trace at the sample points.

Note that, to filter this trace, we can filter each delta function in turn and sum, giving

$$\int_{\infty}^{+\infty} \delta_F(t - j\tau)\, dt = \frac{1}{2\omega_N} \int_{-\omega_N}^{\omega_N} 1 \cdot e^{i\omega(t - j\tau)}\, d\omega = \frac{1}{2\omega_N}\left(-\frac{ie^{i\omega(t-j\tau)}}{t - j\tau} \right)_{-\omega_N}^{\omega_N}$$

$$= \frac{\sin \omega_N(t - j\tau)}{\omega_N(t - j\tau)} \tag{10A.2}$$

and

$$f_F(t) = \sum a_j \frac{\sin \omega_N(t - j\tau)}{\omega_N(t - j\tau)} \tag{10A.3}$$

Thus the trace can now be interpolated by evaluating the expression on the right at the needed sample points. Since $[\sin \omega_N(t - j\tau)]/\omega_N(t - j\tau)$ is an infinite function, however, in the computer process, an approximation limited in time extent must be used.

REFERENCES

Disher, D. A., and Naquin, P. J. (1970), "Statistical Automatic Statics Analysis," *Geophysics*, Vol. 35, No. 4, pp. 574–585.

Dobrin, M. B. (1942), "An Analytic Method of Making Weathering Corrections," *Geophysics*, Vol. 7, pp. 393–399.

Dobrin, M. B. (1951), "Dispersion in Seismic Surface Waves," *Geophysics*, Vol. 16, No. 1, pp. 63–80.

Dusha, L. (1963), "A Rapid Curved Path Method for Weathering and Drift Corrections," *Geophysics*, Vol. 28, pp. 925–947.

Goldin, S. V., and Mitrofanov, G. M. (1975), "Spectral-Statistical Method of Calculating Surface Inhomogeneities of Reflected Waves in an Iterative Tracing System," *Geologiya i Geofizika*, No. 6, pp. 102–111.

Handley, E. J. (1954), "Computing Weathering Corrections for Seismograph Shooting," *World Oil*, Vol. 139, pp. 118–128.

Hileman, J. A., Embree, P., and Pflueger, J. C. (1968), "Automated Static Corrections," *Geophysical Prospecting*, Vol. 16, No. 3, pp. 326–358.

Patterson, A. R. (1964), "Datum Corrections in Glacial Drift," *Geophysics*, Vol. 29, pp. 957–967.

Saghey, G., and Zelei, A. (1975), "Advanced Method for Self-Adaptive Estimation of Residual Static Corrections," *Geophysical Prospecting*, Vol. 23, No. 2, pp. 259–274.

Wiggins, R. A., Larner, K. L., and Wisecup, R. D. (1975), "Automated Residual Status Analysis Using the General Linear Inverse Method," paper presented at the 45th Annual International Meeting of the Society of Exploration Geophysicists, Denver, Colorado, October 1975.

Wine, R. L. (1964), *Statistics for Scientists and Engineers*, Prentice-Hall, Englewood Cliffs, N.J.

ELEVEN

The Interpretation Problem

11.1   INTRODUCTION

At this stage of the book, most of the seismic reflection principles, field acquisition systems, and data processing methods have been outlined. The role they play in reaching the overall goal has been played down previously, but we can now reach out and use any of these tools, hopefully with a good deal of understanding, in our quest for the location of oil and gas fields.

The location of oil and gas fields is, logically, a purely (geo)physical problem. By this we mean that, given the properties of hydrocarbons—that they reside in the pore spaces of rocks and that gas, oil, and water have densities arranged in that increasing order—the entire process of finding them can be regarded as the application of seismic reflection methods only.

Unfortunately for the ego of geophysicists, this is not the case. Geologists, of varying special talents, are also employed in this task:

1. Because the location of fields must be economical.
2. Because of the limited resolving power (ability to see all the detail necessary).

The initial tasks of geologists lie in the direction of evaluating the available information (from surface outcrop information to detailed lithology and geochemical data available from well logs) in order to determine the areas where geophysical work can be done to maximize the chance of finding oil and gas while minimizing the cost of drilling exploratory holes. This geophysical work may include methods other than reflection seismology.

The reflection seismograph was neither the earliest nor the sole method employed; nor is it capable of finding all the hydrocarbon deposits that still remain to be found. A few million holes have been drilled around the world, and the information from these holes, descriptive, geochemical, mineralogical, stratigraphic and physical, has in the hands of geologists played a most important part in the location of oil fields and in producing *leads* which optimize the economical employment of the reflection seismograph. It is difficult, indeed, to assess the fruitfulness of reflection seismic work, since it is so dependent on the state of exploration existing at the time. It has been most fruitful where all the other conditions for hydrocarbon generation and accumulation have been established (or taken for granted) and where the reflection seismograph was responsible for the delineation of the necessary anticlinal structure. While this is by no means its only utility, it is the best known.

11.2 STRUCTURAL INTERPRETATION—THE EARLY YEARS

It was only when areas that could be explored by surface geology were nearing exhaustion—or the projections from surface outcrops were shown to be ambiguous (if not entirely erroneous)—that a need was felt for the reflection seismograph. The assignment was of course to map as accurately as possible the subsurface structure and tectonics, even though these features occurred in several cycles. The cycles were often found to be unrelated because of erosional or depositional periods which gave rise to unconformities.

In some areas, the seismic reflection method of 1940 (say) performed extraordinarily well, and most anticlinal structures found from detailed mapping with a closure of 50 ft had a high probability of being productive. It is no wonder that this led to the adoption of a tool with which, for the expenditure of a few thousands of dollars, an oil field worth millions could be almost confirmed before the first hole was drilled. These areas are chiefly characterized by:

1. Consistent near-surface conditions for which the primary corrections can be made accurately from uphole time and elevation information.
2. Persistent reflections at the requisite depths over large areas—the ability to correlate reflection events from point to point over the area under investigation. This now means invariability of layer thicknesses and rock properties.
3. The presence of gentle anticlinal structures and the (fortuitous) presence of source rocks and other rocks with reservoir characteristics, capped by impermeable layers.
4. The lack of very steep dips and/or major faulting.

At a time when reflection seismograph usage was regarded as an art, and results were obtained only by experience, it was fortunate that areas that had these characteristics were widespread enough. Confidence in the method had to be built up before some of the hazards made their presence known very frequently. It is not necessary to dwell on such halcyon days. It is sufficient to say that simple procedures—correcting the times, transforming to depth through the use of a velocity that matched the well information, and mapping the subsea data for as many different reflections as could be handled—were adequate. The method prospered.

Faulting was inferred either by definite isolated reflection correlations across a more-or-less linear zone on a map or, in many cases, because of such negative evidence as the deterioration of reflection quality. Both structural (anticlinal) traps and traps that relied partially on faulting were drilled and added to the general information for the area, even if they did not produce.

11.3 STRUCTURAL INTERPRETATION—PROCESSING IN A MORE SOPHISTICATED ERA

The time of disillusionment soon came. False structural anomalies—due to time anomalies caused by velocity changes in the near surface or in the subsurface as a whole—were soon found. Deterioration of reflection quality sometimes presented problems and resulted in false structural indications due to miscorrelations. Greater emphasis was then placed on improving record quality—first by the elimination of seismic artifacts or nonreflection events—by patterns of sources and receivers and brute force averaging and then, later, by the invention of methods that achieved some degree of consistency in the source and receiver characteristics.

In retrospect, the methods of structural interpretation changed only in minor ways —institutional prejudices or preferences largely being responsible—even up through these major seismic revolutions. The development of the Vibroseis® method in the early 1950s coupled the use of surface patterns of geophones and source positions with a reproducible source and recording on magnetic tape. By the time the use of the Vibroseis® method was routine in the early 1960s, a further addition—phase control of the vibrators—achieved a degree of consistency that had long been sought. The use of CDP recording, in spite of its unwelcome increase in unit cost, was demanded because of its utility in giving better data quality—but largely for structural interpretation. It is no surprise to practitioners that there are still some areas which, despite the existence of a layered sedimentary system, still refuse to yield good enough seismic reflections on which to base a reliable structural interpretation.

In the CDP and Vibroseis® methods on land and in the use of air or gas guns in marine areas, the seeds of an additional revolution in reflection seismic interpretation were laid. The source output could be made consistent for the first time, and some control could be exercised over the presence of multiple reflections. The innovation of redundant data gathering allowed seismic velocities to be calculated accurately *and often*. Basically, the X^2-versus-T^2 method of velocity determination had been in existence for many years and had been used sparingly to provide velocity information in the absence of well velocity surveys. In marine work, the achievement of consistency was relatively easy because of the constant characteristics of the medium in which the source and the receivers were immersed.

Since our task is to teach the principles of interpretation as practiced today, the combination of a consistent source and CDP methods is taken as the basis for further discussion. We can summarize the *prime needs* for interpretation as:

1. A seismic reflection record section which has established a high signal/noise quality through CDP methods. The degree of redundancy must be determined by the need for such quality.
2. The removal of as many nonreflection (including multiple-reflection) events as possible.
3. The use of consistent source outputs and receiver input and recording characteristics (in other words, changes on the record section, trace by trace, must be due to a change in geology and not to method characteristics).
4. The effects of near-surface abnormalities (at this stage, chiefly time delays) must be removed as accurately as possible.
5. In the first instance, the time section should be produced by carefully following the principal primary data processes. In this context, we list:

   a. Muting
   b. Energy equalization
   c. Expansion
   d. NMO and static corrections (preliminary)
   e. Stacking

f. Predictive deconvolution (before and/or after stack)
g. Filtering
h. Preliminary velocity determinations

as the processes that may all be needed.

The seismic interpreter must assume responsibility for setting up the processing schedule, including determining the processing parameters that need to be known before the digital processing is done. The need is to obtain one specimen section from each area, which will allow determination of the residual problems before finalizing the data processing for the area as a whole. Often this role of the interpreter is the most difficult and time-consuming one that he will have. The prime requisite is to obtain data that, in the time domain, *is truly representative of the geology and minimally affected by artifacts.*

The disturbing events to be removed (if possible) fall in the following categories:

1. *Residual interference* (source-generated noise) which cuts across the reflection events at various angles, depending on the phase velocity along the surface. This should have been removed by patterns of sources and receivers but may, in particular parts of the area under investigation, become very strong because of changes in the weathering or surface. If the events (best seen on the primary or 100% records) can be picked, and therefore aligned by shifting traces with respect to one another, this residual interference can often be removed by two-dimensional filtering. After removal, the traces are returned to their original positions and restacked.
2. *Lack of alignment of reflections* because of inadequate near-surface corrections. The cycle of residual weathering corrections and velocity determination, followed by NMO removal, can be repeated. A final stack may give considerable signal/noise improvement (see Figure 10.10 for a good example).
3. *Noise.* Random noise should have been removed during recording by using an adequate redundancy in the CDP work and by a proper choice of source and receiver patterns. It is often the result of natural sources generating waves which arrive at the receivers from random directions. Radial patterns, although more expensive to use in the field, may be preferable to the linear patterns commonly used to reject source-generated interference. Nevertheless, given the fact that the noise has been recorded on each of the primary records, the following procedures may help to reduce its effect.

   a. Study the effect of successive narrow-band filters on the data. Often the visual coherency of events on these filtered records shows which frequency bands should be retained. Simple frequency filtering can be used to improve the signal/noise ratio.
   b. Some improvement may be obtained by a weighting process in the stacking procedure. For example, if the primary records are *very* noisy, the individual records can be weighted individually in inverse proportion to the total energy (assumed to be noise energy). This process of diversity stacking is often helpful in very noisy areas (e.g., recording near highways or in towns).

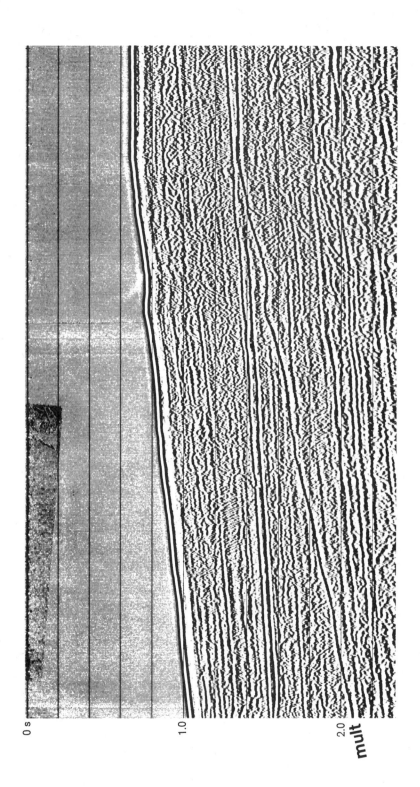

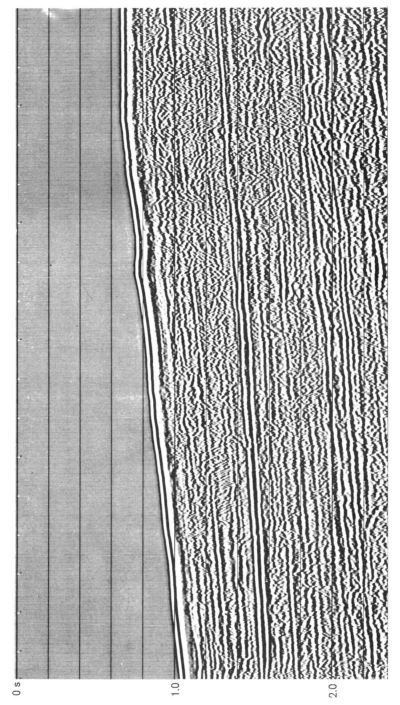

Figure 11.1 The effective removal of a sea bottom multiple by data processing. (Courtesy of Prakla-Seismos. GmbH.)

Events such as diffractions and crossing reflections due to highly curved reflecting surfaces have not been included, as they are indicative of geological features and, in any case, will be minimally disturbing after a later process of digital migration.

4. Multiple reflections, however, are dangerous to interpretation when they are of the long-period type. Discussion has already been given of the usefulness of short-period or peg-leg reverberations which act to improve the signal/noise level of deeper reflections. The periods of the long-period multiples can be ascertained by autocorrelation of the seismic traces, the multiples then showing up as ghosts— subsidiary pulses at multiples of the multiple time delay. If more than one major long-period multiple is present, the ghosts are more complex, consisting of differences in individual multiple delay times as well as occurrences of the multiple times themselves.

The most promising ways of removing such ghosts are:

a. Gapped deconvolution. The operator is obtained by assuming a desired output which consists of a unit pulse followed by a number of zeros (out to the desired time) followed by finite amplitudes out to the length of the operator desired.

b. The Backus (1959) operator. This consists of convolution with the inverse operator to a simple series of pulses, the amplitudes and signs of which are known. If isolated, very prominent multiples are present, these amplitudes and signs will be known from the trace autocorrelation function.

An example of sea bottom multiple removal is given in Figure 11.1 (courtesy of Prakla-Seismos, GmbH.). The final processing desired is thus established. Beyond this, there should always be a relative amplitude section and a migrated section—made with the best velocity information available—provided that the necessary signal/noise ratio has been established. There are areas where the interpreter will decide that it is a waste of time to make the latter two sections.

11.4 THE MECHANICS OF MAKING A STRUCTURAL INTERPRETATION

There is no way, in a summary, to deal with all the idiosyncracies of individual areas. All that is attempted here is to lay out some of the methods, the information needed, some of the problems that will be encountered and, hopefully, some solutions.

The first step is to gather all available information on the area—well logs, correlated and with the major horizons marked and named. Among these logs (if indeed *any* logs are available) should be at least one sonic or well velocity log (preferably of the long-interval type) and one density log. It is obvious, from our previous considerations, that the sonic log(s) should be run over as long an interval of the well as possible, ideally from the base of the weathering to the total depth, although the upper limit usually depends on the surface casing set, hole diameter, and velocity of the sediments near the surface. Synthetic records should be made for all of them, using an input pulse with a frequency spectrum similar to the final playback filter adopted and a phase spectrum determined by the field method and the processing adopted. A deconvolved impulsive seismogram and a correlated Vibroseis® record are both nominal records using zero-phase pulses. SAIL synthetic records should be produced.

The field effort—with which the interpreter should have been coordinated—should have included one field line passing as near as possible to one of these wells.

The identification of reflections and determination of the time lag to use between field and synthetic data is usually accomplished by a straightforward visual correlation between the synthetic SAIL trace and SAIL field traces taken near the well. This correlation is often easy to determine on a zone-to-zone basis. Slight shifts in time are sometimes necessary between zones—because of incorrect well velocity measurement (invaded zones or shale alteration, coupled with an inadequate sonic tool length).

In this manner, a basic grid of geological-seismic arrival time pairs are made for a number of geological horizons over the area. Even though only deep horizons may be required, it is good practice to produce at least one shallow and one intermediate horizon correlation—correlated from trace to trace on the entire set of seismic cross sections and tied from one to the other at line crossing points. This procedure seems simple but can be beset with difficulties. Some of these are now considered.

1. *Faulting.* The continuity of the reflection is interrupted. A note should be made as to whether or not this fault is substantiated by the presence of a diffraction. The picked event must be ascertained by correlation across the fault. It is desirable to be able to correlate between two events a few traces from the fault location on either side. This is done most easily when the bandwidth of the record section is as wide as possible, consistent with a high signal/noise ratio. Any faults picked need to be coordinated from cross section to cross section to make them geologically feasible. Faults are most often clearer on migrated sections, because of the proper placement of the diffracted events. In Figure 11.2, an excellent marine seismic section migrated by Western Geophysical, the author has had no difficulty in marking the discontinuities due to some of the faults. Horizons *A* and *B* can clearly be correlated across the majority of the faults, but what happens to *A* as one proceeds to the right is the subject of dispute. In a complexly faulted area, the interpreter must not lose sight of the fact that dips may not only exist in the plane of the cross section but may come about by reflections from the side of the plane. For this reason, in complex areas, the lines of profile are usually run down the direction of maximum dip, if known. Tie lines should then be placed in locations where the geology appears to be least complex.

It is a mistake to assume that faulting, even of geological strata having a high contrast in acoustic impedance with the neighboring rocks, is invariably noticeable. We have already seen that a discontinuity gives rise to a diffraction which changes sign at the break. Then, for two halves of a broken formation, the positive half from one side is superimposed, with a time delay, on the negative half from the other side. If the delay time (due to the throw of the fault) is small or equal to one period of a particular frequency, the diffraction at that frequency will be nearly eliminated and will be small for neighboring frequencies. The proper frequency analysis of diffracted events may give a clue as to the throw of the fault, although we have no knowledge of any such attempt. In any event, comparison has to be made with the frequency analysis of the reflected event in order to eliminate the effects of stratum thickness. Very often, however, after faulting, one stratum is displaced into a position in which little contrast exists and the diffracted event is eliminated.

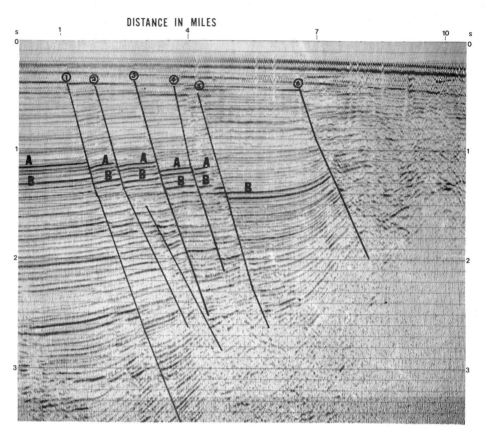

Figure 11.2 A portion of a finite difference-migrated section showing faulting. Even in this excellent example, taken by itself, the question of what happens to some horizons (such as *A*) as they cross a fault cannot be resolved. Is *B* marked in the correct position to the right of fault 5? (© 1975, Western Geophysical Company of America, Inc.)

A question always arises as to the likelihood of reflected events from the fault face. It should really be asked whether or not diffracted events from individual scatterers are persistent enough along the fault plane to organize themselves into a coherent reflection. Usually this means a set of scattering points lying along a break where *one* material is placed opposite a different unique material over a distance of several wavelengths of the seismic wave. This distance can be taken to be hundreds of meters in the usual seismic case, hence a fault of considerable throw is involved. It is clearly better to talk about a distribution of scatterers on a fault surface than about a reflection from the surface. Even if rock distortion and fracturing have resulted in a rubble zone having different fluid contents and acoustic impedance from the rocks laterally in contact with it, "reflection" is probably too precise a description to fit the occurrence, and the distribution of scatterers is to be preferred. Migration of the data, if done with correct velocities, causes the energy of the diffractions to pile up at the fault zone, as long as the scattered points are evenly distributed about the vertical plane of the seismic cross section. Lack of an even distribution, scattering points predominantly on one side of the plane of section, causes the diffracted events to be of an improper geometry to be condensed to the point-scattering pattern, and this results in smearing of the fault zone.

2. *Truncation of strata by unconformities.* An unconformity is a surface of erosion or nondeposition which separates younger strata from older rocks. It is axiomatic that, as far as the seismic reflection method is concerned, this surface will not be distinguishable from a bedding plane if, locally, the planes of deposition of the younger and older rocks are parallel. In a local area where there exists an *angular* unconformity, the upper rock layers are likely to be continuous and (see Figure 11.3) the older ones are truncated by the unconformity. If the lower rocks were subject to peneplanation (erosion to a near-smooth surface) before the younger rocks were laid down, the effects are largely due to the wedges of formations as they are gradually truncated. However, rapid subsidence of an immaturely eroded surface may have resulted in an uneven unconformity surface and consequently the probability of diffracted events.

If the angular unconformity has a small angle, the truncation results in a gradual loss of the younger formations up-dip. By taking the simple view of a constant-velocity formation, gradually reduced in thickness, it is evident that the seismic record will progress from a full rendition of the layer reflection through changes in the shape of the reflection as the layer thins and finally (for a given frequency band) to a gradual loss of amplitude. The more complex cases, of variable velocity zones within a given lithological zone, can be dealt with by the methods of synthetic record construction for a sequence of different velocity functions. There are, however, two important points for the interpreter to recognize:

a. The amplitude of a reflection from a truncated formation diminishes to a level below the noise level (or below the side lobe level for neighboring reflections)

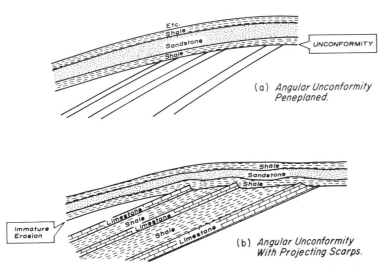

Figure 11.3 Two possible types of angular unconformity: peneplaned, diffraction effects minimized (*a*), and immature erosion, diffraction at the location of the scarps (*b*). There may be some draping of shallower formations over the scarps.

before the formation reaches zero thickness. Thus the mapped zero edge from seismic work is still downdip from the actual zero-thickness line. The distance varies depending on the acoustic impedance contrast of the layer being truncated.

b. If the truncation is gradual, there will be no indication of the zero edge by diffraction phenomena.

Whenever the erosion of the older formation is immature, scarp formation formed by projections of the more resistant rocks may occur. Now we have a case when the reflections from a layer can terminate abruptly and diffractions are formed. The updip limb of the diffracted event must be recognized for what it is; otherwise, the reflection will be continued and show reverse dip. Decisions as to the nature of a reflection-diffraction event are best made in conjunction with all the evidence from the area, that is, by a decision that an unconformity does or does not exist. Modeling programs are later described in which synthetic records can be made for two-dimensional changes in stratigraphy which include the possibility of diffractions. These may be of some help in making what is otherwise a difficult decision.

In mapping of course the reflection terminates at the up-dip limit of the truncated formation. Ties between lines must be made well downdip, where the formation is likely to be at full thickness. One such truncation—of the Hunton formation in Oklahoma—is graphically shown in Figure 11.4.

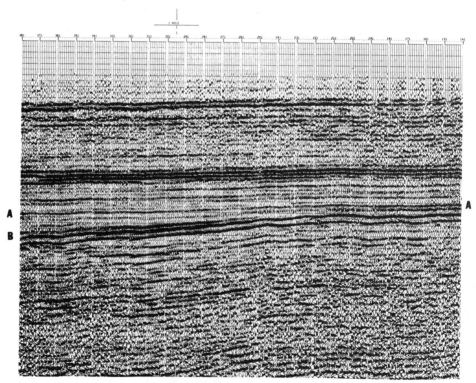

Figure 11.4   Truncation of the Hunton formation (B) by Mississippian rocks (A) as depicted in a modern, unmigrated seismic cross section.

11.5   COMPUTER MODELING—IMPLEMENTATION AND ONE-DIMENSIONAL

The interpretation of a seismic cross section in geological terms (e.g., the sequence of lithologies, thicknesses, and fluids filling the pore spaces) is not possible in an unambiguous sense. It has already been pointed out that, for thin layers compared with the wavelength of seismic waves, it is the product of the layer thickness and the contrast in properties that determines the seismic reflection output. Many possibilities are of course available, and it is the function of the exploration team to decide the most probable combination of parameters. There are some incontrovertible pieces of evidence—the seismic reflection cross sections and the well logs available in the area. Putting the two together and then interpreting the most likely conditions that exist at places where there are no wells are sometimes aided by the ability to forecast what the seismic output should be from a hypothetical set of geological and seismic parameters. This is the sequence of steps now called modeling. It is usually a process of looking at the seismic data in detail—having selected a time zone or a depth zone known to be of interest from the oil production point of view. It may be a truncated zone where the thicknesses and possible fluid contents of the formation change, or (later) we may be considering the stratigraphic trap problem, in which the lithology, the thicknesses, and the fluid contents may change simultaneously.

The basic tool is a synthetic seismogram and the basic equipment is a time-shared computer terminal equipped with a means of rapid input and output of log and reflection data. The methods of making synthetic seismograms have already been discussed (Chapter 4). From the point of view of the explorationist user, the computer terminal must be easy to understand and to use. Figure 11.5 is a schematic of an entire terminal complex. It has an overall function to allow the interpreter to interact with the computer so that he can obtain the best fit possible of the geological input with the known seismic output of the earth.

The terminal consists of the following visible parts:

1. A keyboard and a visual monitor [a cathode ray tube to display graphical (and alphanumeric) output and to retain it long enough for the interpreter to decide if it should be permanently recorded].

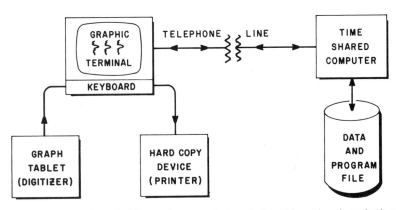

Figure 11.5   Schematic of a typical interactive computer terminal used in exploration seismic modeling.

2. An input graph tablet allowing the easy and comparatively fast input of logs and model descriptions of all kinds.
3. A means of obtaining a permanent record (hard copy) of any results displayed on the screen.
4. A connection device allowing access to a fast digital computer operating in a time-share mode. By this method, the available computer time is allocated to each terminal in small time slices. Return to each user is so rapid that his impression is that he has the computer under his complete control.

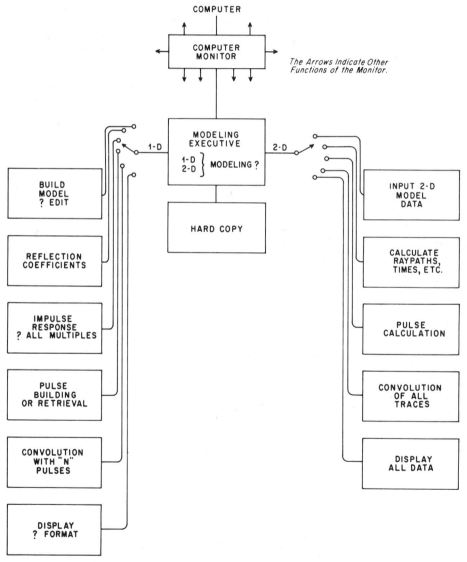

Figure 11.6 Schematic showing the function of the executive module of the modeling system. It allows the interpreter to order the execution of several possible functions without having an exact knowledge of computer requirements.

The invisible part, the programs that control the computer, is complex. It consists of monitor program(s) which control computer action as a whole and user programs to which the computer has access in order to do the special jobs the interpreter requires it to do.

Each institution's modeling system is different, depending on the philosophy of the originator and, to some extent, on the type of main computer and the peripheral equipment available. Usually, however, the system of programs is modular, access to and exit from any particular program being under the control of an executive program. The function of the executive is to gather together initially all the general parameters which are input and communicate them to the individual modules as needed. It facilitates entry into any desired module and, on exit, makes it possible to enter another module, to obtain a hard copy of results, and/or to exit in a tidy manner from the modeling system. All this is concealed from the user, who normally supplies answers to questions posed by the system and displayed on the graphic terminal screen. An ideal modeling system, once the user has been led through it and the terminal controls explained, requires no technical computer knowledge from the user, only geophysical knowledge and a knowledge of the overall capabilities of the system. Figure 11.6 shows how the executive controls the individual modules and isolates the user from technical knowledge of the computer monitor program.

The one-dimensional modeling system needs little comment in addition to the description of synthetic trace construction given in Chapter 4. However, the purpose is to help the interpreter make a decision as to what the characteristics of the velocity and density logs have to be in order to give a seismic output that is already on hand. Thus usually only a small section of these logs is used. Facilities must be available to make modifications quickly (editing) and for displaying the resulting effect on the seismic section. Multiple reflections *within the section being examined* are important only if the section is likely to contain strong reflection coefficients (e.g., a gas-filled sand in conjunction with a shale and/or a water-filled sand). Multiple reflections (including reverberations or peg-leg multiples) occurring in the remainder of the log,

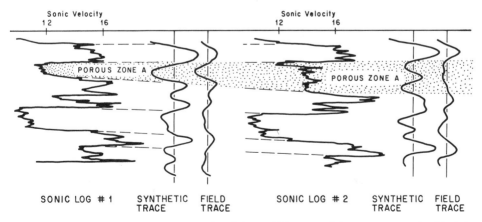

Figure 11.7 Two different sonic logs showing different thicknesses of the porous zone *A*, with their synthetic traces and field recorded traces. The pulse used for modeling was an autocorrelation integrated pulse 9 to 30 Hz. The field traces were integrated and filtered 9 to 30 Hz, and a nominal 30° phase correction applied.

including the near-surface layers, can best be taken into account by changing the input pulse shape. Examples of the output from a one-dimensional modeling system have already been given (Figures 8.5 and 8.6).

These were, however, heuristic models designed to teach a specific point concerning the output of thin beds, and the velocity log concerned was not a realistic one. In Figure 11.7, a more realistic example is shown. Details concerning the log and the pulses making up the section are given in the legend.

One further remark must be made about modeling. The fact that a modeling program is available does not mean that it is possible thereby to obtain a more detailed description of the lithology giving rise to a given seismic output. The amount of detail possible is controlled by the bandwidth and frequency range of the seismic output. The proper use of a modeling program, however, ensures that the lithology determination is *one* possible answer to the question of matching the seismic output. If the interpreter plays the interactive terminal game thoroughly, even when he restricts himself to the confines of the parameter boundaries, several possible answers usually emerge. His choice has to be made based on other factors which can be brought to bear on the problem. One such factor may be the existence of shear reflection records. The modeling program is of course available for use with shear reflections and only the velocity log need be changed. The method of making a fit between P-wave and S-wave reflection data in order to make a better determination of the lithology, porosity, and pore fluid content is in its infancy and is not pursued here.

11.6 TWO-DIMENSIONAL MODELING—THE RAY PATH METHOD

The one-dimensional model of course relies on the assumption of horizontal plane-parallel layering. It gives only an idea of the seismic output from a geological section if structure is of no consequence. That is, if there are no interfering reflections from dipping formations to the side and if no abrupt strata terminations give rise to diffracted events.

In order to gain better insight into the conditions that exist in practice, the next step is to consider a two-dimensional section, such as is shown in two slightly different forms in Figure 11.8. In these idealizations of truncation, it is necessary to compute actual ray paths, since some of them are not vertical (even for a common source-receiver point). In addition, it is desirable to be able to consider diffractions from terminating reflectors. Most two-dimensional modeling programs have facilities for calculating offset traces, with limited facilities for including multiples and thus, by extension, the possibility of creating *n*-fold stacked records. Ray path plots for models A and B are shown in Figure 11.9*a* and *b*. The edge diffraction events have been suppressed in the plot to avoid confusion.

The impulse response for each source-receiver combination is calculated with the two-dimensional program, and it is important that proper consideration be given to the form of the impulse response for different types of events. These are controlled by the type of stationarity of the time path involved. We have already seen that, for a reflection from a flat surface, there is no change in wave characteristic, so that the response to a delta-function input is also a delta function at the appropriate reflection

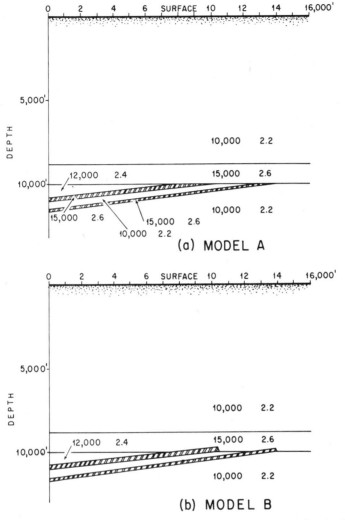

Figure 11.8 Two different models with and without peneplanation of the unconformity. Velocities and densities for models A and B are identical.

time. This is not the case, however, for other types of events. By use of diffraction theory, it is possible to calculate the impulse response (Stolt, 1975).

1. *Travel time minimum* (*normal reflection*). This is the type for reflections from flat, convex, or slightly concave surfaces where the focal point, if any, must lie above the surface observation point. In the model, the form of the output is a delta function whose amplitude is controlled by the radius of curvature and the depth of the reflector. Spherical divergence (or convergence) from the reflecting surface is thereby accounted for.
2. *Travel time maximum* (*buried focus*). If the surface is so concave that the wave reaches a focus below the surface, a travel time maximum will result. This gives rise to a characteristic shape of the form $(t - t_0)^{-1}$ which is an amplitude time

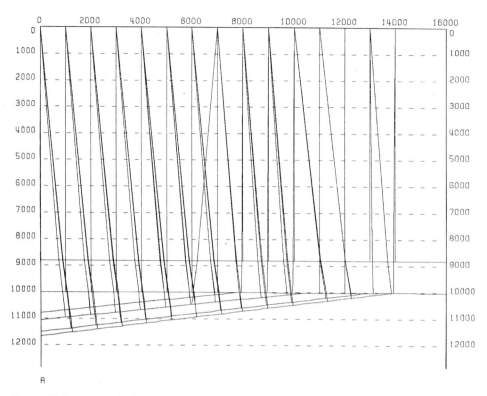

Figure 11.9a Ray paths for model A. Ray paths associated with diffractions are calculated but not included in this display.

shape having a 90° phase. The intensity is also proportional to the inverse of the curvature of the travel time curve, so the deeper the focus, the smaller the amplitude of the event. See Figure 11.10a.

3. *Diffracted event.* The travel time minimum and maximum correspond to points where the slope of the travel time curve, as measured along the reflecting surface, is zero. Higher-order derivatives of the travel time also produce events. The principal such event is caused by a minimum (in absolute value) of the travel time derivative. Its characteristic shape is also shown in Figure 11.10b. Its amplitude depends on the first and third derivatives of travel time at the diffraction point. The event may be expanded or contracted in time depending on the values of the first and third derivatives but again depends most strongly on the first. Thus, at points where the first derivative of travel time is near zero, there is a strong peaked response. The diffractive event can be thought of as a maximum and a minimum occurring at the same point. When moving from trace to trace along the seismic section, one often sees diffractive events split into a maximum and a minimum. While the minima tend to move along the source-receiver surface with the source-receiver midpoint and the maxima tend to move in the opposite direction, the diffraction points tend to remain stationary as long as they exist.

4. *Edge effects.* Whenever a reflecting surface terminates (or abruptly changes slope), it gives rise to an edge effect diffraction. The amplitude of this diffraction varies

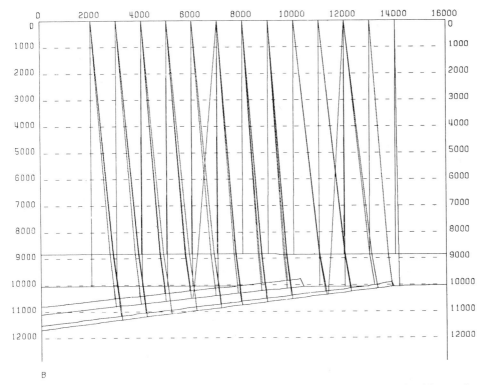

Figure 11.9b Ray paths for model B. Ray paths associated with diffractions are calculated but not included in this display.

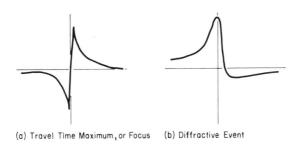

(a) Travel Time Maximum, or Focus   (b) Diffractive Event

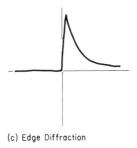

(c) Edge Diffraction

Figure 11.10 Different types of pulses arise from different types of seismic events. Only reflections from flat or convex surfaces (compared with the incident wave front) give a delta-function reflected pulse for a delta-function incident pulse. (After Stolt.)

inversely with the first derivative of the travel time at the edge, and the speed at which it dies out depends on both the first and second derivatives at that point. The general shape shown in Figure 11.10c was derived earlier in Chapter 7. At the edge along which several boundary segments join, the total response is the sum of the responses from each segment.

These effects, which are incorporated in the modeling program, are of no worry to the interpreter. When needed they are all incorporated in the impulse response for each source-receiver combination.

A two-dimensional modeling system normally incorporates NMO removal programs to facilitate stacking the model traces and a convolution program which allows filtering in any manner necessary to try to match the filter characteristics of field traces. This question of the filter characteristics of seismic records is most important in modeling and in the interpretation in geological terms. Besides the question of resolution, however, there is also a question of the representation of a geological boundary and the ease of conversion of the reflection results into geology.

Although this point has been approached before, it is important enough to be reemphasized in the modeling context. As we know well by this time, the seismic record, in a simple no-multiple form, can be envisaged as the sum of pulses, all of the same shape, of different amplitudes and arrival times. If these constituent pulses

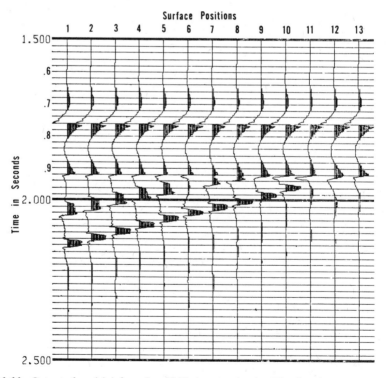

Figure 11.11 Output of model A for a 7 to 90 Hz integrated pulse. The disturbance and confusion of traces 6, 7, 8, and 9 near the 1.95-sec reflection time could have been anticipated because of the juxtaposition of similar materials.

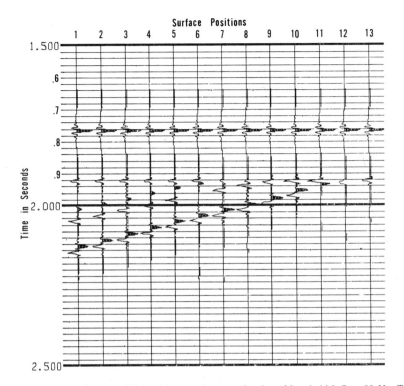

Figure 11.12 Output from model A with a nonintegrated pulse of bandwidth 7 to 90 Hz. The tie-in with geological boundaries is not quite as pronounced as in Figure 11.11, although the positions and senses of the pulses show their positions.

are long and complex in form, the effect of one boundary will be inextricably confused with those of other boundaries. There are two forms of pulse that can be used to the best advantage. Such pulses, in practice, are obtained from Vibroseis ® records after the cross-correlation process. Each symmetrical pulse corresponds to a reflection coefficient or, more exactly, one impulse of the reflection impulse response. Proper deconvolution of other types of records, however, produces pulses of the same shape. Another pulse we have used is the integrated form of the symmetrical pulse as derived in the SAIL process. The latter has the advantage of making a boundary look like a boundary. In the models constructed for illustration of the two-dimensional modeling system, only these two types have been used. Figure 11.11 shows the seismic output from model A using an integrated pulse with a nominal bandwidth of 7 to 90 Hz, with a taper of 4 Hz at the low end and 20 Hz at the high end. Tapers are centered on the bandwidth frequencies given.

Disturbances and confusion in the vicinity of 1.95 sec for traces 6, 7, 8, and 9 could have been anticipated, but the actual form would not have been known without the modeling. At these frequencies, it is possible to discern easily the boundaries of the 200-ft limestone and also the 100-ft limestone beds truncated along the 10,000-ft depth line. The sense of the velocity change at all boundaries is graphically illustrated. While these same things can be seen by the interpreter when a normal, nonintegrated pulse is used for convolution (in Figure 11.12), the tie-in with geological boundaries is not quite as pronounced. In both presentations, the location of diffracted events is unclear.

The composition and mechanics of the two-dimensional modeling system must not be left without a reminder that it is still an artificial model of the earth. The assumption is made that the model cross section is constant in the direction perpendicular to the plane of cross section. A comparison with field records can be made only if such field records were obtained on a line directly down-dip. In areas where dipping events exist, such a unique direction (for shallow and deeper zones) may not exist, and it is indeed fortuitous if a seismic line runs in the required direction. The effect of reflections coming from out of the plane of cross section must therefore be considered. As far as is known, no three-dimensional modeling system is used routinely, although it is seen later that one exists for special purposes.

11.7 STRATIGRAPHIC TRAP LOCATION—
TRANSFER FUNCTION USAGE

Application of the seismic reflection method to the location of stratigraphic oil traps has been a favorite subject for discussions and for papers. Unfortunately, there appears to be no common understanding of what constitutes a stratigraphic trap. It is usually defined from the point of view of geologists; that is, a stratigraphic trap is one that does not rely on structural or tectonic processes to trap hydrocarbons. As a corollary, the process involved must involve changes in the permeability and porosity of the host sediments achieved either by sedimentation or by metamorphic processes. Examples usually given are those in which porosity in a sandstone body silts up until the pore spaces are insufficiently large to allow the passage of hydrocarbons; or a change from limestone to dolomite by the action of magnesium-bearing salts in groundwater, thereby causing porosity because the specific volume of dolomite is lower than that of calcite.

If the stratigraphic trap does not involve structure or faulting, from the seismic point of view it is not possible to find a stratigraphic trap simply by measuring changes in the arrival time of the *associated* reflection. Note that this does not preclude indirect evidence of a stratigraphic trap possibility by time interval changes in which the interval includes the trapping formation. While this definition undoubtedly has some holes and is specialized to information the geologist does not always have, it is adopted here because it leads to a clear-cut break between structural and stratigraphic traps. With the latter, we simply abandon one piece of evidence the seismograph easily provides. What seismic parameters are left? The easy ones of course are time differences and comparative reflection wave shapes. The cross section, which has been flattened using a strong, consistent shallow marker as a reference time—made constant by shifting all traces—is a seismic analog of the isopach cross section of the geologist and is just as useful. Emphasis must be placed on the criterion of consistency for the reference reflection. If it is known, or can be assumed, to have been derived from a constant-thickness, constant-lithology-type layer, then changes in its character or time of arrival must be due to near-surface layering. In the critical analysis of seismic data for changes due to changes in lithology or bed thickness, effects of the surface and near surface must be removed. This has been done by the use of transfer functions.

These transfer functions are simply filters, usually digital filters, by the use of which the character of a particular zone of one trace A can be made to look *exactly* like the same zone of a second trace B. We simply convolve trace $A(t)$ with the transfer function $T(t)$ to arrive at $B(t)$:

$$A(t) * T(t) = B(t) \tag{11.1}$$

Leaving, for the moment, the question of how $T(t)$ is determined, we can see that, if the shallow portions of two records are made to look alike and if in fact they *are* geologically similar, the records have to have been produced with input pulses that are alike. If the early portion of trace A had to be treated with the transfer function T to make it look like B, it is plausible that the later portion of A must also be treated with T to be able to compare A with B in the later zones. This filter takes into account both phase and amplitude spectrum deviations of the source pulse A from that of B and therefore corrects any time errors as well as amplitude spectrum errors.

The determination of T is made directly from the sequences $A(t)$ and $B(t)$ early in both records. It is easiest to visualize in the frequency domain, so that it is assumed that

$$A(\omega)T(\omega) = B(\omega) \tag{11.2}$$

Therefore

$$T(\omega) = \frac{B(\omega)}{A(\omega)} \tag{11.3}$$

where A, B, and T = complex functions of ω.

There is a corresponding matrix method for the time domain determination of a convoluting pulse.

The operation is particularly necessary when several different vintages and methods of shooting have been employed in the area under investigation. It should be noted that the two traces are brought into conformity only within the common bandwidth so that, at best, the bandwidth will be left the same with considerable likelihood that it will be reduced. In the determination of $T(\omega)$, any values of $A(\omega)$ that approach zero cause instability, and $T(\omega)$ must be (artificially) set to a number determined by other considerations. In calculating $T(\omega)$, it is only the smoothed amplitude and phase spectra that are to be made similar, since any rapid variations in amplitude and phase are due to the relative positions of reflected events and are not a function of the primary pulse.

11.8 STRATIGRAPHIC TRAP LOCATION— GENERAL METHODOLOGY

Additional seismic evidence for stratigraphic trap determination is provided by measurements of the seismic interval velocity and by comparison of consecutive seismic traces through the various methods (Section 7.5) of measuring trace similarity

(Waters and Rice, 1975). Both of these parameters are reduced in effectiveness by the fact that they are averaged over a substantial time interval compared with the time thickness of most stratigraphic oil traps. If the record section has adequate bandwidth, the role of *local* (interval) velocity indicator is taken over by the SAIL-type trace for, as previously pointed out, the trace deflection is, within limits imposed by the frequency spectrum, roughly proportional to the product of the velocity, density, and thickness. The answers sought about changes in lithology, porosity, and pore fluids are all in the SAIL section—as far as they can be determined by the seismic method with this spectrum. Thus the similarity coefficients and statistically measured interval velocities can be plotted, and their gradients will fulfill the role of attention getters. However, the absolute value of velocity for a portion of the section plays an important role in deciding on the type of section, for example, clastic versus carbonate or evaporite, or normal-pressured versus overpressured.

Since the stratigraphic traps so far discovered have little in common geologically, it is difficult to describe a set of invariant methods to discover them. The general methodology is nevertheless the same, with variations in interpretation caused by different geological conditions. The general procedures are:

1. Obtain a set of seismic cross sections with maximum bandwidth and a maximum signal/noise ratio through the use of highly redundant CDP procedures.
2. Process these records in such a manner that reflection events are symmetrical pulses of maximum bandwidth. It is important that, at least, a set of relative amplitude–preserved sections is obtained, if these are not the only sections available.
3. Integrate the processed records to obtain SAIL-type sections which may also be relative-amplitude.
4. If on inspection it is found that there is enough structure to influence section complexity (e.g., synclines which would give a focus below the surface), then the best migrated sections possible should be made from the relative amplitude–processed and SAIL cross sections. These form the first part of the seismic evidence.
5. Prepare a contoured section of the subsurface interval velocity of the proper scale to overlay the seismic cross sections. These are often filtered in two dimensions to remove noise due to low-accuracy velocity determinations.
6. Prepare a contoured similarity coefficient overlay of the correct scale to overlay the seismic cross sections.

The initial data are now at hand. Since stratigraphic trap location is of the same order of difficulty as looking for a needle in a haystack, there needs to be a preliminary winnowing of this material to determine which zone offers the best chance for finding a stratigraphic trap. If anything, other than the seismic data, is known about the area under investigation, this is normally a task for geologists, who can examine well logs and samples to grade the different depth zones for source and reservoir potential. If this is not possible, the sieving procedure devolves on the seismic interpreter.

The most obvious evidence of source rock potential is the indication that gas gives away its presence in a reservoir by enlarging considerably the amplitude of the reflection associated with the reservoir in which it occurs (see Figure 11.13). The change in velocity caused by the gas replacing water in the reservoir sand (Section 7.3) gives rise to such secondary indicators as delays in the deeper reflections and sudden

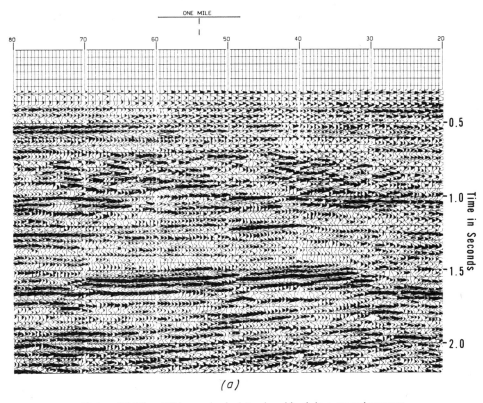

Figure 11.13a Offshore seismic data played back in a normal manner.

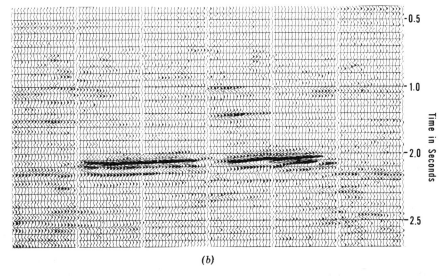

Figure 11.13b Offshore seismic data relative amplitude–processed. (Courtesy of the Continental Oil Company.)

attenuation of energy on these reflections. Diffractions and changes in reflection character near the boundary of the reservoir provide additional evidence for gas occurrence. Even within this small subset of hydrocarbon location problems, there exists some danger. Domenico (1975) showed that even a small (5%) amount of gas in the reservoir is sufficient to lower the mixture velocity suddenly. Thus the sudden increase in reflection amplitude, or bright spot, and associated phenomena may come from a small amount of gas in the sand. Such gas may be biogenic, that is, derived from the direct decomposition of plant material, or it may even be carbon dioxide. Economically, the use of bright spots may be fraught with the danger of overestimation but, as a source rock indication, there could be little better evidence. Some care is still necessary, however, because of the possibilities of noncommercial gas such as the examples listed. In basins where direct hydrocarbon indicators (DHI) are not seen on seismic sections, the presence of abundant shale can be regarded as the best indirect indicator. Shales are usually indicated by low seismic velocities. The presence of large sections with interval velocities in the range 2000 to 3500 m/sec (6500 to 11,000 ft/sec) points to clastic sections which may have both source beds and reservoirs.

For the likelihood of reservoirs within the clastic section, however, there are zones wherein the seismic traces have little similarity because of the coming and going of sands whose acoustic impedance is different from the average. Location of stratigraphic traps in carbonates can be eliminated by this procedure, but they have to be dealt with separately in any case.

As a result of the foregoing qualification work, some traps may have been discovered —they are evidenced directly by the location of bright spots. The possibility of others, occurring in small geographic areas and in smaller time zones on the record sections, have been indicated and the more detailed work of obtaining a corresponding geological model can begin.

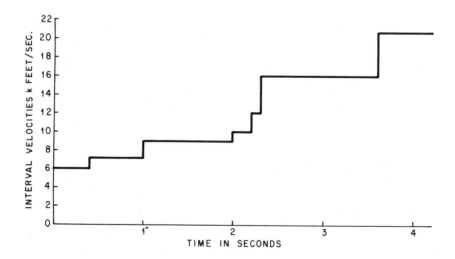

Figure 11.14 Interval-velocity plot from statistical analysis of CDP field data. It is the low-frequency components of this curve that are needed to supplement the (higher) frequencies determined directly. In this manner. SAIL traces can be made to give pseudovelocity logs.

There appear to be two possibilities, a direct and an indirect way of obtaining this geological model.

1. The direct method is to use seismic traces converted to acoustic impedance logs—with the important qualifications that the latter be made by incorporating both a knowledge of the statistical velocities obtained and that the acoustic impedance logs so obtained be calibrated in terms of actual velocities obtained in one well.

The use of statistical velocities involves the following steps:

a. Make a steplike log of interval velocities. These are determined at times of good reflections (see Figure 11.14).
b. This velocity log must be Fourier-analyzed over the same time interval as the seismic trace. This gives amplitudes and phases at the same intervals of frequency as those obtained for the seismic trace.
c. Compare the amplitudes and phases for the velocity log and for the integrated seismic record, over the overlapping frequency band, and determine a scaling factor. It is hoped that this will be a real number. Otherwise, a complex scaling factor can be used to adjust the velocity log amplitude and phase—and these values, for the low-frequency interval, have to be added in.
d. The completed Fourier analysis is then synthesized and shifted by a constant velocity, to take care of the dc shift which has not been determined.
e. If necessary, the velocity values can be inverted to interval time values (e.g., microseconds per foot). This has been done by Lindseth (1975).

Note that the band-limited acoustic impedance log involves reflection coefficient contributions by both changes in velocity and changes in density. Before incorporating the statistical velocity information, it is best to remove this duplication on some physical basis. One basis that has some support is that the density and velocity contributions to the reflection coefficients are equal. If this is the case, there will be no problem, because there is an arbitrary scaling factor involved in step c. However, if the contributions are not linearly related, the compensation has to be made before step c is undertaken.

Thus each integrated seismic trace is forced into becoming an approximate velocity log, and these are used in the same manner as other well logs, for correlation. The advantages are that:

a. The changes in lithology and/or porosity can be seen in detail *between* wells.
b. The calibration is in velocity.

It must be borne in mind, however, that even with the broadest obtainable seismic section, the details of the change from one lithology to another are blurred. The existence of long-period multiple reflections is an ever-present danger.

Some pseudodiagraphies or approximate acoustic impedance logs obtained by Grau, Hemon, and Lavergne (1975) are given in Figure 11.15, where their excellent correlation qualities can be seen. These traces were 50 m (164 feet) apart.

2. The indirect method involves modeling in which an assumption is made as to the layering thicknesses, densities, and velocities; either the two-dimensional or the

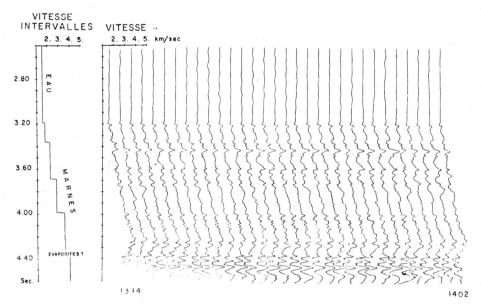

Figure 11.15 Pseudovelocity logs spaced 50 m apart. Interval velocities are shown on the left, to the same scale. [After Grau, Hemon, and Lavergne. Courtesy of the World Petroleum Congresses (1975).]

one-dimensional interactive modeling system is used to predict the seismic record(s) coming from a defined section, and alterations are made until the model fits the field trace(s). Given a starting point at a well or, better still, a tie into a well at each end of a line, the modeling can be reasonably fast—but the direct model from the acoustic impedance logs may, even then, be a help in providing a first model at each position along the line.

Two-dimensional modeling, because of its greater complexity, is probably only used when steeper dips or structural complications affect the ability of the interpreter to predict a model under very complex conditions. The effect of diffractions is difficult to predict, and they can sometimes easily be misinterpreted as lithofacies changes.

In both the direct and the inverse methods, it is important to remember that only one model is obtained that fits the seismic data. Other inputs are necessary to decide if this is the best model that can be obtained.

11.9 TWO-DIMENSIONAL MODEL USAGE IN COMPLEX VELOCITY PROBLEMS

The reader has been alerted several times in this book that false structural anomalies can arise through lateral changes in velocity, either in the near surface or in the deeper subsurface. Some excellent examples of false anomalies have been given by Tucker and Yorston (1973).

The anomalies range all the way from multiples due to formations of uneven thickness near the surface to the simple turnover into a fault plane—which may be due to slower velocities on the downside of the fault. In the past, such pseudotraps could be tested only by drilling, but today CDP velocity information can very readily be obtained in great detail. Of course, such velocities are subject to experimental error but, to avoid too gross errors, the practice of smoothing the velocities (either for a constant depth or to a constant reflection) can be used to minimize experimental error. A smooth curve of known algebraic form can always be fitted to the observations by the method of least squares. Suitable curves may be single low-order polynomials to fit the data over its entire range, or a series of cubic splines may be constructed to show the relation between the velocities at different intervals along the profile. Each piece of the curve has a cubic relation of the form

$$y_i = a_{0i} + a_{1i}x_i + a_{2i}x_i^2 + a_{3i}x_i^3$$

where x_i lies between x_{iL} and x_{iH}, and at the join

$$x_{(i+1)L} = x_{iH} \qquad \left(\frac{dy_i}{dx_i}\right)_H = \left(\frac{dy_{i+1}}{dx_{i+1}}\right)_L$$

so that the *slopes* as well as the curve values are constrained to be continuous going from one spline to the next.

With modern two-dimensional modeling processes, it is possible to specify horizontal gradients of velocity. Thus, in cases such as interpreting structure under overthrusts, it is possible to construct the estimated model and, after a short time, decide whether or not it gives the required seismic output. In many cases, this procedure is complex and time-consuming—chiefly in setting up the model. Nevertheless, the problem today is tractable, whereas 10 years ago it would have been unthinkable in a very detailed form.

11.10 THE USE OF STATISTICAL METHODS IN STRATIGRAPHIC TRAP DETERMINATION

In maturely drilled areas, more geological knowledge is available. In some cases, this may consist of sonic logs in producing wells and/or dry holes. These sonic logs can be used for comparison with field records, and some discrimination between the types of record where sands are present and those where they are not can be obtained by eye. A more quantitative method of combining geological knowledge with detailed seismic information was reported by Mathieu and Rice (1969). The procedure makes use of a seismic time zone referenced to a geological marker depicted in the seismic cross section. Each seismic trace provides a suite of variables which are later analyzed using multivariate statistical procedures. The quantitative variables used in the example were directly measured by taking amplitudes of the seismic trace at specified time intervals referenced to a reflection event. Other quantifiable variables relating to the acoustic properties of the suspected anomaly can also be used in place of or in addition to the direct digital values. Examples include the spectral properties of the zone, measured times of the zero crossings from the reference time, and interval thickness.

It must be remembered that the data are band-limited, and therefore there exists a sampling interval, determined by the highest frequency present. If the sampling frequency is increased above this Nyquist sampling rate, the samples taken are not independent variables, and this will cause instability problems in further statistical work.

The method assumes that a set of such data can be measured for all wells in the area or collected from seismic data in the immediate vicinity of the wells.

Linear discriminant analysis is used to establish classification functions for groups of traces defined a priori. In the example given, two groups (sand and no sand) were constructed from synthetic seismograms. Linear functions of the variables were defined in the following manner:

$$Z^k = \lambda_1^k x_1 + \lambda_2^k x_2 + \lambda_3^k x_3 + \cdots + \lambda_n^k x_n + \lambda_{n+1}^k \qquad (11.4)$$

where $x_1, x_2, x_3 \ldots, x_n$ = measured variables
$\lambda_1, \lambda_2, \lambda_3 \ldots, \lambda_{n+1}$ = estimated coefficients that maximize the following function of the two groups
k = 1 or 2, indicating the sand and no-sand conditions

$$D = \frac{(\bar{Z}^1 - \bar{Z}^2)^2}{\sum_{i=1}^{ns} (\bar{Z}^1 - Z_i)^2 + \sum_{j=1}^{nn} (\bar{Z}^2 - Z_j)^2} \qquad (11.5)$$

where ns and nn = number of sand and no-sand examples

This statistical process optimally separates the two groups from one another by maximizing differences between group means while minimizing differences among samples of the same group. The mode of action can be most easily visualized by taking the sample case of dependency on only three variables. In this case, the individual samples can be plotted in three dimensions as shown in Figure 11.16a. If these are looked at in a plane parallel to the X-Y axis, the projections of the samples are all intermixed. However, by the proper choice of a plane, the projections from the different samples are optimally separated and, at the same time, the distance apart of samples within the same group is overall as small as possible. In Figure 11.16b, this optimal plane was chosen to be perpendicular to the X-Z plane, making an angle θ with the X axis. The results of using this plane are shown. In general of course the plane necessary to effect an optimal separation is achieved only by rotation of the original axes through two different angles. The important point is that *rotation* of axes is involved. In the more general (n-dimensional) case, the rotations necessary are achieved by matrix manipulations. Tests can be made, subject to the usual statistical distribution assumptions, which will determine if the group means are distinguishable from one another at the desired confidence level. Nonlinear discriminants (Rao, 1967) are available but are more difficult to use.

Essentially, the characteristics of this method are:

a. The known samples are used to determine a statistical discriminant function whose power to distinguish between two (or more) groups can be ascertained.
b. These groups must be decided before calculation is made. This is relatively simple in the sand–no-sand problem but may be much more difficult for more complex problems.

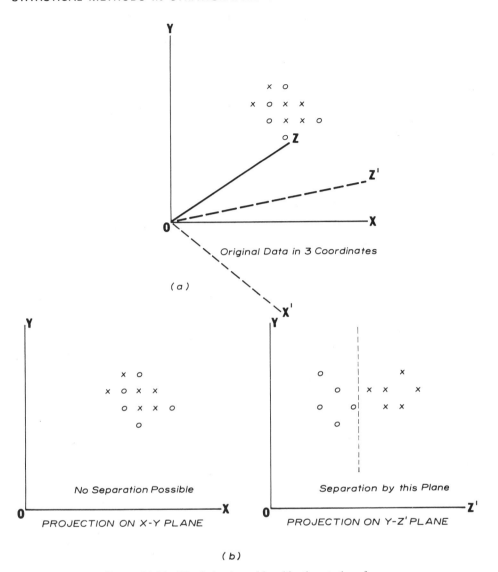

Figure 11.16 Discrimination achieved by the rotation of axes.

c. New and unclassified samples—that is, new seismic traces—can now be classified into one or another of the established groups.

It can be said that although the method was successful (see Figure 11.7) in testing preconceived ideas, other significant stratigraphic trends may not be detected.

It now seems to be desirable to formulate a method that removes some of the restrictions imposed in the early method.

The new technique identifies underlying patterns in a fixed window of seismic data and, wherever possible, associates a geological meaning with each pattern. The identification of basic components in a set of quantitative data has been discussed by Griffith, Pitcher, and Rice (1969). It is a problem to be tackled by factor analysis—a

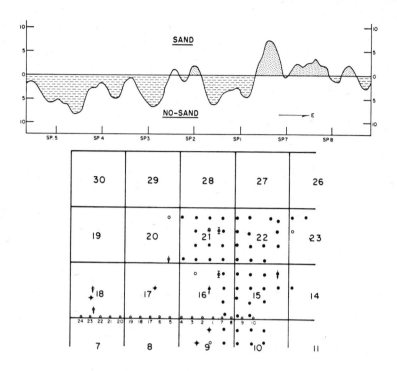

Figure 11.17 Classification of field data into one of two groups (sand, no-sand) from discriminatory analysis. Classification is corroborated by wells shown in the map at the bottom of the figure. [After Mathieu and Rice, (1969). Reprinted with permission from *Geophysics*.]

multivariate statistical technique. These investigators used factor analytic techniques to delineate the number of environmental facies that existed in the area of analysis and associated each sample with the proper (geological) facies. Similar techniques are applicable for quantitative variables derived from a zone of seismic data and can be used effectively to outline the underlying patterns mentioned earlier. Other clustering techniques, such as those described by Anderberg (1973) can be used to augment factor analysis methods.

The original data are digitized at the usual sample interval, and these amplitude sequences corresponding to the set of trace segments constitute the original data. We first search for underlying constituents which are orthogonal to one another (see Section 10.4) and which explain a large fraction of the data. The technique is one of determining the characteristic components and the weights to be given to each one, so that a major portion of the energy of the traces is explained. The characteristic components (sometimes called eigenvectors) and their weights (eigenvalues) are determined by standard matrix operations. Each trace can then be constructed (almost) by combining the characteristic functions in the proper proportions. The number of characteristics is determined largely by the degree to which the interpreter wishes to approximate the data. The factor analysis, given this degree of approximation, determines how many eigenvectors are needed and what their form is. It is important that each eigenvector be entirely independent of another, and this is what is meant by orthogonality. In mathematical terms,

$$E_1(x_i) = a_1 x_1 + a_2 x_2 + a_3 x_3 + \cdots + a_n x_n$$
$$E_2(x_i) = b_1 x_1 + b_2 x_2 + \cdots + b_n x_n$$

(11.6)

Then, $a_1 b_1 + a_2 b_2 + \cdots + a_n b_n = 0$ for orthogonality. In the sense of n-dimensional vectors, the eigenvectors are perpendicular to each other.

By using these eigenvectors as axes, each individual trace can be expressed in terms of them. For example,

$$T_1 = A_{11} E_1 + A_{12} E_2 + \cdots + A_{1m} E_m$$
$$T_i = A_{i1} E_1 + A_{i2} E_2 + \cdots + A_{im} E_m$$

(11.7)

One can imagine the traces being displayed in n-dimensional space, with the eigenvectors as coordinates, just as the samples were shown in three-dimensional space in Figure 11.16.

After the basic patterns have been established in the data, it is desirable to be able to associate some physical or geological meaning with them. A more rigorous procedure than visual comparison is needed for cases that require simultaneous consideration of several complex variables. One possible approach is the application of discriminant analysis techniques to groups previously established by factor analysis techniques.

The resulting functions can be used to classify new seismic data or synthetic seismograms derived from the acoustic impedances found from well logs. Corresponding tests of significance can also be computed to give a quantitative assessment of the uniqueness of the groups that have been defined.

Because of the sensitivity of this method to reflection character, which we know to be controlled by the type of input pulse (as seen after filtering by the recording system), it is essential that all data either be of the same vintage, collected in the same manner with identical instruments, or the different data types undergo treatment with transfer functions to convert them to one basic type. The choice of standard is optional except that the transfer functions cannot convert data with one bandwidth to data with a wider bandwidth. The choice is open, but it invariably leads to degradation of all data to the characteristics of the lowest acceptable set. But one always has the choice of throwing some away.

The results of applying this more general method have been given by Waters and Rice (1975). The original data, with the zone of interest (A-B) marked, is given in Figure 11.18. Below the main data is shown the interval concerned, flattened on the zero crossing above the lower reflection. Five different characteristic waveforms were found to be present, and the proportion of S_1 (the factor found to be chiefly associated with gas production) is shown in the lower part of the figure, along with some shading which identifies the chief constituent of each trace segment.

Figure 11.19 shows the association of the characteristic waveforms (end member), the average waveform for a group, and the synthetic waveforms from producing wells and dry holes. The upper characteristic waveform is most indicative of production.

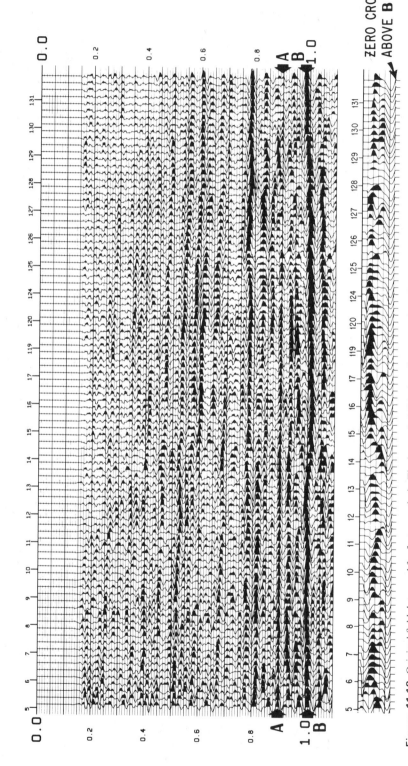

Figure 11.18 Original field data used for factor analysis. The interval *A-B* is the relevant portion of the cross section, reproduced below the seismic section by flattening on the zero crossing above *B*. [After Waters and Rice. With permission from the Ninth World Petroleum Congress (1975).]

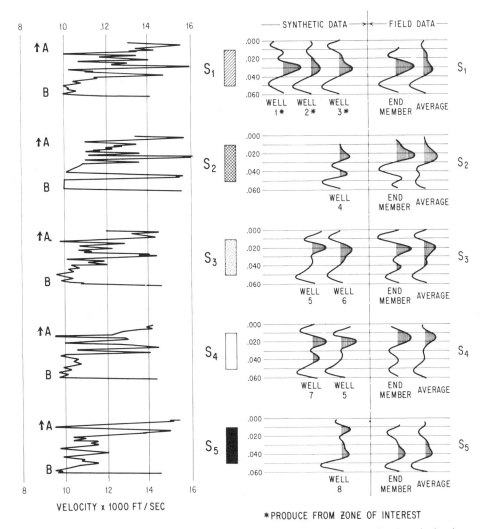

Figure 11.19 Average and end member waveforms for the factor types and associated synthetic seismograms classified by discriminant analysis. Type logs for each group are shown on the left. [After Waters and Rice. Courtesy of the Ninth World Petroleum Congress (1975).]

Finally, we must remark that this *method* does not, by itself, constitute a method of increasing resolution—rather it organizes the task of producing the best information that can be obtained from the bandwidth available.

The effect of bandwidth is significant in the test example and can be quantified through statistical procedures. Since three vintages of seismic data were available in the area, with different bandwidths, groups derived from dynamite coverage were processed with band-pass filters equivalent to the Vibroseis® data. Figure 11.20 illustrates the waveform differences for the five groups.

Some loss of ability to distinguish between the groups is associated with loss of high-frequency content. The loss is not proportional to bandwidth reduction and, in the example, the loss of the 45 to 52 Hz band is much more critical than the loss of

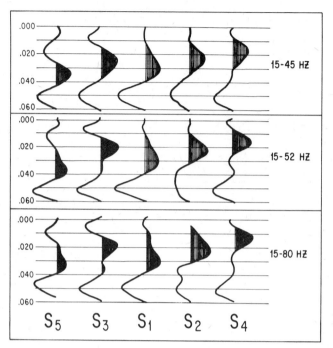

Figure 11.20 Waveform differences associated with different filters for the five groups shown previously. [After Waters and Rice. Courtesy of the Ninth World Petroleum Congress (1975).]

the 52 to 80 Hz band. When using a bandwidth of 15 to 52 Hz, average probabilities of misclassifying a sample are approximately 1.3 times as great as when the 15 to 80 Hz data are used. The 15 to 45 Hz data increase the probability of misclassification by a factor of 3 over the 15 to 80 Hz data. Thus a substantial improvement in successful classification can result from a relatively small bandwidth increase.

11.11 SUMMARY AND CONCLUSIONS

In this chapter, the methods of interpreting seismic reflection data are examined. Although the task of finding hydrocarbons should be a purely geophysical one, the resolving power of the tool has yet to be proved sufficient to determine exactly enough of all the physical data necessary. Geologists of varying special disciplines are needed to make exploration by seismograph economical and to supply a knowledge of likely values of other parameters at present beyond the reach of the geophysicist.

Structural interpretations, although relatively straightforward, are still subject to errors due to unsuspected velocity variations, inadequate migration procedures, or basic limitations in resolution at truncations.

Stratigraphic trap interpretations are basically limited by resolution, both in the thickening and thinning of thin beds and by changes in their other physical characteristics. Imaginative construction of one-dimensional models, together with interactive modeling programs, can check whether or not the models can be related to the seismic reflection results. At best, however, a single answer is found for what, in view of limited resolution, is probably a multianswer problem.

In cases of structural complexity, resort has to be made to two-dimensional modeling, in which all types of seismic events are included. Such programs are slow and expensive to run and of course are valid only when the earth structure under the line of section really is two-dimensional. Cases of reflections coming from the side of the line (side swipe) are not dealt with. One would have to rely on two-dimensional data gathering with some discrimination in the processing to remove, or display separately, reflections from outside the line of section.

In all types of interpretation, but particularly in stratigraphic trap modeling, reliance is placed on consistent reflection character where consistent geology is present. In order to achieve this, if more than one vintage of seismic record is present, resort has to be made to achieving consistency through the use of transfer functions. Unfortunately, although these put all seismic reflection data on an equal footing, the level is that of the set of records with the least bandwidth.

An important clue to traps containing gas is the relative amplitude–processed section, in which the gas zones stand out by virtue of the large change in acoustic impedance between normal and gas-filled sands. The presence of large-amplitude anomalies, evidence of high attenuation, and change in phase at the ends of the high-amplitude reflection are not necessarily, however, indications of commercial concentrations of gas.

The direct method of obtaining a "first guess" geological model is to use the SAIL trace modified to include low frequencies by making use of statistical velocity information. Some success has been achieved in contouring the values of inverse velocity (in microseconds per feet) to show up low-velocity zones, indicating porosity, in carbonates.

Statistical methods have now been developed for use in maturely drilled areas, by which characteristic wave shapes can be calculated from the seismic data for the zone of interest. These eigenvectors can be used to express the waveforms of all traces, and a discriminant analysis is used to divide them up into identifiable groups, for a given confidence level. Association can then be made between groups so defined and certain properties of the section—such as oil production or not, and so on. This process works best in areas where consistent, high-bandwidth records are available and where numerous wells, some producers and some dry holes, have been drilled. It is really an exploitation rather than an exploration tool. At present, the tool is limited to considering zones of constant-time interval.

APPENDIX 11A: LOCATION OF DEPTH OF OVERPRESSURED ZONES

In all geological sections, the internal pressure of the fluids increases with depth. This increase is brought about by the fact that part of the weight of the sediments above a particular depth is borne by the fluids and part by the rock matrix. If the fluid in a porous rock is connected, even tenuously, with the surface through the permeability of the rocks or through faults and fissures, this pressure can be somewhat relieved and can be reduced to a pressure consistent with a column of water equal to the depth being considered.

In some areas, however, thick sections of clastics may have been laid down very quickly and may have retained water within an impermeable matrix. Under these conditions, the water in the shales helps to bear the overburden load and may have considerably greater pressure than normal. It is of considerable importance to be able to predict before drilling a hole the depth to the top of any overpressured zones, so that a warning can be given to the drilling crew that additional care is necessary when these depths are approached. Producing sands do occur within overpressured zones and, when found, are often prolific producers. They, and possible sands below overpressured shales, are the economic incentive for continuing to drill.

While reflection seismology cannot provide exact depths to the top of overpressured shales, the accurate prediction of average or interval velocities as a function of depth can often provide a depth within a few hundreds of feet. Figure 11A.1 shows diagrammatically the form of the average velocity-versus-depth curve for a location in which overpressured shales are present. The shales themselves have low P-wave velocities, and this causes a rapid decrease, even of average velocity, with depth. However, the "knee" of the velocity depth curve is gradual and can, at best, be determined to within 100 m (300 ft).

Some care is necessary, since many statistically determined velocity depth curves show zones where the average velocity appears to be constant, or decreases slightly. Such zones may be due to causes other than overpressure, such as multiple reflections and diffracted events, and so on.

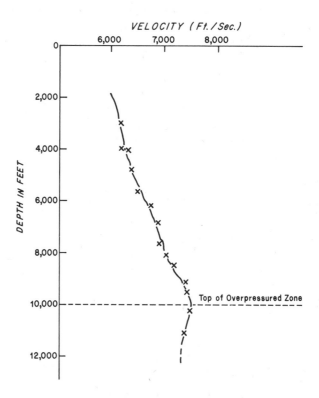

Figure 11A.1 A velocity-depth curve, showing the effect of an overpressured zone.

There is reason to believe that some overpressured zones, giving rise as they do to gradual changes in acoustic impedance (transition zones), should be observable from reflection data. Transition zones are characterized (Berryman, Goupillaud, and Waters, 1958) by having a higher reflection coefficient at low frequencies. In addition, any slow-velocity medium delays reflections coming from layers deeper in the section.

Overpressured zones have been found at markedly different depths on opposite sides of some faults, and a contoured, partially smoothed, interval velocity section superimposed on the normal seismic section can provide excellent clues to their presence and depth.

REFERENCES

Anderberg, M. R. (1973), *Cluster Analysis for Applications*, Academic, New York.

Backus, M. M. (1959), "Water Reverberations, Their Nature and Elimination," *Geophysics*, Vol. 24, No. 2, pp. 223–261.

Berryman, L. H., Goupillaud, P. L., and Waters K. H. (1958)," Reflections from Multiple Transition Layers—Theoretical Results," *Geophysics*, Vol. 23, pp. 223–243.

Cruz, R. B. (1975), "Direct Hydrocarbon Detection: Modeling and Interpretation," paper presented at the 45th Annual International Meeting of the Society of Exploration Geophysicists, Denver, Colorado.

Domenico, S. N. (1975), "Effect of Brine-Gas Mixture on Velocity in an Unconsolidated Sand Reservoir," *Proceedings of the 9th World Petroleum Congress, Tokyo, Japan*.

Grau, G., Hemon, C., and Lavergne, M. (1975), "Possibilities Nouvelles pour la Sismique Stratigraphique," *Proceedings of the 9th World Petroleum Congress, Tokyo, Japan*.

Griffith, L. S., Pitcher, M. G., and Rice, G. W. (1969), "Quantitative Environmental Analysis of a Lower Cretaceous Reef Complex," SEPM Special Publication 14, Tulsa, Oklahoma.

Harms, J. C., and Tackenberg, P. (1972), "Seismic Signatures of Sedimentation Models," *Geophysics*, Vol. 37, No. 1, pp. 45–58.

Lindseth, R. O. (1975), "Interpretation of Seismic Data from Derived Velocity Logs," paper presented at the 45th Annual Meeting of the Society of Exploration Geophysicists, Denver, Colorado.

Mathieu, P. G., and Rice, G. W. (1969), "Multivariate Analysis Used in the Detection of Stratigraphic Anomalies from Seismic Data," *Geophysics*, Vol. 34, No. 4, pp. 507–515.

May, B. T., and Hron, F. (1975), "Synthetic Seismic Sections of Typical Petroleum Traps," paper presented at the 45th Annual International Meeting of the Society of Exploration Geophysicists, Denver, Colorado.

Po-Hsi Pan and de Bremaecher, J. Cl. (1970), "Direct Location of Oil and Gas by the Seismic Reflection Method," *Geophysical Prospecting*, Vol. 18, Supplement, pp. 712–727.

Stolt, R. L. (1975), personal communication.

Taner, M. T., Cook, E. E., and Neidell, N. S. (1970), "Limitations of the Reflection Seismic Method—Lessons from Computer Simulations," *Geophysics*, Vol. 35, No. 4, pp. 551–573.

Trorey, A. W. (1970), "A Simple Theory for Seismic Diffractions," *Geophysics*, Vol. 35, No. 5, pp. 762–784.

Tucker, P. M., and Yorston, H. J. (1973), "Pitfalls in Seismic Interpretations," Society of Exploration Geophysicists, Monograph No. 2.

Waters, K. H., and Rice, G. W. (1975), "Some Statistical and Probabilistic Techniques to Optimize the Search for Stratigraphic Traps on Seismic Data," *Proceedings of the 9th World Petroleum Congress, Tokyo, Japan*.

Wilson, J., and Embree, P. (1975), "Interactive Modeling of Direct Hydrocarbon Indicators," paper Presented at the 45th Annual International Meeting of the Society of Exploration Geophysicists, Denver, Colorado.

Yacoub, N. K., Scott, J. H., and McKeown, F. A. (1970), "Computer Ray Tracing through Complex Geologic Models for Ground Motion Studies," *Geophysics*, Vol. 35, No. 4, pp. 586–602.

BIBLIOGRAPHY

Anderson, T. W. (1957), *An Introduction to Multivariate Statistical Analysis*, John Wiley, New York.

Collins, R. E. (1968), *Mathematical Methods for Physicists and Engineers*, Reinhold, New York.

Dixon, W. J. (1968), *BMD Biomedical Computer Programs*, University of California Press, Berkeley and Los Angeles, Calif.

Dixon, W. J. (1969), *BMD Biomedical Computer Programs—X Series Supplement*, University of California Press, Berkeley and Los Angeles, Calif.

Harman, H. H. (1960), *Modern Factor Analysis*, University of Chicago Press, Chicago.

Marriott, F. H. C. (1974), *The Interpretation of Multiple Observations*, Academic, New York.

Press, S. J. (1972), *Applied Multivariate Analysis*, Holt Rinehart and Winston, New York.

TWELVE

New Tools in the Making

12.1 INTRODUCTION

Reflection seismology has never been a static science, and it is no surprise that, given a fixed time for the termination of any discussion, there are new concepts and new methods in an active stage of development. These methods do not form coherent material for an additional chapter, except that some coherence arises from the ever-present need to increase the amount of detailed information derived about the sub-surface rocks. However, the reader who has arrived at this point should be able to place these anomalous subjects in the proper place to allow each facet of reflection seismology to have its own developing phase which fits in with the overall goal. This is not a comprehensive catalog of research and development being carried on in industry, government, and universities—such as is published every three years in *Geophysics*. Instead, we present the main advancing phases of which we have some knowledge and place them in context so that their future probable importance can be assessed. If this means a discussion of the problems still to be solved, then this is done.

12.2 THE USE OF SHEAR WAVES TO PROVIDE SUPPLEMENTARY DATA

In spite of the progress made in the interpretation of seismic data in areas when thin beds produce, it is well known that *thin* beds produce a response that is proportional to the product of the reflection coefficient and the thickness. Consequently, as the thickness is not known, the acoustic impedance contrast is not either. Second, even given the acoustic impedance (or velocity) of a formation, this is not a perfect discriminant as far as rock type is concerned. Thus a long-range goal of exploration seismology—to be able to predict lithology—is not yet feasible. The addition of another factor could make discrimination between various rock types, rock porosities, or fluid types much easier. One such factor is the shear wave velocity or the acoustic impedance contrast.

Methods of producing shear wave records with good signal/noise ratios of a quality comparable to P-wave records have long been sought (Ricker and Lynn, 1950; Jolly, 1956; White, Heaps, and Lawrence, 1956; White and Sengbush, 1963; Cook, 1964; Cherry and Waters, 1968; Erickson, Miller, and Waters, 1968; Geyer and Martner, 1969) in the United States. There are indications that in the U.S.S.R. shear waves have been produced using explosive sources (White, 1974). However,

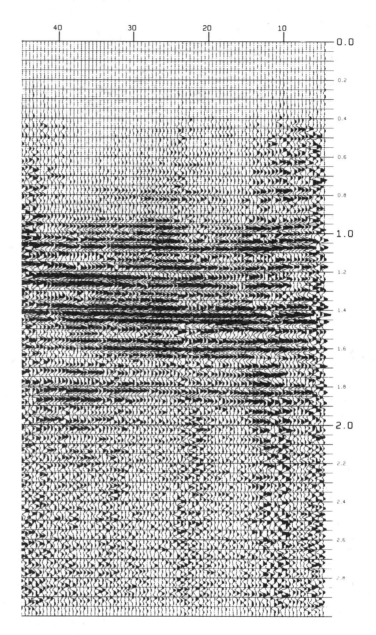

Figure 12.1 A compressional wave reflection record is shown on the right, taken with a vertical vibrator and vertical geophones. The compressional wave reflection record on the left was made with a horizontal vibrator operating at half-frequencies, but the output on vertical geophones is correlated with the original control signal. This is conclusive evidence that the rocking motion of the SH vibrator generates compressional waves of double frequency. (Courtesy of the Continental Oil Company.)

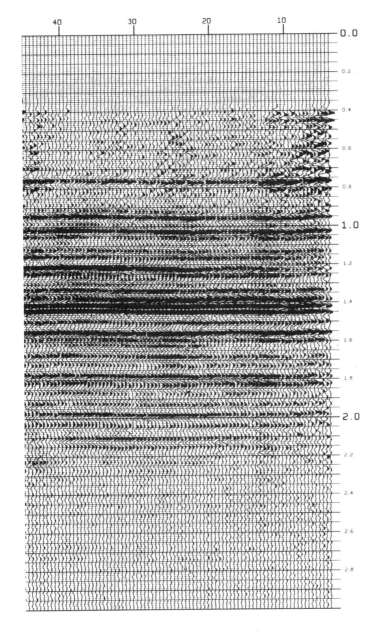

Figure 12.1   (*continued*)

the use of vibrational sources has not yet begun there. Their method of producing SH waves from explosions is ingenious but time-consuming. Essentially, SH motion is obtained by exploding dynamite in a hole near another one, which relieved the pressure in that direction. Of course, the SH waves are accompanied by compressional waves. The next step is to use another, nearby, pair of holes but to reverse their roles so that this time the explosive hole is on the other side. SH waves, out of phase with those previously produced, are obtained, accompanied by P waves of the same polarity. The procedure then continues in the data processing domain by normalizing one record with the other—to produce P waves of equal power. The two records are then subtracted to eliminate the P-wave energy and to reinforce the SH-wave energy.

Horizontally polarized shear waves (SH) are used in most of the experiments and appear to be most suitable because of the lack of conversion at horizontal boundaries. SH waves can be produced relatively easily by suitable vibrators (see Chapter 3, Appendix B) but are accompanied at low level by some P-wave energy at double the frequency of SH signals. Although this had been suspected for a few years, because of the rocking nature of the horizontal vibrator, it has recently been proved by Hopkins (personal communication). Figure 12.1 shows, on the right, a normal P-wave Vibroseis® record taken using vertical vibrators and vertical geophones. On the left is a record made using SH vibrators operating at half-frequencies. The geophones were vertical, and this record section was obtained by correlating the geophone outputs with a SH double-frequency sweep. The similarity between the two sections is evidence that P-wave energy is generated, and the loss of signal/noise ratio draws attention to the low level of the signal input.

Modern shear wave Vibroseis® techniques employ CDP methods and, for optimum comparison with the P-wave records in the same area, use a frequency band which is one-half the frequency band used for the P-wave records. The rationale for this is

1. Seismic reflection records of any energy type are responsive to rock layers on a wavelength basis; that is, the comparison of the wavelength spectrum in the rock itself to the thickness of the rock layer. Since wavelengths are velocities divided by frequency, if the velocities decrease by a certain factor ($V_s = kV_p$), the frequencies have to decrease by the same factor in order to keep the record character the same.
2. In recent years, shear wave reflection work has suggested that $k \simeq \frac{1}{2}$ for a clastic section.

Other types of rocks have different V_s/V_p ratios. It is reasonable that the shear wave swept frequency signal should cover a little more than the half-bandwidth of the P-wave sweep signal. Later processing can then, if needed, trim up the filter band to fit the V_s/V_p ratios found from the SH- and P-wave sections themselves.

Most of the problems of obtaining good SH-wave reflection data appear either to be solved or are capable of being solved with the means at our disposal. The deployment of an additional field crew to work over the same areas as P-wave reflection crews requires much economic justification, however. It is necessary to examine what, at best, can be achieved from this increased expenditure.

In the first place, all the velocities with which SH waves are propagated are different from those for P waves. This applies to weathering, subweathering, and subsurface layers, and the ratios are unlikely to be the same. One can therefore expect to see a

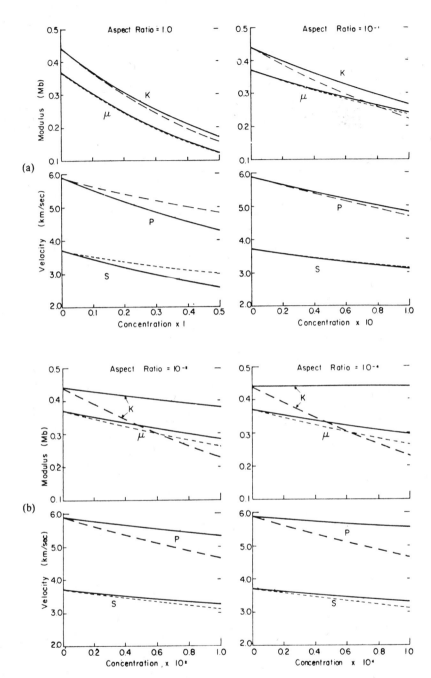

Figure 12.2 Variation in elastic moduli and compressional (*P*) and shear (*S*) velocities in a composite medium as a function of the volume of inclusions. The solid curves are water-saturated and the dashed curves are air-saturated. $\alpha = 1.0$ for spherical inclusions; $\alpha = 10^{-4}$ for very flat oblate spheriods corresponding to fine cracks or grain boundaries. Note the different multipliers on the concentration scale. [After Kuster and Toksoz (1974). Courtesy of Society of Exploration Geophysicists.]

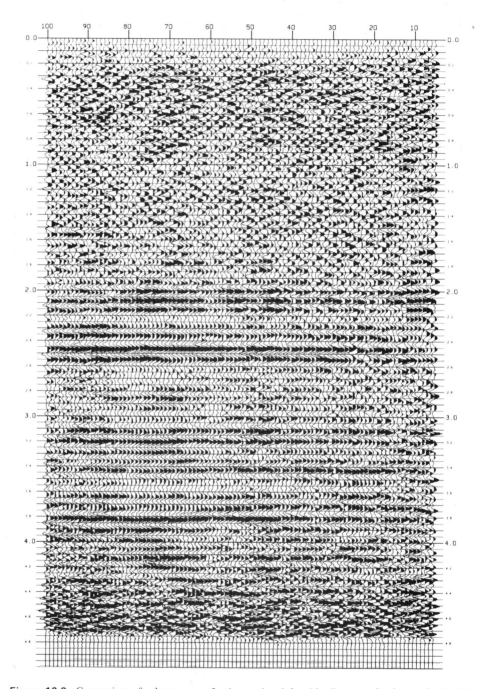

Figure 12.3 Comparison of a shear wave reflection section (left) with a P-wave reflection section (right). The vertical time scales and frequencies used for P-wave work differ from those of SH by a factor of 2. This leads to a better match but may not be exact. (By permission of the Continental Oil Company.)

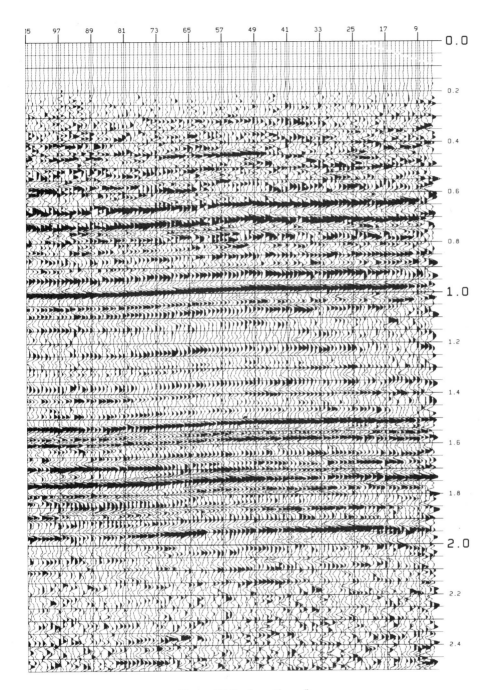

Figure 12.3 (*continued*)

349

different structure map. There are different weathering and elevation corrections to be made, which should not correlate exactly with the P-wave corrections. A comparison of the seismic cross sections obtained may offer a clue to false structure caused by velocity anomalies. The amount of experience with duplicate P and SH lines of data is insufficient to indicate whether or not this comparison is a useful one.

Second, SH-wave velocities should behave differently from P-wave velocities when one pore-filling fluid is replaced by another. This fact is derived from both experimental and mathematical treatments of velocities in porous rocks (Kuster and Toksoz, 1974). Such results deal with ellipsoidal pores which can be varied in aspect ratio (to simulate pores from spherical to thin cracks). On the assumption that sedimentary rocks have, in the main, pores that can be assumed to be spheroidal, it is shown that P-wave velocities are lower with water infilling than with air infilling, and S-wave velocities have the same characteristics, but to a lesser degree. As the aspect ratio decreases (the cracks become thinner), this situation reverses. The P-wave velocity with air infilling becomes less than with water, while the S-wave velocity is relatively independent of pore fluid. Finally, for very fine cracks, both P-wave and S-wave velocities increase when the fluid is water rather than air, but the S-wave change is much smaller than the P-wave change. Graphically, these theoretical results are shown in Figure 12.2.

From the practical point of view, shear wave results should throw new light on the process of direct location of gas deposits, on the amount and content of fracturing, and on variations in porosity within the same general matrix rock.

Third, one can examine the relation between the generation of surface waves and reflections, since this is an entirely new problem using SH-wave energy. At first sight, it appears to introduce a more difficult separation problem. Surface SH waves (Love waves) travel with approximately the same velocity as pseudo-Rayleigh waves which constitute interference for P-wave reflections. Moreover, the reflection SH waves travel with approximately half the velocity of P waves and, for the same spread and reflector geometry, the phase velocity along the surface is one-half that of P-wave reflections. Thus the separation between phase velocities for the signal and interference is more difficult to effect. However, it must not be forgotten that shear wave reflections arrive later than P-wave reflections, so the interference has a better chance to die away before the relevant S-wave reflections arrive.

Fourth, the contrast in velocities between layers is only, accidentally, the same for P waves and S waves. The elastic moduli controlling the velocities are not necessarily proportional to one another. Hence the reflections are not the same. This alone is sufficient reason for investigating the method in a large number of areas. The unraveling of the geological reasons for the variation in one reflection pattern with respect to the other is a difficult task but may be a very rewarding one in terms of both hydrocarbon and coal exploration. Some examples of SH-P seismic section comparisons were given in Erickson, Miller, and Waters (1968), but a more recent comparison, using CDP methods, is given in Figure 12.3 (courtesy of the Continental Oil Company).

The shear wave (SH) record section is on the left, and the P-wave section on the right. As usual, the vertical time scales as well as the frequency bands used differ by a factor of 2. While this is convenient for obtaining a *better* match, it must not mislead the interpreter into expecting an exact match with this time scale and frequency ratio. For example, on an isolated correlation basis alone, there is a possibility that the S-wave reflection near 2.5 sec is caused by the same lithological change as the good P-wave reflection near 1.0 sec.

This problem of correspondence cannot be taken lightly. The use of CDP recording methods has alleviated this to some extent, since the data themselves allow velocities to be calculated, which permit approximate calculations of depth on each of the sections for any chosen points. This facility, combined with identification of P-wave reflections in terms of geology by synthetic records, allows an approximate correspondence to be established. Additional methods, based on similarity functions between traces having a variable time expansion factor between reflections, exist but are largely untested.

The potential of shear wave reflection methods is great and, within a few years, may bring interesting new methods to light for determining the presence of possible stratigraphic traps for oil and gas.

12.3 INCREASED RESOLUTION THROUGH THE USE OF HIGHER FREQUENCIES

For several years, in marine prospecting, impulses having a higher frequency content than normal have been generated by special equipment in order to obtain much higher resolution between the near-water bottom layers than could be obtained with conventional marine seismic equipment. A survey of such equipment was made by Saucier (1970). The sound energy in the water is produced by a variety of means— piezoelectric or magnetostrictive transducers (relying on a change in volume of a material when an electric or magnetic field is applied), electromechanical types, sparkers, arcers, gas guns, air guns, and others. The type of source and the power used have to be chosen based on the discrimination and the depth of penetration required, since it is rare that a single type of transducer is efficient over a wide range of frequencies. In these discussions, we consider only devices developed for obtaining geological information below the sea floor, and we therefore disregard an enormous amount of work done on the initiation of pulses in the water for military purposes.

The generation of surface waves is negligible enough (because of the zero velocity of shear waves in a liquid) that it is rare that an array of sources or receivers is used specifically for the cancelation of such waves. Arrays are used, however, both for increasing the input energy by using multiple sources and for decreasing the ambient noise level. With single sources and receivers, results such as those shown in Figure 12.4 can be obtained. In this example, an input of several hundred joules of electric power can give resolution of near-bottom sediments down to about a 1 ft thickness, which is equivalent to the reflection of frequencies up to 5000 Hz. The penetration, however, is limited to about 100 ft. Although these are reflection means of obtaining information, they do not use most of the data enhancement principles and methods developed for reflection seismology. However, the principle of using an extended, unique signal and a correlation receiver was employed by Clay and Liang (1962, 1964) and, even though not always an integral part of the equipment, the principle can always be used if the outgoing pulse shape is known (e.g., recorded on magnetic tape) and the reflected signals are recorded also.

An increase in power and a lowering of the useful frequency spectrum but still staying with the single sparker source and receiver system lowers the resolution but

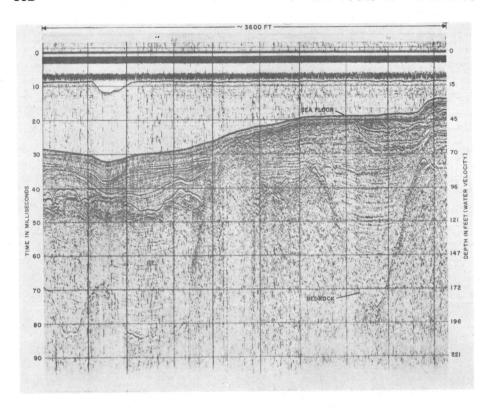

Figure 12.4 Example of higher-resolution sparker results in thin sediments. (By permission of E. G. and G. International.)

increases the depth of penetration, as shown in Figure 12.5. Frequencies of 700 to 800 Hz can be identified down to penetration depths of 1000 ft, and the clarity of faulting and erosional unconformities in near-surface sediments is quite evident. Clearly these devices are extremely useful for gaining information about the sediments near the sea floor, and examples are known that show gas-filled sediments angling upward to the sea bottom and even gas seeping into the water.

For depths of penetration useful in exploring for oil and gas deposits offshore, however, a further decrease in the upper limit of the frequency spectrum and a further increase in high-frequency power is required. In some areas, the evidence for a gain in resolution compared to the conventional 10 to 45 Hz records can be spectacular. Figure 12.6 shows such a record section which was made with a 62 KJ sparker array and recorded on several channels so that a 24-fold stacked section could be obtained. The clarity of the faulting is striking, and clear evidence can be obtained for coherent 100-Hz events at a record time of 1.5 sec, which corresponds roughly to 5200 ft.

As with all marine records, some care has to be taken with multiple reflections from the ocean bottom. Some are clearly evident here, with double the dip and double the reflection time of the water bottom. However, it is clear that much progress has been made with the use of higher frequencies in marine oil and gas prospecting. The general problem of resolution versus penetration has been indicated and is put

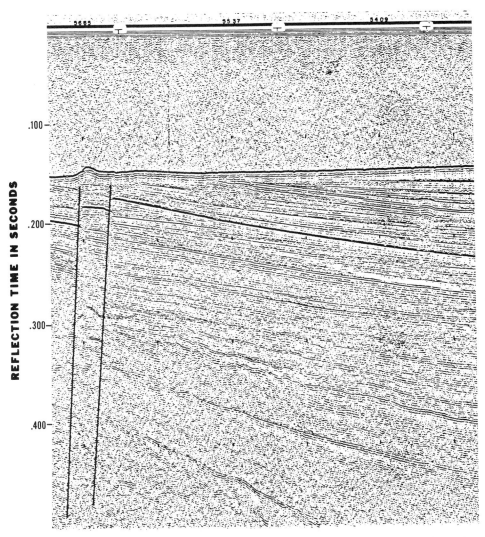

Figure 12.5 High-frequency marine data from marine sparker. Frequencies of 700 to 800 Hz can be identified at reflector depths of 1000 feet. (Courtesy of the Teledyne Exploration Company.)

on a firmer footing later. We must first, however, consider the progress that has been made on land.

Seismic records with usable P-wave reflection energy with frequencies up to 130 Hz have been made by Vibroseis® methods. In addition to the one shown in Figure 8.22, over a coal mine, high-frequency P-wave and S-wave reflection cross sections have been made for other purposes (Dunster and Miller, 1968). Two of these appear in Figure 12.7a and b. The structure shown has not been corrected for surface weathering thickness or for elevation changes. These were taken in north central Oklahoma, and the decrease in average frequency with depth is quite pronounced, even though the waves travel through Paleozoic, Permian, and Pennsylvanian formations.

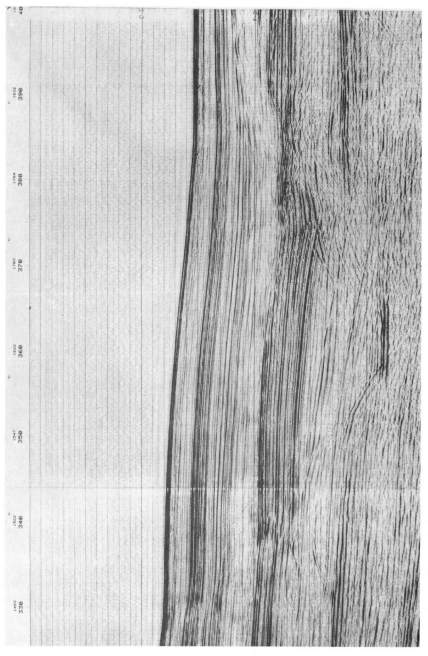

Figure 12.6 High-resolution marine system (62 KJ sparker array), 24-fold-stacked. Note the clarity of faulting. (Courtesy of Digicon, Inc.)

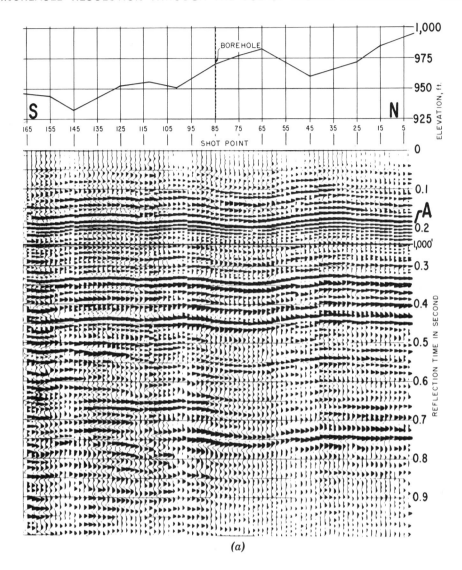

(a)

Figure 12.7a High-frequency P-wave Vibroseis® section. The frequency range is 32 to 130 Hz, and the data are fivefold-stacked. No corrections have been made for near-surface variations. (Courtesy of the U.S. Corps of Engineers.)

In areas where the water table is very close to the surface, some success with high frequencies has recently been reported (Figure 12.8). The system uses explosives and special hydrophones in shallow (4 ft deep) holes filled with water. Explosives follow the hydrophones in hole usage, and CDP methods can be used for data enhancement. The fault shown has been confirmed by drilling, and the throw varies from 12 ft at a depth of 200 ft to 36 ft at a depth of 800 ft. It is evident that the success of this method, in these areas, stems from the fact that the weathered layer is saturated and evinces

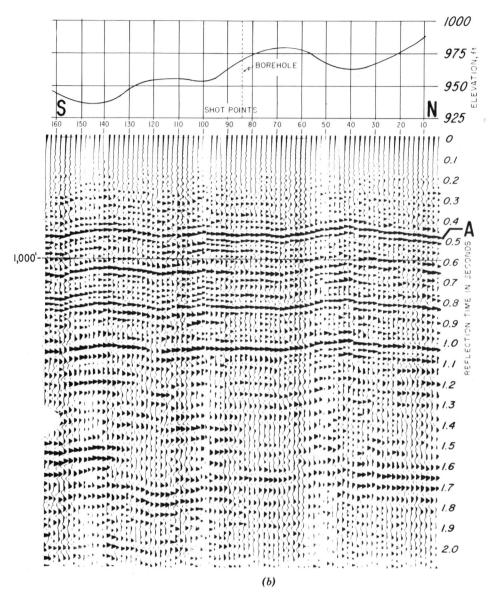

(b)

Figure 12.7b High-frequency SH-wave Vibroseis® section over the same line as in Figure 12.7a. Frequency range is 15 to 50 Hz. (Courtesy of the U.S. Corps of Engineers.)

neither the variability nor the attenuation for seismic waves that is a feature of a normal weathered layer.

Apart from the weathered layer which, for higher frequencies, can destroy coherence and scatter and give very rapid, unknown delays, which are all difficult to counter, the problem of pursuing the goal of higher resolution to greater depths is of course attenuation—as explained in Section 7.4. The attenuation due to primary reflection scattering from thin beds is not frequency-sensitive. In order to understand this, it is

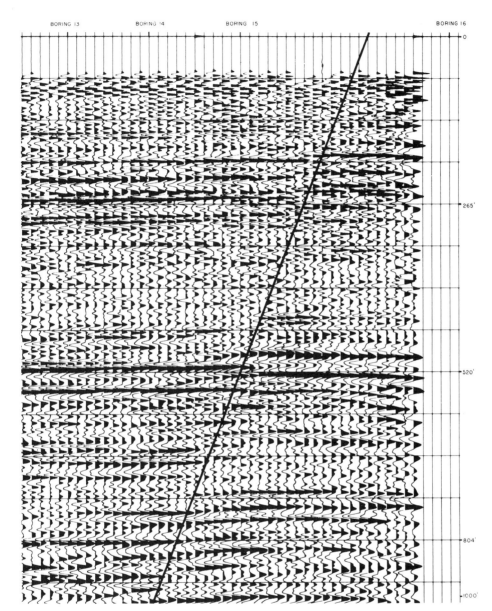

Figure 12.8 High-frequency (65 to 500 Hz) land record section showing a subsidence fault whose existence is supported by drilling. (© 1975, Western Geophysical Company of America.)

simply necessary to remember that the amplitude loss in passing through a series of boundaries (in both directions) is

$$\frac{A_F}{A_0} = \prod_{j=1}^{N}(1 - R_j^2)$$

where A_F = final amplitude
$\quad\;\; A_0$ = initial amplitude

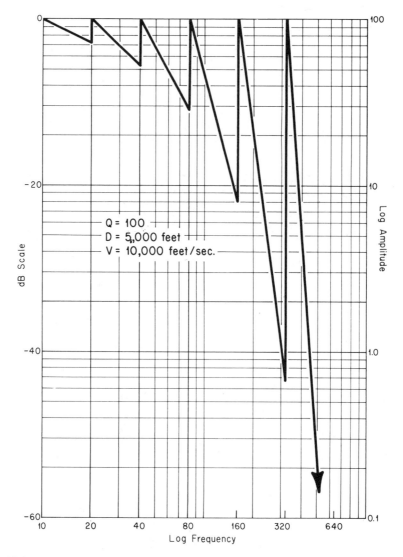

Figure 12.9 Variation in amplitude of reflection across successive frequency octave bands. $Q = 100$, depth = 5000 ft, $V = 10,000$ ft/sec.

This is a function of the reflection coefficients only, which are not frequency-dependent —and the high frequencies pass through the same boundaries as the low frequencies.

The effect of peg-leg multiples should not, however, be forgotten, and it has been shown that they can cause a loss of high frequencies with depth. However, for simplicity, they are lumped in with material attenuation, where the loss of energy to heat is an exponential function. The exponent is proportional to both distance traveled and (approximately) the first power of the frequency.

To try to record a very wide-frequency band during a single experiment involves a dynamic range problem as well as the problem of the signal/noise ratio for the higher frequencies. To solve the dynamic range problem, it is possible to record the required frequency bandwidth in several sequential bands. For convenience, we use octaves so that, for each octave, the resolution increases by a factor of 2. During each octave recording, the signal input can be increased so that the amplitude stays at the same level for some reference point (say the starting frequency of each band). Figure 12.9 shows the situation graphically. An end point for this process is provided either by the exponential increase necessary in input power or by the very rapidly increasing rate of decay in amplitude returned across each octave band. The decibel (log amplitude) range across the octave bands is proportional to the frequency bandwidth in hertz. There is a point at which the higher frequencies of an octave eventually dip below the noise level.

In practice, this method of recording different bands and adding the results together, after inverse filtering to flatten the contributions of each spectrum, necessitates an overlap from one octave to the next in order to be able to judge the weights with which the spectra are to be added.

It is to be expected that the barriers of obtaining wide-band, high-frequency records will gradually be forced back but, as can be seen from the preceding analysis, the problems rise exponentially with frequency, and an economic limit is soon reached.

A final note should be added. Figure 12.9 was computed using a Q of 100, a depth of 5000 ft (two-way distance 10,000 ft), and a velocity of 10,000 ft/sec. The decibel loss across an octave is proportional to both the linear frequency range and the depth. It is obvious that a greater extension of the frequency range can be expected to be feasible for shallow reflection targets than for very deep ones. In a general sense therefore frequency increase is apt to be more useful in coal applications than in seismic prospecting for oil and gas.

12.4   THREE-DIMENSIONAL SEISMIC DATA

In spite of the great strides made in the quality of linear seismic cross sections, such isolated profiles are at times limited in their usefulness, for two main reasons.

First, events are recorded from out of the vertical plane of the cross section, thereby yielding noise as far as geological interpretability is concerned. In order to sort out and eliminate these anomalous events, some form of three-dimensional control is needed. For this purpose, the most widely publicized method is the Wide Line Profiling System® (Michon, 1972) previously mentioned in Section 5.3.

Second, unless the type of sedimentation responsible for the accumulation of hydrocarbons is known before the accumulation itself is found, an unlikely supposition, there is no way in which the trend of the accumulation can be known. This makes the disposition of the available seismic effort, and the assignment of line positions, difficult. A sandbar can cross over the seismic line at almost any angle (many oil fields are controlled by fluvial sands in meandering ancient stream channels), and it is difficult to make an assessment of the value of such an isolated anomaly. Thus the detail of seismic reflection work along the line should be matched by equal detail across the line. Other methods exist, such as the seismic swath method, which give good CDP coverage for this purpose. However, in order to preserve the high redundancy needed for good data, it must be possible to have a recording system capable of recording a large number of surface positions simultaneously—a greater number than is common at present. In addition, the arrangement on the surface of the earth is awkward for coaxial cabling or other cabling methods. Moreover, if radio communication is considered, the number of separate channels is prohibitive and, in many land areas where seismic work is done, the line-of-sight (recorder to receivers) conditions demanded by ultrahigh-frequency radio communication are just not present.

The technology gap is capable of being filled in within a year or two by the development of recorders that are self-contained units having large but inexpensive memories (magnetic casette, solid-state, or bubble memories) and which are controlled by low-frequency, pulse-coded radio initiation signals. Accurate internal clocks can control digitization of the incoming signals picked up by a single group of geophones. None of these technologies need development, but the mass production effort to make them inexpensive may hold up application.

A real gap exists in the condensation of information, data processing methods, and in the presentation of data in a form that allows easy assimilation. Anyone who has worked with three-dimensional data knows that the last of these should not be taken lightly. However, significant advances can be made within a few years by computer calculations of changes in patterns recorded on two-dimensional cross sections as these sections are moved perpendicular to their planes. It is essential to remove from consideration geological indications that remain the same or change very slowly.

12.5 FINITE DIFFERENCE MODELS—FURTHER UNDERSTANDING OF THE SEISMIC REFLECTION PROCESS

In addition to advances in instrumentation and field methods, there is a need for advancement in our understanding of the wave propagation process in elastic media that have complex distributions of the elastic parameters and the density. This problem exists not only in the field of seismic exploration, for which some specific problems are listed later, but also in the field of seismic measurements of nuclear bomb and earthquake effects. Some progress has been made, particularly in the latter two fields, largely through the use of a finite difference approximation to the elastic wave propagation process (Cherry and Hurdlow, 1966). The original TENSOR code, developed at the Lawrence Radiation Laboratory at Livermore, California, allows various states of rock materials—plasma, gaseous, plastic flow, cracked rocks, and linear elastic conditions—to be simulated, most of which are of limited usefulness in seismic applications.

Developments in the field have been rapid, both for earthquakes and for exploration purposes. Boore (1972) and Alterman and Loewenthal (1972) have established a firm mathematical basis for finite difference approximations. A more limited code, in the sense of being suitable only for linear elastic conditions, is available from Systems, Science and Software, of La Jolla, California. It is called SAGE (seismic analysis for geologic environments) and allows investigation of two-dimensional Cartesian or axisymmetrical models. Both these basic systems are shown in Figure 12.10. In the plane of the diagram, all grids are rectangles, but the size can vary separately in each direction. Both surface loading or interior pressure loading are possible. The duration, several pulse forms, and whether the loading is pressure or velocity controlled can be specified. Each grid can, if necessary, have its own elastic parameters and densities specified separately, although at present only simple models are being investigated.

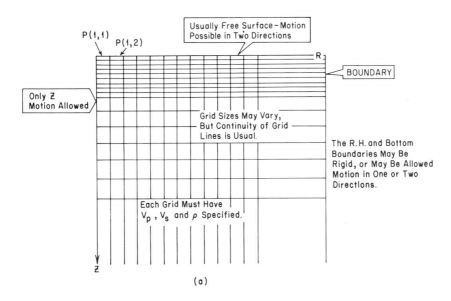

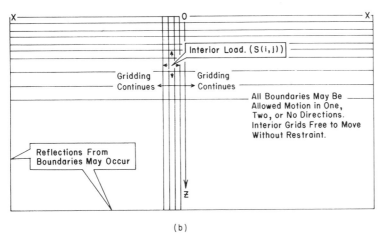

Figure 12.10 Illustrations of the basic arrangements for the radial model (*a*) and the *X-Z* model (*b*) of the SAGE elastic modeling program.

The types of permissible motion at the boundaries of the model can also be specified; that is, the cells can be allowed to move in a particular direction as, for example, vertical motion only along the radial model axis or motion in any direction in the plane for the free surface and other boundaries. As Boore (1972) showed, there are potential instabilities associated with the finite difference model, and some (small) damping factors can be specified that effectively damp out high-frequency oscillations which can be falsely generated.

There are two interrelated, troublesome aspects of the elastic modeling system. One of these is that, even in the two-dimensional system, computer time consumption can very rapidly become prohibitive. Associated with this are the very large arrays

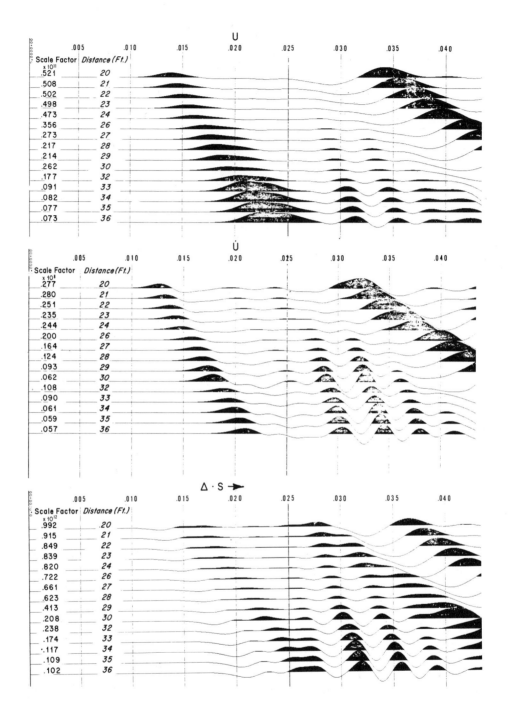

Figure 12.11 Results of an elastic modeling program described in the text. (Courtesy of the Continental Oil Company.)

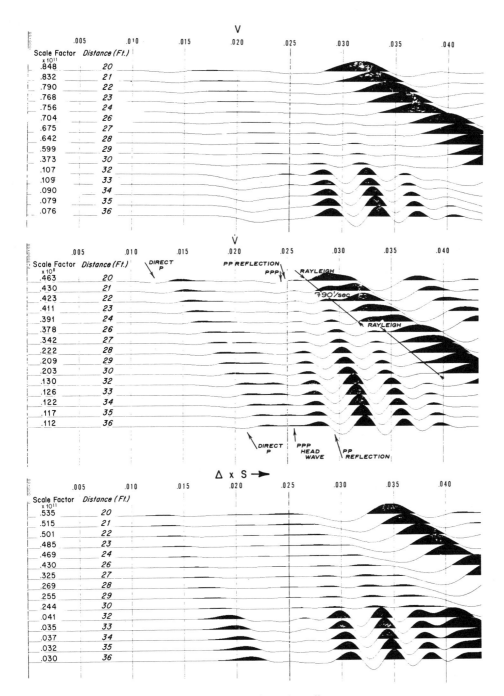

Figure 12.11 (continued)

needed. If these are to be held in core memory, the sizes are too large, while if the parameters are held on disk memories, the transfer times in and out of the computer become a controlling factor.

At first sight, there appears to be no need for overwhelmingly large arrays. However, the stipulations, first of the method requirements and then of the results needed, soon cause the size to become large. An example may make this clear. It is necessary that at least seven grids be present in the shortest compressional wavelength present. If conventional frequencies are used and the upper frequency is 50 Hz, while the lowest P-wave velocity is 2000 ft/sec, the grid size must not exceed 5 ft.

Then we must make sure that the model is large enough to ensure that the reflections from any of the edges do not come back and interfere with the desired results. For example, if the results are needed out to 0.2 sec and the major portion of the medium has a P-wave velocity of 8000 ft/sec, the model should be at least 1600 ft long and deep and, allowing a safety factor of 1.5, the number of grids in each direction should be 480. Five parameters are stored for each grid, making the total storage requirement 1,152,000.

The cycle time (rate of progression of the waves in the time domain) must not exceed three-eights of the time taken for a compressional wave to travel through the fastest grid. In our sample case, the transit time is 0.625 msec, and a cycle time of 0.0002 sec would be wise. Thus, for 0.2 sec of data, 1000 cycles would be required. Even on very fast machines, this sample problem would take several hours of computation time.

Nevertheless, these are models that simulate the elastic wave equation and can answer some hitherto unanswerable questions. They can be used to simulate actual seismic experiments if done in three dimensions, and much valuable information can be obtained from the more restricted two-dimensional or axisymmetrical systems (Cherry, 1973; Kelly et al., 1976). To some extent, if the restrictions are borne in mind, they may replace some forms of experimental field tests. While the ultimate test for any new equipment, or new idea, is a field test, accurate testing by computer simulation offers an alternative which will make the expensive field test necessary only when the idea has crystallized, all unsuspected "bugs" have been removed, and it shows definite promise.

While this tool may seem esoteric to the practicing field geophysicist and remote from the realities of exploration, consideration of the following two unsolved problems may convince him that research of this nature can well bear fruit. The first problem is that of the influence of the conversion of compressional waves to shear waves, or vice versa, in the long-offset traces now commonly employed in the CDP reflection method. It is not logical to assume that the extension to longer and longer offsets will continually increase the accuracy of velocity determinations or the quality of the reflection P-wave data. At some stage, as the angle of incidence of the P wave increases, the amount of conversion from one wave type to another increases, and of course refractions or head waves of both types (P and SV waves) are possible. Thus it is a matter of practical importance to decide what the limit of the spread length should be to optimize the desirable factors compared with the deleterious effects.

A second possibility is the use of finite difference programs to investigate the influence of a complex weathered zone on the character of the reflection record. Other examples with graphic displays have been given by Kelly et al. (1976). Figure 12.11 shows the results of the input of a 200-Hz pulse at the surface of a radial model incorporating a weathered zone 20 ft thick and having velocities $V_p = 2000$ ft/sec and

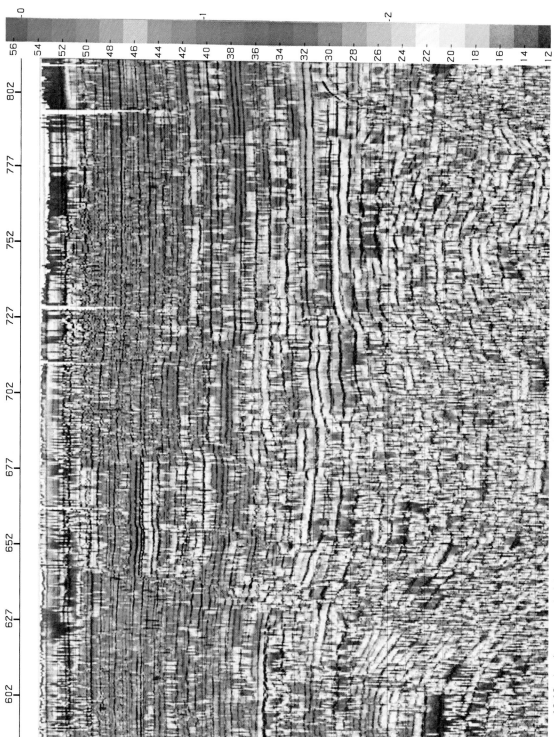

Figure 12.12 A color portrayal of the average frequency associated with a particular seismic section as a function of depth (reproduced from a Seis-Chrome display with the permission of Seiscom Delta, Inc.)

$V_s = 800$ ft/sec and specific gravity 1.6 superimposed on a half-space having the parameters $V_p = 8000$ ft/sec, $V_s = 4000$ ft/sec, and specific gravity $= 2.2$.

The output is taken from groups of geophones at every foot on the surface from intervals 20 to 24 ft, 26 to 30 ft, and 32 to 36 ft. Six different quantities are displayed, the horizontal and vertical displacements (U and V), the horizontal and vertical particle velocities ($\dot{U}$ and $\dot{V}$), and the compressional and shear strains ($\mathbf{V} \cdot \mathbf{S}$ and $\mathbf{V} \times \mathbf{S}$). The arrival times of the different theoretical arrivals [theoretical impulse time plus one-half of the pulse width (0.0025 sec)] are shown only on the $\dot{V}$ record. Detailed information of this type helps to solve the problems of making measurements close to a surface source.

All the records were played out using a program that scales the largest event so that its amplitude is just equal to the trace spacing. The program also makes available the scale factors, so that absolute amplitudes and their rates of decay with distance are calculable.

In order to reduce the complexity of computer codes and consequently the computer cost of solutions, some finite difference programs have been written that restrict wave propagation to the acoustic type only. These undoubtedly have a place in a research and development program but, with recent advances in computer technology, in which extremely fast programmable arithmetic modules (fast boxes) have been produced which can be interfaced to a small general-purpose computer, the outlook is already rosier for substantial, relatively low-cost usage of finite difference programs.

An industry-sponsored research program at Stanford University, under the direction of J. F. Claerbout, has already produced some noteworthy successes with P waves only. There is no doubt that this effort, combined with the effort of industry groups, will lead to innovations in data processing and in the understanding of the seismic reflection method.

12.6   NEW OR DEVELOPING DISPLAY METHODS AND METHODS FOR INTERPRETATION

There has been an emergence during the past year or two of new color displays for seismic cross sections (Anstey, 1973). As noted by Levin et al. (1975), "Color presentations were tried nearly 20 years ago, but advances in technology have eliminated difficulties that delayed the use of color as a practical aid to the interpreter." Two examples, one of a color presentation of amplitude and the other of average frequency, are given in Figures 1.3 and 12.12. The chief appeal (for the noncolorblind) seems to be a combination of immediate standout of the color difference and the fact that the normal eye is able to distinguish more changes in color than it can discriminate different densities in the grey scale. The use is subjective and economic justification probably has to be made on an individual basis, since the addition of color to structure hunting adds very little, but it may well be very valuable in exploration for stratigraphic traps.

While the deduction of geology is the standard practical use of reflection seismic data, it is rare that analyses between data in different basins are attempted. Recently Vail, Mitchum, and Todd, 1975, have shown that some geological features, notably unconformities caused by worldwide sea-level changes, can be inferred from seismic record data. This represents a startling advance in seismic reflection interpretation and is undoubtedly the beginning of a considerable expansion in seismic data use by geologists.

REFERENCES

Alterman, Z., and Loewenthal, D. (1972), "Computer Generated Seismoforms," *Methods in Computational Physics*, Vol. 12, pp. 35–164.

Anstey, N. A. (1973), "The Significance of Color Displays in the Direct Location of Hydrocarbons," paper presented at the 43rd Annual International Meeting of the Society of Exploration Geophysicists, Mexico City.

Boore, D. M. (1972), "Finite Difference Methods for Seismic Wave Propagation in Heterogeneous Materials," *Methods in Computational Physics*, Vol. 11, pp. 1–37.

Cherry, J. T. (1973), "Computer Simulation of Stress Wave Propagation in One and Two Dimensions," paper presented at the 43rd Annual International Meeting of the Society of Exploration Geophysicists, Mexico City.

Cherry, J. T., and Hurdlow, W. R. (1966), "Numerical Simulation of Seismic Disturbances," *Geophysics*, Vol. 31, No. 1, pp. 33–49.

Cherry, J. T., and Waters, K. H. (1968), "Shear-Wave Recording Using Continuous Signal Methods, Part I—Early Development," *Geophysics*, Vol. 33, No. 2, pp. 229–239.

Clay, C. S., and Liang, W. L. (1962), "Continuous Seismic Profiling with Matched Filter Detector," *Geophysics*, Vol. 27, pp. 786–795.

Clay, C. S., and Liang, W. L. (1964), Seismic Profiling with a Hydroacoustic Transducer and Correlation Receiver, *Journal of Geological Research*, Vol. 69, No. 16, pp. 3419–3428.

Cook, J. C. (1964), Progress in Mapping Underground Solution Cavities with Seismic Shear Waves: Transactions of the *Society of Mining Engineers*, March, pp. 26–32.

Dunster, D. E., and Miller, D. E. (1968), "A Research Study to Determine the Occurrence and Structural Competency of Deep Rock Formations by the Use of Surface Vibrators," U.S. Army Engineer Waterways Experiment Station, Vicksburg, Mississippi, Contract No. DACA-39-67-C-0053.

Erickson, E. L., Miller, D. E., and Waters, K. H. (1968), "Shear-Wave Recording Using Continuous Signal Methods, Part II—Later Experimentation." *Geophysics*, Vol. 33, No. 2, pp. 240–254.

Geyer, R. L., and Martner, S. T. (1969), "SH Waves from Explosive Sources," *Geophysics*, Vol. 34, No. 6, pp. 893–905.

Jolly, R. N. (1956), "Investigation of Shear Waves," *Geophysics*, Vol. 21, No. 4, pp. 905–938.

Kelly, K. E., and Alford, R. M. (1974), "Accuracy of Finite Difference Modeling of the Acoustic Wave Equation," *Geophysics*, Vol. 39, No. 6, pp. 834–842.

Kelly, K. R., Ward, R. W., Treitel, S., and Alford, R. M. (1976), "Synthetic Seismograms, a Finite Difference Approach," *Geophysics*, Vol. 41, No. 1, pp. 2–27.

Kisslinger, C., Mateker, E. J., and McEvilly, T. V. (1961), SH Motions from Explosions in Soil, *Journal of Geophysical Research*, Vol. 66, pp. 3487–3496.

Kuster, G. T., and Toksoz, M. N. (1974), "Velocity and Attenuation of Seismic Waves in Two-Phase Media, Parts I and II," *Geophysics*, Vol. 39, No. 5, pp. 587–618.

Levin, F. K., Bayhi, J. F., Dunkin, J. W., Lea, J. D., Moore, D. B., Warren, R. K., and Webster, G. H. (1975), "Developments in Exploration Geophysics, 1969–1974," paper presented at the 45th Annual International Convention of the Society of Exploration Geophysicists, Denver, Colorado.

Michon, D. (1972), "Wide Line Profiling Offers Advantages," *Oil and Gas Journal*, Vol. 70, No. 48, pp. 117–120, November 27, 1972.

Vail, P. R., Mitchum, R. M., and Todd, R. G. (1975), "Global Unconformities from Seismic Sequence Interpretation," paper presented at the 45th Annual International Meeting of the Society of Exploration Geophysicists, Denver, Colorado.

Saucier, R. T. (1970), "Acoustic Sub-Bottom Profiling Systems—A State-of-the-Art Survey," U.S. Army Engineer Waterways Experiment Station, Technical Report S-70-1.

Ricker, N., and Lynn, R. D. (1950), "Composite Reflections," *Geophysics*, Vol. 15, No. 1, pp. 30–49.

White, J. E., Heaps, S. N., and Lawrence, P. L. (1956), "Seismic Waves from a Horizontal Force," *Geophysics*, Vol. 21, No. 3, pp. 715–723.

White, J. E., and Sengbush, R. L. (1963), "Shear Waves from Explosive Sources," *Geophysics*, Vol. 28, pp. 1001–1019.

White, J. E. (1974), personal communication.

Epilogue

At the time of writing, this book has been brought up to date as much as possible. Although the lack of details is evident, the reader can obtain most of those by reading the literature cited. However, it is the writer's present intention, after an appropriate period of rest and rehabilitation, to supply some of the deficiencies, possibly in the form of additional appendices for those chapters in need of expansion or possibly in the form of a supplementary book.

These, however, are intentions, and there is a well-known road paved with them, so it behooves the individual reader not to wait but to fill in for himself as and where he can. The author wishes him well in this endeavor.

Index